ADVANCES IN THE BIOSCIENCES

Volume 74

ALLERGY AND MOLECULAR BIOLOGY

ADVANCES IN THE BIOSCIENCES

Latest volumes in the series:

Volume 56: RECENT TRENDS IN MEDICAL GENETICS
Editors: K. M. Marimuthu and P. M. Gopinath

Volume 57: ADAPTIVE PROCESSES IN VISUAL AND OCULOMOTOR SYSTEMS
Editors: E. L. Keller and D. S. Zee

Volume 58: BRAIN TUMORS: BIOPATHOLOGY AND THERAPY
Editors: M. A. Gerosa, M. L. Rosenblum and G. Tridente

Volume 59: LEARNING & MEMORY: MECHANISMS OF INFORMATION STORAGE IN THE NERVOUS SYSTEM
Editors: H. Matthies

Volume 60: DISORDERS OF EATING BEHAVIOUR: A PSYCHONEUROENDOCRINE APPROACH
Editors: E. Ferrari and F. Brambilla

Volume 61: DYNAMIC PROPERTIES OF GLIA CELLS II: CELLULAR AND MOLECULAR ASPECTS
Editors: T. Grisar, G. Franck, L. Hertz, W. T. Norton, M. Sensenbrenner and D. M. Woodbury

Volume 62: RESEARCH IN RETINITIS PIGMENTOSA
Editors: E. Zrenner, H. Krastel and H-H. Goebel

Volume 63: RETINITIS PIGMENTOSA: PATIENTS' FIGHT FOR SIGHT
Editors: F. Brunsmann and R. von Gizycki

Volume 64: ADVANCES IN AGE PIGMENTS RESEARCH
Editors: E. Aloj Totaro, P. Glees and F. A. Pisanti

Volume 65: PEPTIDES AND PROTEASES: RECENT ADVANCES
Editors: R. L. Schowen and A. Barth

Volume 66: THE BIOLOGY OF PHAGOCYTES IN HEALTH AND DISEASE
Editors: C. Mauri, S. C. Rizzo and G. Ricevuti

Volume 67: RED BLOOD CELLS AS CARRIERS FOR DRUGS: POTENTIAL THERAPEUTIC APPLICATIONS
Editors: C. Ropars, M. Chassaigne and C. Nicolau

Volume 68: IMMUNOMODULATORS & NONSPECIFIC HOST DEFENCE MECHANISMS AGAINST MICROBIAL INFECTIONS
Editors: K. N. Masihi and W. Lange

Volume 69: ADVANCES IN PITUITARY ADENOMA RESEARCH
Editors: A. M. Landolt, J. Girard, P.U. Heitz, E. del Pozo and J. Zapf

Volume 70: NEURON, BRAIN AND BEHAVIOUR
Editors: M. Bajić

Volume 71: ALCOHOL TOXICITY AND FREE RADICAL MECHANISMS
Editors: R. Nordmann, C. Ribière and H. Rouach

Volume 72: SMALL CELL LUNG CANCER
Editors: M. Marangolo and G. Fiorentini

Volume 73: TRENDS IN CHRONOBIOLOGY
Editors: W. Th. J. M. Hekkens, G. A. Kerkhof and W. J. Rietveld

ALLERGY AND MOLECULAR BIOLOGY

Proceedings of the DPC First International Symposium on Allergy and Molecular Biology
11–12 April 1988, Laguna Niguel, California, USA

Editors:

A. SAID EL SHAMI

Diagnostic Products Corporation, Los Angeles, California, USA

and

T. G. MERRETT

Allergy Unit, Benenden Chest Hospital, Cranbrook, Kent, UK

DPC

PERGAMON PRESS OXFORD · NEW YORK
BEIJING · FRANKFURT · SÃO PAULO · SYDNEY · TOKYO · TORONTO

U.K.	Pergamon Press plc, Headington Hill Hall, Oxford OX3 0BW, England
U.S.A.	Pergamon Press, Inc., Maxwell House, Fairview Park, Elmsford, New York 10523, U.S.A.
PEOPLE'S REPUBLIC OF CHINA	Pergamon Press, Room 4037, Qianmen Hotel, Beijing, People's Republic of China
FEDERAL REPUBLIC OF GERMANY	Pergamon Press GmbH, Hammerweg 6, D-6242 Kronberg, Federal Republic of Germany
BRAZIL	Pergamon Editora Ltda, Rua Eça de Queiros, 346, CEP 04011, Paraiso, São Paulo, Brazil
AUSTRALIA	Pergamon Press Australia Pty Ltd., P.O. Box 544, Potts Point, N.S.W. 2011, Australia
JAPAN	Pergamon Press, 5th Floor, Matsuoka Central Building, 1-7-1 Nishishinjuku, Shinjuku-ku, Tokyo 160, Japan
CANADA	Pergamon Press Canada Ltd., Suite No. 271, 253 College Street, Toronto, Ontario, Canada M5T 1R5

First edition 1989

ISBN: 0–08–036883–2

ISSN: 0065–3446

In order to make this volume available as economically and as rapidly as possible the authors' typescripts have been reproduced in their original forms. This method unfortunately has its typographical limitations but it is hoped that they in no way distract the reader.

Printed in Great Britain by A. Wheaton & Co. Ltd., Exeter

Preface

When Dr. Terry Merrett first proposed that DPC sponsor a technical meeting on potential roles for recombinant DNA technology in allergy studies, we had visions of its being attended by just a small group of specialists. We soon discovered that the topic was of great interest to a very large number of the world's leading researchers, because of the general consensus that genetic engineering techniques would lead to major improvements in allergy diagnosis and treatment. Many of these experts were able to attend the symposium, which took place on April 11-12, 1988 at the Ritz Carlton Hotel in Laguna Niguel, Southern California.

We hope that the presentations and discussions at the DPC-sponsored 1st International Symposium on Allergy and Molecular Biology will help to channel further research in directions ultimately beneficial to the estimated 15 to 20 percent of the world's population suffering from allergenic disorders.

We extend our gratitude for a job well done to Dr. Merrett and to Said El Shami, who were instrumental in organizing this very meeting and who gave generously of their time to insure its success. Special thanks are likewise due to Dr. James D. Watson, Dr. Jack Pepys and all the others who acted as chairmen or contributors throughout the intensive, two-day symposium. We are indebted also to the many attendees who helped to promote sharp debate during the discussion periods, and who saw to it that even more questions were raised than answered. Finally, we must thank Bob Smith and Casper De Clercq for their untiring efforts in assembling the proceedings volume and Jean Smith for typing the manuscripts.

Sigi Ziering, Ph.D.
Chairman of the Board
Diagnostic Products Corporation
June 27, 1988

Participants at DPC First International Symposium on Allergy and Molecular Biology

Laguna Niguel, April, 1988

Abrahamsson, Lars - Sweden
Alaba, Sola - USA
Ansari, Aftabasari - USA
Bahna, Sami - USA
Baines, David - United Kingdom
Baldo, Brian - Australia
Baxter, Barrie - United Kingdom
Belin, Lars - Sweden
Bonini, Sergio - Italy
Bragg, Alice - USA
Braun, Jonathan - USA
Brüggemann, Marianne - United Kingdom
Burr, Michael L. - United Kingdom
Carter, Bob - United Kingdom
Chapman, Martin D. - USA
Chng, Hiok - USA
Condado, Enrique - Spain
Cuesta, Emilio - Spain
De Clercq, Casper - USA
Diaz, Teresa Carrillo - Spain
Donta, Sam T. - USA
Duncan, Alex - USA
El Shami, A. Said - USA
Esch, Robert - USA
Fergeson, A.C. - Canada
Grey, Howard - USA
Hamburger, Robert N. - USA
Hatton, T.W. - Canada
Helm, B.A. - United Kingdom
Hernandez, Jose - Spain
Hockett, Bill - USA
Impellitteri, Diane - USA
King, Te Piao - USA
Kingo, Kaj - Denmark
Klapper, David - USA
Knox, Bruce - Australia
Kumano, Masanobu - Japan
Lau, Susanne - Germany
Longbottom, Joan - United Kingdom
Lorenzotti, Fabio - Italy
Mansmann, Herbert C. - USA
Marsh, David G. - USA
Merrett, Greg - United Kingdom
Merrett, Terry - United Kingdom
Miller, Jeff - USA
Mitchell, E. Bruce - Ireland
Pantin, Charles - United Kingdom
Pavuk, Jack - USA
Pepys, Jack - United Kingdom
Perelmutter, Louis - USA
Rafnar, Thorunn - USA
Reese, John - United Kingdom
Riva, Antonio - Spain
Saxon, Andy - USA
Schwartz, Lawrence B. - USA
Shakib, Farouk - United Kingdom
Shiova, Rakujj - Japan
Smith, Bob - United Kingdom
Spector, Sheldon - USA
Spiegelberg, Hans L. - USA
Stephan, Volker - Germany
Steppich, Hermann - Germany
Stewart, Geoff - Australia
Sullivan, Tim - USA
Thomas, Wayne R. - Australia
Tovey, Euan - Australia
Van Asten, Mark - Australia
Warner, John - United Kingdom
Watson, James - USA
Witte, Owen - USA
Wyatt, Jeremy - United Kingdom
Ziering, Marilyn - USA
Ziering, Sigi - USA

Participants

Contents

Session One

ALLERGENS

Chairman: J. Pepys

Allergenic Extracts

Robert E. Esch

Greer Laboratories, Inc., P.O. Box 800,
Lenoir, NC 28645, USA

ABSTRACT

Allergenic extracts are prepared from hundreds of environmental source materials. Their composition is complex and only a few allergens have been characterized. At present, the only way to assure extracts of consistent quality is to control the source material quality and manufacturing methods. As our knowledge of the identity and nature of allergens increase, more refined approaches to producing allergenic extracts will be possible.

KEYWORDS

Allergenic extracts; source materials; potency; stability.

INTRODUCTION

Allergenic extracts are prepared from a wide variety of sources including pollens, fungi, arthropods, foods and dusts. Crude extracts containing all the extractable components of the source material have been used for diagnosis and immunotherapy in allergic patients since the turn of this century [1-2]. The methods of preparing and using diagnostic and therapeutic extracts have not significantly changed since that time. Notable exceptions are the "allergoids" or chemically modified allergenic extracts [3-5]. These preparations are manufactured in such a way to reduce the allergenic activity while retaining the desirable immunogenic properties.

Our knowledge of the chemical composition of allergenic extracts is very limited: They are complex mixtures of proteins, glycoproteins, and polysaccharides. They may contain unknown quantities of poorly defined substances having various biological activities.

PURITY AND IDENTITY

The assurance of identity and purity of allergenic extracts requires the proper collection, identification, handling and processing of the allergen

Advances in the Biosciences Vol. 74

sources. Pollen extracts from more than a hundred species of grasses, weeds, trees, shrubs, and cultivated agricultural plants are commercially available for the diagnosis and treatment of pollinosis. Pollens from properly identified plants can be harvested by several methods including vacuum collection, water setting and cut-dry-and-sieving [6-8]. Vacuum collection and water setting methods yield pollens that are relatively pure and fully developed. Contamination by plant parts are usually less than 1%. Cut-dry-and sieving yields pollens contaminated with plant parts at a level of 5-10% and thus require additional separation steps in solvents such as perchloroethylene or carbon tetrachloride. Botanical classification of the plant source and microscopic examination of the pollen sample is routinely used to confirm the identity and purity of the source material prior to extraction. Although pollens have unique structures that aid in the identification, many pollens from related plants are morphologically indistinguishable. In these cases, techniques such as analytical isoelectrofocusing and reactivity to a panel of antigen specific monoclonal antibodies of sample extracts can be used to confirm their identity.

Purity and identity assessments of fungal source materials are more difficult because fungi display a high degree of genetic variability among strains of the same species and culture conditions can affect the morphological and biochemical characteristics of fungi [9-13]. Information on the culture conditions required for optimal allergen production and the relative importance of the spores, mycelia and spent culture media as source materials is scarce and often contradictory. The maintenance of relatively stable strains without repeated subculturing and the use of standard culture conditions are minimum requirements for batch-to-batch uniformity. The purity and identity of fungal source materials can be confirmed by examination of the cultural and morphological characteristics under varying growth conditions, metabolic characteristics as determined by their ability to assimilate various substrates, and crossed-immunoelectrophoretic analysis of sample extracts.

Allergens of animal origin can be isolated from their dander, skin, saliva, serum, and urine [14-19]. The most suitable source for preparing diagnostic extracts varies according to the animal. Salivary allergens appear to be more important in reactions to the cat and rabbit. On the other hand, urine allergens appear to be the major allergens of the rat and mouse. Due to obvious reasons in using some of these source materials, commercial extracts are prepared from the hair, dander and pelts of animals which frequently contain varying amounts of the relevant allergens [20-21].

Based on the volume of house dust extracts sold each year, it is the single most important and used extract in the United States. House dust is a heterogeneous mixture of animal proteins, skin scales, hair, insects, mites, cotton and synthetic fibers, vegetable and food particles, and other household debris. Studies by a number of investigators have confirmed the importance of mites of the genus Dermatophagoides in house dust allergy [22-24]. Animal protein concentrations in home environments with pets can reach clinically significant levels [20,25] and frequently the dusts from such homes gives higher yields of animal allergens than the animals themselves. Pollen allergens can be a major component of house dust from homes during the summer months [26]. In urban house dust, cockroach allergens can be a major constituent [27-28]. Thus, the composition of house dust extracts prepared from dusts collected from different homes or from the same home at different times of the year could vary significantly. The characterization and definition of house dust remains one of the most complex problems in allergen chemistry.

Insects are potential sources of injectant, inhalant or even ingestant allergens [29-30]. Hymenoptera venoms constitute the most important insect allergens. Venom allergens from the honeybee (*Apis melifera*), yellow jacket (*Vespula spp.*) and wasp (*Polistes spp.*) have been isolated and characterized [31-34]. Whole body and venom extracts have been used in diagnosis and therapy. In nearly all cases, the venoms are more efficacious in this regard [35-36]. Isolated cases of inhalant sensitivity to bee body dust among beekeepers have been described [36-37] and in these cases, it appears that whole body extracts are the extract of choice for diagnosis.

POTENCY

Allergenic extracts are used for both diagnosis and therapy. The intended use of the extract dictates the potency assay to be employed. Diagnostic extracts for skin tests are ideally tested by quantitative skin testing or an appropriate *in vitro* correlate. The potency of therapeutic extracts can only be fully assessed by measuring their efficacy in controlled clinical trials in humans. Once the efficacy of a therapeutic extract is established, the problem of assaying subsequent production batches still remains. In contrast to many vaccines such as diptheria and tetanus, allergenic extracts do not have an appropriate animal model which permits a laboratory assay. The immunogenic properties of allergenic extracts include the induction of allergen specific IgG antibody production and stimulation of T suppressor cells; and the determinants to which these responses are directed vary from individual to individual. There is presently no simple potency assay which precisely measures these properties of extracts.

A major problem of extract manufacturers as well as the regulatory agencies is to develop appropriate laboratory potency assays for the hundreds of allergenic extracts currently being used clinically. Several approaches are being used to standardize extracts (see Table 1). Methods based on weight of source material per volume of extraction fluid (w/v) and protein nitrogen content are most widely used but are inadequate since they do not necessarily correlate with allergenic activity. Quantitative skin testing, RAST or ELISA inhibition potency and the measurement of specific allergen content are the most recent favored approaches being employed. The use of a combination of these methods could prove to be the best approach in view of the current technology [38-39]. When an allergenic component of an extract is known and has been purified, specific immunoassays can be used to measure its content. Assays for the major allergen of short ragweed (*Amb a I*) and ryegrass (*Lol p I*) pollens, cat dander (*Fel d I*), and the house dust mites *Dermatophagoides farinae* (*Der f I*) and *D. pteronyssinus* (*Der p I*) are examples. The list of allergens available in a purified state is growing at a rapid rate [40] and thus standardization of a larger number of allergenic extracts based on the direct measurement of allergen content appears more feasible. The list of hybridoma derived monoclonal antibodies is also growing and these reagents will allow for the standardization of reagents used in allergen quantitation [39,41,48].

STABILITY

The production of stable allergenic extracts has long been a goal of extract manufacturers. Conditions that promote the deterioration of potency include storage in aqueous solution with phenol added as a preservative, at higher temperatures and at greater extract dilutions [42-45]. Thus, storage of extracts in glycerine or in a lyophilized state, at reduced temperatures and

Table 1. Methods used to standardize allergenic extracts

Weight to volume ratio [2]
Protein Nitrogen content [51-52]
Leukocyte histamine release [53]
Radial immunodiffusion [54-55]
Crossed immunoelectrophoresis [56-57]
RAST inhibition [58]
Gel isoelectric focusing [59]
Complement consumption [60]
Quantitative skin testing [61-62]
Monoclonal immunoassay [39,48]

lower dilutions has been employed to delay the reduction of allergenic potency. The use of human serum albumin and polysorbates (eg. Tween 80) have been used to increase the stability of diluted extracts presumably by inhibiting adsorption of the allergens to the wall of glass containers [46].

Both thermostable and thermolabile allergens are found in allergenic extracts. For example, boiling short ragweed pollen extracts for 1 hour leads to the destruction of *Amb a I* (AgE), *Amb a II* (AgK) and *Amb a III* (Ra3) immunoreactivity. This treatment had no affect on the immunoreactivity of the *Amb a V* (Ra5) component [42].

The stability of allergens vary over a wide pH range. The major allergen of perennial ryegrass pollen, *Lol p I*, remains stable between pH 2 - 10. In contrast, the major allergen of short ragweed pollen, *Amb a I*, is stable only between pH 6 - 8.5.

The presence of endogenous enzymes in many crude pollen and fungal extracts may contribute to the instability of aqueous preparations. Table 2 shows enzyme profiles of representative fungal and pollen extracts as determined by the Api-Zym system. Enzyme activities can be measured colorimetrically in this system and the levels were graded semi-quantitatively. High glucosidase activity was characteristic of fungal extracts while high protease activities were characteristic of the pollen extracts. Phosphatase activity can be found in nearly all extracts examined. The role of these enzymes in extract stability has not been clearly documented, although the presence of the proteases especially in grass pollen extracts could explain some of the effects on the yield of grass allergens during purification (unpublished observations). The presence of a trypsin-like enzyme in short ragweed pollen extract is responsible for the generation of the two chain form of the allergen *Amb a I* (AgE) during fractionation of the pollen extract [47]. It can also be proposed that much of the heterogeneity in molecular size and charge observed with "purified" allergens is due to the action of endogenous proteolytic enzymes.

Thus, the stability of allergens under various conditions vary widely and this poses a problem for establishing optimal extraction and storage conditions of crude allergen extracts. This problem is compounded by the fact that individual components within an extract differ in stability and extraction procedures devised to preserve the activity of one allergenic component may be deliterious for another.

Table 2. Enzyme profiles of various allergenic extracts[1]

	Phosphatases		Proteases					Glucosidases	
Extract	Alk	Acid	Leu	Val	Cys	Tryp	Chym	Alph	Beta
Moulds:									
A. tenuis	4	5	5	1	0	1	0	5	5
A. fumigatus	3	1	1	0	0	0	0	5	5
H. hordei	0	1	1	0	0	0	0	5	5
Pollens:									
June grass	5	3	5	2	1	5	0	0	0
Timothy grass	5	5	4	2	1	4	0	0	0
Short ragweed	5	5	5	0	0	2	0	0	0
Plantain	2	5	5	0	0	1	0	0	1

[1]Enzyme activity was measured in the ApiZym system (Analytab Products, Plainview NY). Readings were made according to the manufacturers instructions using diluted (1:100 w/v) crude extracts. Alk = Alkaline phosphatase; Acid = Acid phosphatase; Leu = Leucine aminopeptidase; Val = Valine aminopeptidase; Cys = Cystine aminopeptidase; Tryp = Trypsin; Chym = Chymotrypsin; Alph = alpha-glucosidase; Beta = beta-glucosidase.

FUTURE TRENDS

Achievements made in the field of protein chemistry and the development of immunological assays based on the detection of antigen specific IgE have led to a rapid growth of our understanding about allergenic extracts. Purified allergens and allergen-specific antibodies are becoming more commonly available. These preparations are not only useful in the standardization of allergenic extracts but will most likely contribute to our basic understanding of allergic diseases. More recent developments in hybridoma and recombinant DNA technologies are beginning to have an impact on this knowledge base. Allergen-specific monoclonal antibodies are now being used in the development of sensitive and specific immunoassays for allergens [48], the study of allergenic cross-reactivity [49] and the immune structure of allergens [50]. The cloning of allergens raises our hopes of producing allergens more efficiently and cleanly. Finally, the potential of manipulating the structure of allergens and thus their function may have profound effects on immunotherapeutic strategies.

REFERENCES

1. Blackley, C.H. (1880). Hay fever: its causes, treatment and effective prevention. In: Experimental Researches, 2nd edition, Balliere, Tindall and Cox, London.
2. Noon, L. (1911). Prophylactic inoculation against hay fever. Lancet, 1, 1572.
3. Marsh, D.G., Lichtenstein, L.M., Campbell, D.H.(1970). Studies on allergoids prepared from naturally-occurring allergens. I. Assay of allergenicity and antigenicity of formalinized rye group I component. Immunol., 18, 705.
4. Marsh, D.G., Norman, P.S., Roebber, M., Lichtenstein, L.M. (1981). Studies on allergoids from naturally-occurring allergens. III. Preparation of ragweed pollen allergoids by aldehyde modification in two steps. J. Allergy Clin. Immunol., 68, 449.
5. Patterson, R., Suszko, I.M., McIntire, F.C. (1973). Polymerized ragweed antigen E. I. Preparation and immunologic studies. J. Immunol., 110, 1402.
6. Paetzel, M.M. (1967). Hints for pollen collectors. Greer Laboratories, Lenoir NC.
7. Nelson, R., Townsend, B. (1972). Guidelines for pollen collectors. Hollister-Stier Laboratories, Spokane WA.
8. Wodehouse, R.P. (1935). Collecting pollen in large amounts. In: Pollen Grains (ed. R.P. Wodehouse), New York, Hafner Publishing Co., Inc., pp.101-105.
9. Aukrust, L. (1979). Crossed radioimmunoelectrophoretic studies of distinct allergens in two extracts of *Cladosporium herbarum*. Int. Arch. Allergy Appl. Immunol., 58, 371.
10. Thurston, J., Richard, J., McMillen, S. (1973). Cultural and serological comparison of ten strains of *Aspergillus fumigatus*. Mycopathol. Mycol. Appl., 51, 327.
11. Reed, C. (1979). Variability of antigenicity of *Aspergillus fumigatus*. J. Allergy Clin. Immunol., 61, 227.
12. Foster, J.W., McDaniel, L.E., Woodruff, H.B., Stokes, J.L. (1945). Microbiological aspects of penicillin. V. Condiospore formation in submerged cultures of *Penicillium notatum*. J. Bacteriol., 50, 365.
13. Flaherty, D., Burrell, R. (1970). Further environmental factors affecting the antigenicity of *Trichophyton rubrum*. Mycopathol. Mycol. Appl., 42, 165.
14. Siraganian, R.P., Sandberg, A.L. (1979). Characterization of mouse allergens. J. Allergy Clin. Immunol., 63, 435.
15. Lowenstein, H., Markussen, B., Weeke, B. (1976). Isolation and partial characterization of three major allergens of horse hair and dandruff. Int. Arch. Allergy Appl. Immunol., 51, 48.
16. Wallenbeck, I., Einarsson, R. (1987). Identification of dander-specific and serum-specific allergens in cat dandruff extract. Ann. Allergy, 59, 131.
17. Longbottom, J.L. (1980). Purification and characgerization of allergens from the urines of mice and rats. In: Advances in allergology and applied immunology (eds. A. Oehling, I. Glazer, E. Mathov, C. Arbesman) Oxford, Pergamon Press, pp.483-490.
18. Anderson, M.C., Baer, H. (1981). Allergenically active components of cat allergen extracts. J. Immunol., 127, 972.
19. Ohman, J.L., Lowell, F.C., Bloch, K.J. (1974). Allergens of mammalian origin. III. Properties of a major feline allergen. J. Immunol., 113, 1668.
20. Ohman, J.L., Lorusso, J.R., Lewis, S. (1987). Cat allergen content of commercial house dust extracts: comparison with dust extracts from cat-

containing environment. J. Allergy Clin. Immunol., 79, 955.
21. Ohman, J.L., Baer, H., Anderson, M.C., Leitermann, K., Brown, P. (1983). Surface washes of living cats: an improved method of obtaining clinically relevant allergen. J. Allergy Clin. Immunol., 72, 288.
22. Miyamoto, T., Oshima, S., Ishizaki, T., Sano, S. (1968). Allergenic identity between the common floor mite (*Dermatophagoides farinae* Hughes 1961) and house dust as a causative agent in bronchial asthma. J. Allergy, 42, 14.
23. Voorhorst, R., Spieksma, F.T.M., Varekamp, N., Leupen, M.J., Lyklema, A.W. (1967). The house dust mite (*Dermatophagoides pteronyssinus*) and the allergens it produces: identity with the house dust allergen. J. Allergy, 39, 325.
24. Platts-Mills, T.A.E., Chapman, M.D. (1987). Dust mites: immunology, allergic disease, and environmental control. J. Allergy Clin. Immunol., 80, 755.
25. Swanson, M.C., Agarwal M.K., Reed, C.E. (1984). An immunochemical approach to indoor aeroallergen quantitation with a new volumetric air sampler: studies with mite, roach, cat, mouse, and guinea pig antigens. J. Allergy Clin. Immunol., 73, 191 (Abstr.).
26. Platts-Mills, T.A.E., Hayden, M.L., Chapman, M.D., Wilkins, S.R. (1987). Seasonal variation in dust mite and grass-pollen allergens in dust from the houses of patients with asthma. J. Allergy Clin. Immunol., 79, 781.
27. Kang, B. (1976). Study on cockroach antigen as a probable causative agent in bronchial asthma. J. Allergy Clin. Immunol., 58, 357.
28. Bernton, H.S., Brown, H. (1967). Cockroach allergy. II. The relation of infestation to sensitization. South Med. J., 60, 852.
29. Shulman, S. (1968). Insect allergy: biochemical and immunological analyses of allergens. Progr. Allergy Vol.12, pp.246 (Karger, Basel/New York)
30. Wirtz, R.A. (1984). Allergic and toxic reactions to non-stinging arthropods. Ann. Rev. Entomol., 29, 47.
31. Hoffman, D.R., Shipman, W.H. (1976). Allergens in bee venom. I. Separation and identification of the major allergens. J. Allergy Clin. Immunol., 58, 551.
32. King, T.P., Sobotka, A.K., Kochoumian, L., Lichtenstein, L.M. (1979). Allergens of honey bee venom. Arch. Biochem. Biophys., 172, 4350.
33. King, T.P., Kochoumian, L., Joslyn, A. (1984). Wasp venom proteins: phospholipase A1 and B. Arch. Biochem. Biophys., 230, 1.
34. Hoffman, D.R. (1985). Allergens in Hymenoptera venom. XIII: Isolation and purification of protein components from three species of vespid venoms. J. Allergy Clin. Immunol., 75, 599.
35. Lichtenstein, L.M., Valentine, M.D., Sobotka, A.K. (1974). A case for venom treatment in anaphylactic sensitivity to Hymenoptera sting. N. Engl. J. Med., 290, 1123.
36. Yunginger, J.W., Jones, R.T., Leiferman, K.M., Paull, B.R., Welsh, P.W., Gleich, G.J. (1978). Immunological and biochemical studies in beekeepers and their family members. J. Allergy Clin. Immunol., 61, 93.
37. Ellis, R.V., Ahrens, H.G. (1931-32). Hypersensitiveness to airborne bee allergen. J. Allergy, 3, 247.
38. Brede, H.D., Steven, E. (eds.) (1985). Regulatory Control and Standardization of Allergenic Extracts. Third International Paul-Ehrlich Seminar. New York, Gustav Fischer Verlag.
39. Baldo, B.A. (1983). Standardization of allergens. Allergy, 38, 535.
40. Marsh, D.G., Goodfriend, L., King, T.P., Lowenstein, H., Platts-Mills, T.A.E. (1987). Allergen nomenclature. J. Allergy Clin. Immunol., 80, 639.
41. Tovey, E.R., Baldo, B.A. (1988). An on-line computer data bank of monoclonal antibodies that recognize allergens: invitation for entries. J. Allergy Clin. Immunol., 81, 611.

42. Baer, H., Anderson, M.C., Hale, R., Gleich, G.J. (1980). The heat stability of short ragweed pollen extract and the importance of individual allergens in skin reactivity. J. Allergy Clin. Immunol., 66, 281.
43. Anderson, M.C., Baer, H. (1982). Antigenic and allergenic changes during storage of a pollen extract. J. Allergy Clin. Immunol., 69, 3.
44. Nelson, H.S. (1979). The effect of preservatives and dilution on the deterioration of Russian thistle (*Salsola pestifer*), a pollen extract. J. Allergy Clin. Immunol., 63, 417.
45. Augustin, R. (1959). Grass pollen allergens. III. Their differentiation from the other pollen antigens by immunoelectrophoretic studies in relation to skin reactivity, enzymic digestions, heat and cold stabilities. Immunology, 2, 230.
46. Norman, P.S., Marsh, D.G. (1978). Human serum albumin and Tween 80 as stabilizers of allergen solutions. J. Allergy Clin. Immunol., 62, 314.
47. King, T.P., Alagon, A., Kochoumian, L., Kuan, J., Sobotka, A.K., Lichtenstein, L.M. (1981). Limited proteolysis of antigens E and K from ragweed pollen. Arch. Biochem. Biophys., 212, 127.
48. Chapman, M.D., Heymann, P.W., Wilkins, S.R., Brown, M.D., Platts-Mills, T.A.E. (1987). Monoclonal immunoassays for the major dust mite (*Dermatophagoides*) allergens, *Der p I* and *Der f I*, and quantitative analysis of the allergen content of mite and house dust extracts. J. Allergy Clin. Immunol., 80, 184.
49. Esch, R.E., Klapper, D.G. (1987). Cross-reactive and unique grass group I antigenic determinants defined by monoclonal antibodies. J. Allergy Clin. Immunol., 79, 489.
50. Olson, J.R., Klapper, D.G. (1986). Two major human allergenic sites on ragweed pollen allergen antigen E identified by using monoclonal antibodies. J. Immunol., 136, 2109.
51. May, J.C., Sih, J.T.C., Miller, J.R., Seligmann, E.B. (1979). Optimization of parameters in protein nitrogen unit precipitation procedures for allergenic extracts. J. Allergy Clin. Immunol., 63, 87.
52. Stull, A., Cooke, R.A., Tennant, J. (1933). The allergen content of pollen extracts: its determination and its deterioration. J. Allergy, 4, 455.
53. Siraganian, R.P. (1976). Histamine release and assay methods for the study of human allergy. In: Manual of Clinical Immunology (eds. N.R. Rose, H. Friedman), Washington DC, American Society for Microbiology, pp.603-615.
54. Baer, H., Godfrey, H., Maloney, C.J., Norman, P.S., Lichtenstein, L.M. (1970). The potency and antigen E content of commercially prepared ragweed extracts. J. Allergy, 45, 347.
55. Mancini, G., Carbonara, A.O., Heremans, J.G. (1965). Immunochemical quantitation of antigens by single radialimmunodiffusion. Immunochemistry, 2, 235.
56. Lowenstein, H. (1978). Quantitative immunoelectrophoretic methods as a tool for the analysis and isolation of allergen. Prog. Allergy, 25, 1.
57. Axelsen, N.H., Kroll, J., Weeke, B. (1973). A manual of quantitative immunoelectrophoresis. Methods and applications. Scand. J. Immunol., 2 (suppl.), 1.
58. Gleich, G.J., Larson, J.B., Jones, R.T., Baer, H. (1974). Measurement of the potency of allergy extracts by their inhibitory capacities in the radioallergosorbent test. J. Allergy Clin. Immunol., 53, 158.
59. Brighton, W.D. (1975). Profiles of allergen extract components by isoelectric focusing and radioimmunoassay. Develop. in Biol. Standard., 29, 362.
60. Berrens, L., Guikers, C.L.H. (1975). Value of fluid phase complement consumption as a means of *in vitro* allergen standardization. Develop. in Biol. Standard., 29, 235.

61. Aas, K., Backman, A., Belin, L., Weeke, B. (1978). Standardization of allergen extracts with appropriate methods. The combined use of skin prick testing and radio-allergosorbent tests. Allergy, 33, 130.
62. Turkeltaub, P.C., Rastogi, S.C., Baer, H. (1986). Quantitative intradermal test procedure to determine relative potency of allergic extracts using parallel line bioassay. In: Methods of analysis: allergenic products, Bethesda, DBB/OBRR, Center for Drugs and Biologics, FDA.

Toward a Definition of the 'Complete' Spectrum and Rank Order of Importance of the Allergens from the House Dust Mite: *Dermatophagoides Pteronyssinus*

B. A. Baldo[*†], S. A. Ford[†] and E. R. Tovey[†]

[*]Department of Medicine, University of Sydney, Syndey;
[†]Kolling Institute of Medical Research, Royal North Shore Hospital, St. Leonards, NSW 2065, Australia

ABSTRACT

Despite the introduction of an international reference preparation for the common house dust mite *Dermatophagoides pteronyssinus* and continuing research on the allergenic nature of this important allergen source, there remains an urgent need to identify the full spectrum of *D. pteronyssinus* allergens, relate this to the allergen recognition spectra of a large number of mite-allergic subjects and rank the allergens in order of clinical importance. By employing sera from 96 house dust mite-allergic subjects together with a whole body extract of *D. pteronyssinus*, 32 different IgE-binding components were identified by protein blotting in the MW range ~6 to >110 kD. Most sera recognized an often unique combination of 6-10 components. The MWs and frequencies of binding of the six most frequently recognized components were: 16 kD (*Der p II*) 88%; 30 and 32 kD (pair) (*Der p III*) 68%; 26 kD 50%; 15 kD 49%; 25 kD (*Der p I*) 43%. Other components bound by more than 30% of the sera had MWs of 54 and 56 (pair), 95 and >110 kD. The above mentioned components also bound the most IgE as detected by the uptake of ^{125}I-anti-human IgE. *Der p I* is most likely the 25 kD component. It is not clear if it also comprises the 26 kD component.

KEYWORDS

House dust mite allergens; *D. pteronyssinus*; *Der p I*; *Der p II*; *Der p III*; protein blotting; mite allergy.

The Importance of Identifying the Full Spectrum of *D. pteronyssinus* Allergens

Of all the allergens sources that provoke immediate hypersensitivity reactions in man, few are as important in terms of their widespread distribution and frequency of involvement in allergies as the common house dust mites *D. pteronyssinus* and *D. farinae* (Voorhorst *et al.*, 1964; Maunsell *et al.*, 1968; Pepys *et al.*, 1968; van Bronswijk and Sinha, 1971; Sarsfield, 1974; Platts-Mills and Chapman, 1987). In view of this, it is unfortunate that there is still no widespread agreement on the number and rank order of importance of the allergens from these species. The major reason for this is that the important initial task of identifying the complete spectrum of allergens has been leap-frogged in efforts to pursue and promote individual research on one or only a few allergens and this has led, in our view (Krilis *et al.*, 1984a; Tovey and Baldo, 1987a), and in the view of some other research groups (Yman, 1985; Maasch *et al.*, 1987), to the premature acceptance of an inadequate international standard for *D. pteronyssinus* allergenic extract (A. Ford *et al.*, 1985a,b). The first step in standardizing any allergen extract involves the adequate definition of the source material. This is achieved by employing reliable analytical methods designed to reveal the number and nature of the antigens and allergens in the extract being studied (Baldo, 1983; Tovey and Baldo, 1984). This information has not been available for *D. pteronyssinus* and *D. farinae* extracts nor has it been adequately pursued.

Allergen Identification by Protein Blotting

Because of its sensitivity, resolving power, ease of use, high capacity and need for the employment of small quantities of extract, protein blotting (Towbin *et al.*, 1979) is the method of choice for identifying IgE-binding components in complex allergen extracts (Sutton *et al.*, 1982; Tovey *et al.*, 1987). The method permits the rapid side-by-side examination in the same experiment of large numbers of patients' sera and, since employment of an heterologous antiserum is unnecessary, the procedure is easy to standardize and results from different laboratories are more likely to be comparable than with, for example, crossed radioimmunoelectrophoresis. The value of protein blotting in allergen identification and standardization has now been established in a large number of different studies (Tovey and Baldo, 1987b; Baldo and Tovey, 1988) and, in view of the power of the method for providing reliable qualitative definitions of a variety of complex extracts, we set out to obtain the most detailed, accurate and reliable definition of *D. pteronyssinus* allergenic extract that it is practical to achieve with the available methodology. Before outlining these findings, however, attention is drawn to two important points in relation to allergen-specific IgE antibodies.

i. Heterogeneity of the IgE antibody response. Because large numbers of sera from different allergic patients can be studied side-by-side, the unique allergen sensitivity spectrum of individual patients can be rapidly identified. Results from studies in which large numbers of sera were employed to probe transferred components from a number of different important allergen sources, particularly pollens and the mite *D. pteronyssinus*, often revealed an extraordinary degree of heterogeneity in patient IgE responses to individual allergenic proteins. While some response patterns observed so far have sometimes proved relatively simple, for example, the sera recognition spectra obtained with *Parietaria judaica* extract (Ford *et al.*, 1986), with some other sources such as *D. pteronyssinus* (Tovey and Baldo, 1987a), *Lolium perenne* (Ford and Baldo, 1986), *Dactylis glomerata* (Ford *et al.*, 1985) and *Cynodon dactylon* (Ford and Baldo, 1987), identical allergen recognition profiles by two or more different sera were not often seen despite the often striking occurence of one or more dominant IgE-binding components. On reflection, and in view of the known heterogeneity of the humoral immune response at the level of antibody class and subclass and the antibody combining site (Kabat, 1976), we should not be too surprised by these results. Such clear demonstrations of unique allergen recognition patterns by individual patients' sera have not been obvious in the past simply because the techniques used to identify allergens have not lent themselves to the sort of rapid, precise and individual analysis that is achievable with protein blotting followed by probing with sera from large numbers of allergic subjects.

ii. IgE-binding *in vitro* and allergenic activity *in vivo*. Methods that simply detect IgE-binding provide no data on the biological properties of the identified antigens and we should therefore always be seeking to establish, one way or the other, the clinical relevance of the identified reactants. Only when this has been done can we be confident of the allergenic activity of a particular IgE-binding antigen. In relation to this point, however, two other important points should be emphasized. Identification of the complete spectrum of allergens in any source material is essential for any subsequent attempts to standardize the extract and ultimately for successful diagnosis and therapy with the extract. The easiest and most practical method of identifying allergens is by undertaking an initial screen for the identification of those components in an extract that bind IgE antibodies. All of the components found to bind IgE antibodies, especially those that show a low frequency of binding in an allergic population, may not prove to be clinically important but, on the other hand, those components that do not bind IgE or bind it infrequently, are almost certainly going to lack allergenic activity or be of minor clinical importance when tested *in vivo*.

Comparison by Electroblotting of IgE-binding Components in *D. pteronyssinus* Bodies and Culture Extracts

All mite extracts are not the same; they vary depending upon their source, the extraction times and procedures employed and the conditions of storage. Figure 1 shows protein-stained patterns of heated but unreduced mite proteins following separation by SDS-PAGE and transfer to nitrocellulose. An extract of *D. pteronyssinus* whole bodies yielded a complex pattern (Fig. 1a) of a large number of bands distributed throughout the MW range from ~6 to >110 kD. The *D. pteronyssinus* culture extract was noticeably less complex and less well resolved with three main components visible (Fig. 1b). The pattern seen in Fig. 1 is similar to that obtained with mite faeces extract (Tovey and Baldo, 1987a).

When both whole body and culture extracts were probed with sera from mite-allergic subjects, the patterns of IgE-binding components seen were similar for the two different extracts (Fig. 2) (Tovey and Baldo, 1987a). Both extracts were shown to contain IgE-binding proteins with MWs 95, 53, 32, 28, 25 and 15 kD, although more components could be resolved in the body extract.

Influence of Two Technical Aspects on the Detection of Important Mite Allergens by Protein Blotting

Fig. 3 outlines the general procedure we employ to identify IgE-binding components in crude allergen sources. While attempting to identify *D. pteronyssinus* allergens, two aspects of the general scheme, namely, the employment of reducing conditions and the blocking step carried out with the nitrocellulose transfer, yielded interesting and unexpected results.

i. Optimum blocking conditions. In probing blotted mite proteins it became apparent that the nature of the blocking agent used influenced the subsequent detection of those components that bound IgE antibodies. Tween 20 proved

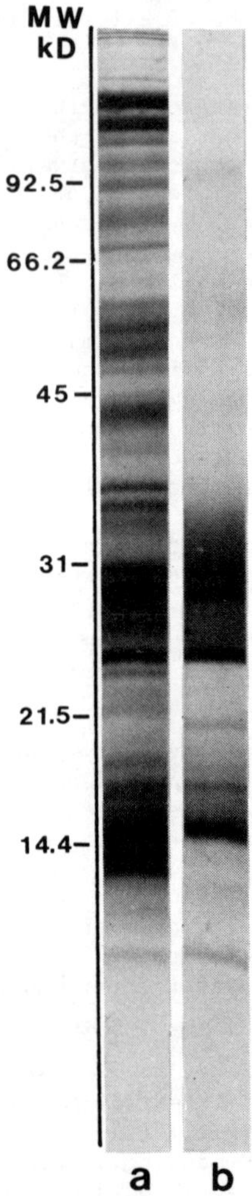

Fig. 1. Separation by SDS-PAGE of mite *D. pteronyssinus* (a) body extract and (b) culture on 9-27% gradient gels.

superior to protein blocking agents (bovine serum albumin, milk powder, gelatin, serum etc.) but, in addition, we also found that the longer the incubation time employed, the stronger the IgE-binding became, especially for the detection of *Der p II*. The apparent enhanced IgE-binding to *Der p II* mediated by Tween was not due to 'non-specific' effects since probing of mite transfers with sera containing IgE antibodies to pollens and/or other non-mite allergens did not result in binding to mite proteins. The enhancement of IgE-binding of *Der p II* induced by Tween was strikingly shown in

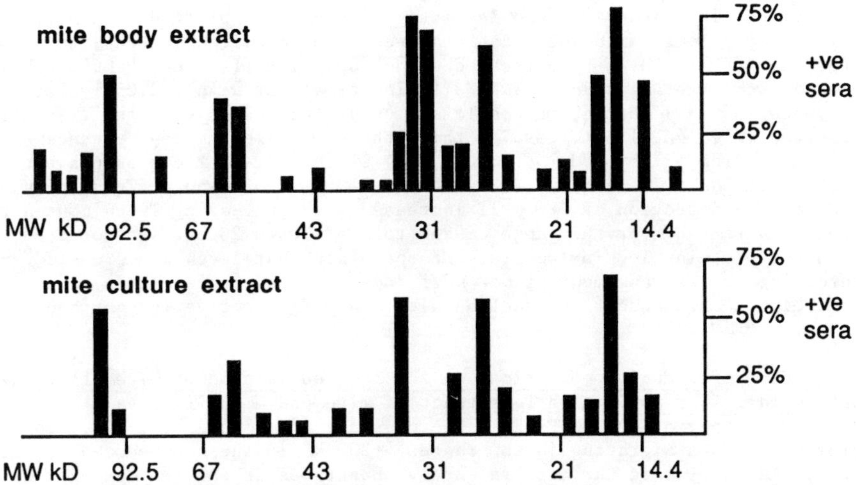

Fig. 2. Histogram showing the frequency of binding of patients' sera (n=22) containing specific IgE antibodies to each of the SDS-PAGE-separated allergenic components of mite body and culture extracts (data adapted from Tovey and Baldo, 1987a).

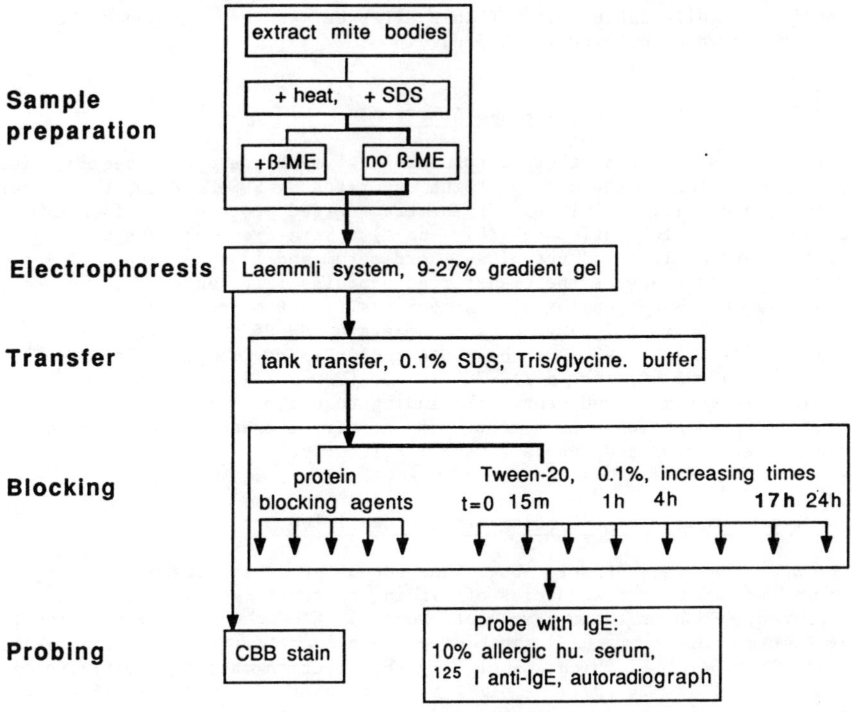

Fig. 3. Procedures employed to identify IgE-binding components in a crude extract of mite bodies.

experiments in which ^{125}I-labelled mite proteins were blotted onto nitrocellulose, blocked with Tween and compared to unlabelled blotted mite proteins that had also been treated with a solution of Tween 20 before being probed with mite-allergic sera. Following treatment with a 0.05% solution of Tween 20 for 17 hours, marked losses of protein bands from the transfer resulted (Fig. 4a,b) but, despite this, there was a concurrent enhancement of IgE-binding (Fig. 4c,d), probably only slight for most components but marked for *Der p II* (Fig. 4d; location of *Der p II* arrowed). This enhancement of the detection of *Der p II* increased with increasing Tween concentration when tested over the range 0.025% to 2.5% Tween 20 but was not seen with the protein blocking agents. We speculate that Tween may promote refolding or rearrangement of *Der p II* to a 'native' conformation which is conducive to subsequent interaction with *Der p II*-reactive IgE antibodies.

ii. Effect of reducing conditions on the immunodetection of mite allergens, particularly *Der p I*. A whole extract of mite and a purified sample of *Der p I* were prepared for SDS-PAGE in the presence or absence of 2-mercaptoethanol. Bands migrating in the range 14-200 kD in the non-reduced samples (Fig. 5Aa,b) stained far more intensely than those in the reduced extract (Fig. 5Ac), but this pattern was reversed in the MW region <14 kD. The MW of *Der p I* under non-reducing and reducing conditions was found to be 25 and 30 kD respectively (Fig. 5Ad,e). Fig. 5B shows the corresponding IgE-binding patterns of the same four samples. Protein bands from the non-reduced preparations showed stronger binding (compare Fig. 5Bb and c). Again, the electrophoretic mobility of *Der p I* increased in the absence of reducing conditions while IgE binding of this allergen was reduced when reducing conditions were employed (Fig. 5Bd,e).

The Spectrum of *D. pteronyssinus* Allergens

Sera from 96 patients allergic to house dust mites were examined individually for IgE-binding to the mite proteins separated by SDS-PAGE and then electroblotted onto nitrocellulose. Of the transferred proteins, 32 in the MW range 6-140 kD bound IgE antibodies (Fig. 6c-t). Sera from subjects highly allergic to pollens but not mites, cord serum and ^{125}I-anti-human IgE alone did not bind to any of the transferred proteins (Fig. 6u-w). The components that gave the bands on the autoradiographs (ie. the components that presumably bound the most IgE) were band numbers 4 (MW 95 kD), 8 (56 kD), 9 (54 kD), 14 (32 kD), 15 (30 kD) (bands 14 and/or 15 may comprise *Der p III*), 18 (26 kD), 19 (25 kD) (*Der p I*), 26 (16 kD) (*Der p II*) and 28 (15 kD). An occasional serum showed dense IgE-binding to one or more of the other identified IgE-binding components, for example, strips o and q in Fig. 6 revealed heavy staining of components 12 and 7 respectively.

The Frequencies of IgE-binding of the Identified Mite Components

The 9 components mentioned above that bound IgE antibodies most strongly also showed the highest frequencies of binding to the large panel of sera examined. Each of these components bound ≥34% of the sera tested with the 16 kD component (*Der p II*) exhibiting the highest frequency of binding (88%). Apart from *Der p II* which was clearly the allergen most frequently recognised, component 18 (MW 26 kD), *Der p III* and component number 28 (MW 15 kD) were the only other proteins that reacted with IgE antibodies in the sera of approximately 50% or more of the patients' sera examined (Fig. 7).

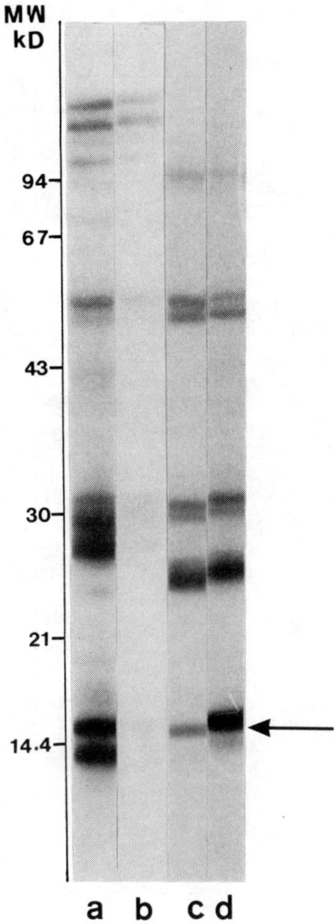

Fig. 4. Effect of Tween on the detection of blotted *D. pteronyssinus* proteins. a,b: Autoradiographs of ^{125}I-labelled mite whole body proteins separated by SDS-PAGE and transferred to nitrocellulose. Strips were blocked with 0.1% Tween 20 for 1 and 17 h respectively. c,d: Autoradiographs of nitrocellulose strips with transferred unlabelled mite proteins, blocked with 0.1% Tween for 1 and 17 h respectively and probed with sera from mite-allergic subjects followed by ^{125}I-anti-human IgE. Arrow indicates the position of *Der p II*.

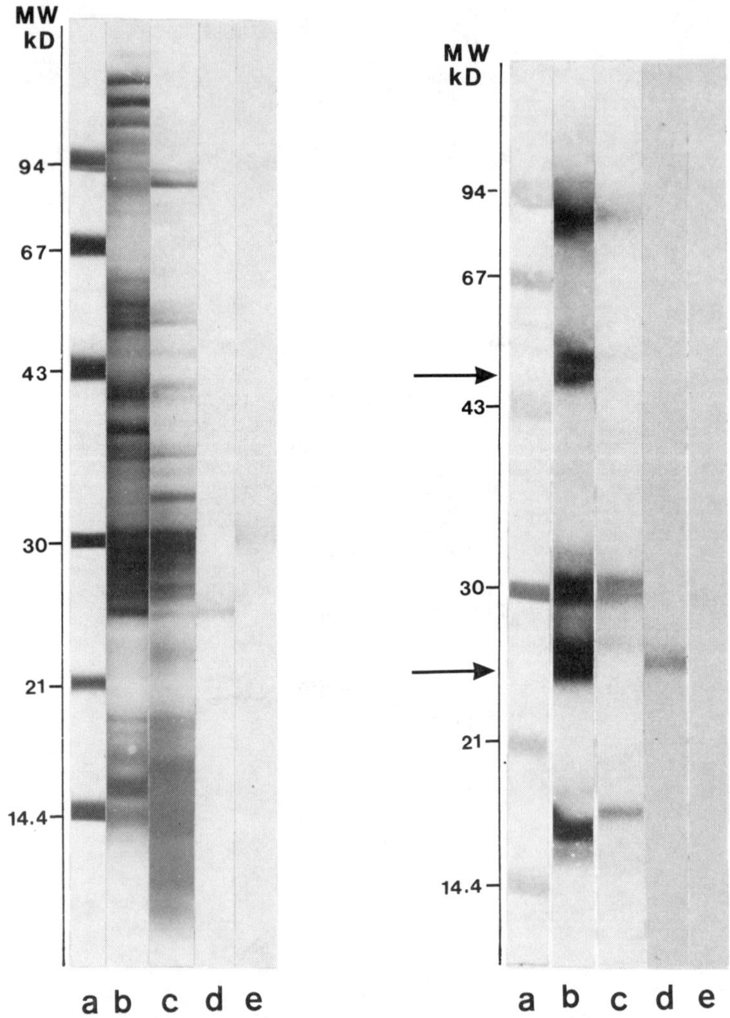

A B

Fig. 5.

Fig. 5. Effect of reduction (2-mercaptoethanol) on the electrophoretic mobility of *Der p I*. Mite body extract and purified mite allergen *Der p I* (the latter provided by Professor T.A.E. Platts-Mills, Charlottesville Va. USA) separated by SDS-PAGE and (A) stained for protein with Coomassie brilliant blue and (B) transferred to nitrocellulose and probed with sera from mite-allergic subjects followed by ^{125}I-anti-human IgE and autoradiography. a: Low MW markers phosphorylase b (94 kD), bovine serum albumin (67 kD), ovalbumin (43 kD), carbonic anhydrase (30 kD), soybean trypsin inhibitor (21 kD) and lysozyme (144 kD). b,c: Separated mite body components from an extract prepared in the absence or presence of 2-mercaptoethanol (5%), respectively. Arrows indicate the positions of 2 mite components that disappear after reduction. d,e: Purified *Der p I* prepared in the absence, or presence, of 2-mercaptoethanol respectively.

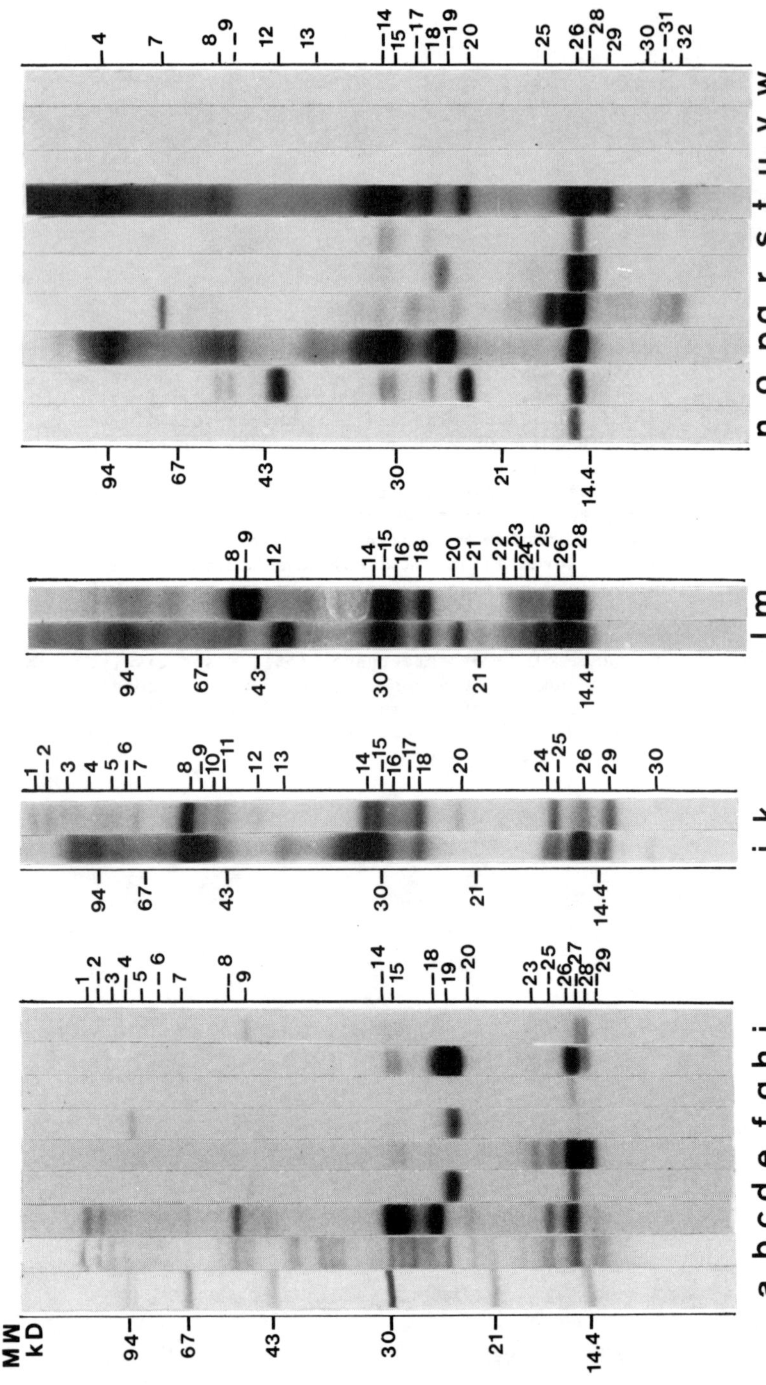

Fig. 6.

Fig. 6. Identification by autoradiography of the IgE-binding mite body components transferred to nitrocellulose after SDS-PAGE. a: Low MW markers. b: Amido black protein stain of separated mite body extract after transfer to nitrocellulose. c–t: Strips incubated with individual human sera from mite-allergic patients. u: Strip incubated with human sera from a subject not allergic to mite but highly allergic to ryegrass. v: Strip incubated directly with ^{125}I-anti-human IgE.

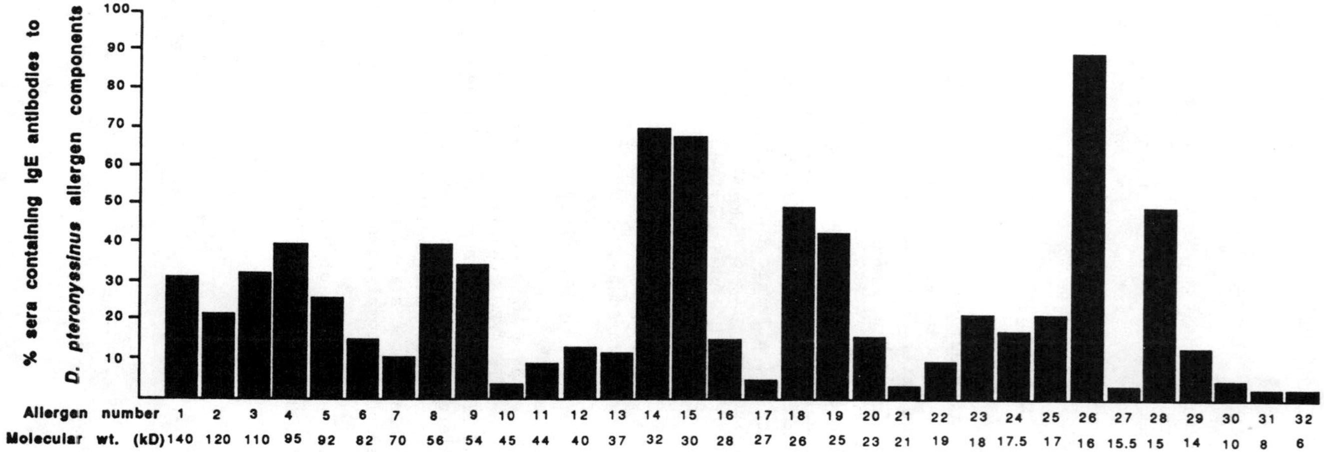

Fig. 7. Histogram showing the frequency of binding of patients' sera (n=96) containing specific IgE antibodies to each of the SDS-PAGE-separated allergenic components of mite body extract.

Detailed Analysis of IgE-binding Components in the Region 6-32 kD

Figs. 8 and 9 comprise a selection of probed nitrocellulose strips demonstrating previously unreported complexity in the MW regions where the important allergens *Der p I*, *Der p II* and *Der p III* occur. In the range 23-32 kD, 7 IgE-binding components were identified (Fig. 8b-i) while 8 closely-located bands were observed in the range 14-19 kD (Fig. 9b-j). By employing antisera specific for DpX (*Der p II*) it was possible to identify with confidence band number 26 (MW 16 kD) as *Der p II*. *Der p I* and *Der p III* were not quite as easy to identify. The latter is either band 14 (MW 32 kD) or band 15 (MW 30 kD) or both of these components since in all but 1 of the 65 positive sera, both components were recognized. The identification of *Der p I* was aided by comparing electrophoretograms of mite extract with the protein band obtained when a purified sample of *Der p I* (supplied by T. Platts-Mills) was included in the same SDS-PAGE gel, transferred and probed with human allergic sera. Even so, the close proximity of bands 18 and 19 made it difficult to say just which of these two bands is *Der p I*. However, from results of experiments employing reducing and non-reducing conditions, we concluded that component 19 (MW 25 kD) is *Der p I*. The IgE-binding protein we have recently cloned is component number 28 (MW 15 kD) (Tovey, Coban, Johnson and Baldo, this volume).

While some sera showed clear differences in their IgE-binding patterns in these two regions, for example, compare Fig. 8 strips d and e and Fig. 9 strips d and e, there was also recognition of common groups of allergens. Table 1 shows the frequencies of recognition of allergens *Der p I*, *p II* and *p III* only, that is, the frequencies obtained when one or more of these three allergens were recognized. This table emphasizes the dominance of *Der p II* in the recognition profiles of the 96 sera examined. If the presence of *Der p I*, *II* or *III* reactivity was used as the basis for determining allergy to *D. pteronyssinus* (for example, in a RAST assay) with or without associated IgE-binding to one or more of the other 29 identified components, then allergy would have been identified in 43, 88 and 68% of the 96 subjects respectively. If pairs of these major allergens were used (without excluding reactivity with the other components) the overall rates of detection would have been 92% for *Der p I* and *II*, 76% for *Der p I* and *III* and 95% for *Der p II* and *III*. If all three allergens had been used in combination, the frequency of detection would have risen marginally to 96%. Of course, detection of IgE-binding to these three allergens may not reflect the quantity of IgE directed against the total number of mite allergens.

Five sera showed monospecific recognition of *Der p II* (RAST uptakes 6.1, 21.4, 12.9, 41.1 and 23.9%). The binding pattern of one of these is shown in Fig. 9e.

Significance of these Findings

The difference in MW of *Der p I* determined by SDS-PAGE under reducing and non-reducing conditions and the decrease in IgE-binding capacity of the allergen on reduction, have important implications for the identification of house dust mite, and perhaps other, allergens by protein blotting. Non-reducing conditions preserves the IgE-binding capacity of *Der p I* and allows it to be resolved from the other IgE-binding components with MWs of ~30 kD, in particular, *Der p III*. Employment of reducing conditions, on the other hand, highlights the presence of a major IgE-binding component of MW 26 kD. The allergenic importance of this component and its relationship, if any, to

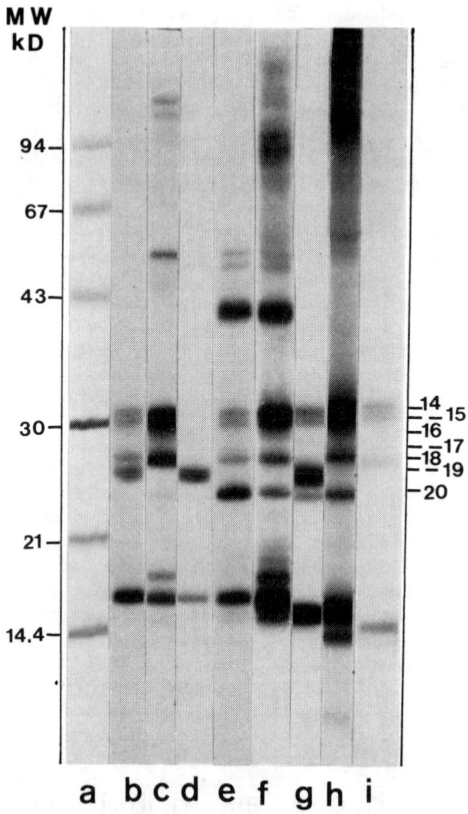

Fig. 8. Identification by autoradiography of the IgE-binding mite body components present in the 24-32 kD MW region. a: Low MW markers. b-i: Autoradiographs of nitrocellulose strips incubated with individual human sera from mite allergic patients.

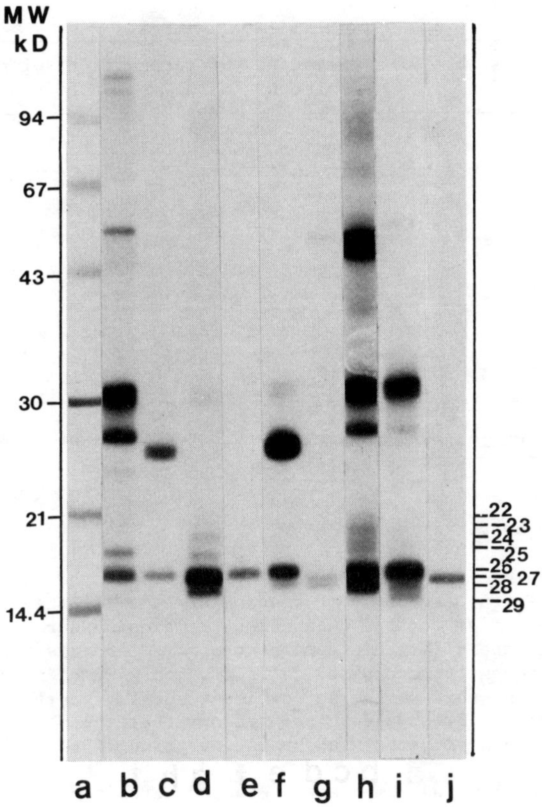

Fig. 9. Identification by autoradiography of the IgE-binding mite body components present in the 14-19 kD MW region. a: Low MW markers. b-j: Autoradiographs of strips incubated with individual human sera from mite-allergic patients.

Table 1. Frequencies of recognition by IgE-antibodies of *Dermatophagoides pteronyssinus* allergens *Der p I*, *Der p II* and *Der p III* in sera from 96 mite-allergic subjects

Allergen(s)	Frequency of recognition* of allergen patterns in the same patients (%)
p I	0
p II	20
p III	4.2
p I, *p II*	7.3
p I, *p III*	2.1
p II, *p III*	27.4
p I, *p II*, *p III*	32.7

*Frequencies obtained when only one or more of these 3 allergens were recognised.

Der p I need to be clarified.

The peculiar enhancement by Tween 20 of the detection of IgE-binding to *Der p II* enabled the true frequency of binding of this allergen to be determined although the mechanism of this enhancement remains obscure. It is now clear that *Der p II* is the allergenic component of *D. pteronyssinus* body extract that is most frequently recognised, a fact which was not previously generally appreciated but one that we can be confident about given the large sample size of sera we studied. It should be pointed out that the allergenic importance of *Der p II* has been previously emphasized by Lind (1981, 1985).

The number of IgE-binding components detected here is greater than the number detected in any previous study of *D. pteronyssinus* (Lind and Lowenstein, 1983; Le Mao *et al.*, 1983; Krilis *et al.*, 1984b; Lind *et al.*, 1984; Tovey and Baldo, 1987a). This may be explained by the much larger number of sera studied and recognition of the effects of reduction and blocking on IgE-binding undertaken during the process of identifying allergens by protein blotting.

We conclude that *Der p II* is the ~16 kD component, *Der p I* is most likely the 25 kD component and *Der p III* is probably the 30 or 32 kD component or, perhaps, both. Hence, in terms of the frequency of recognition, *Der p II* > *Der p III* > *Der p I* and components 18 (MW 26 kD) and 28 (MW 15 kD) are each more frequently recognized than *Der p I*.

So far, only the *D. pteronyssinus* allergens *Der p I* and *Der p II* have been isolated and subjected to at least some biochemical analysis although monoclonal antibodies to *Der p III* have been prepared (Platts-Mills, personal communication). The MW of *Der p I* has been variously reported as ranging from 24 to 30 kD (Chapman and Platts-Mills, 1980; Stewart, 1982; Lind, 1986) and to focus in the pI range 4.7 - 7.1. Recently, the isolation of a cDNA clone coding for *Der p I* has led to a derived MW of 25,371 for this allergen (Chua *et al.*, 1988).

The MW of *Der p II* has been reported to be 12 kD (Lind, 1981) and 18-20 kD (Lind, 1985), although a more commonly quoted, but unsubstantiated, value in the allergen standardization literature is 15 kD. We found that the MW of *Der p II* is close to the latter value on SDS-PAGE under non-reducing conditions and a slightly higher value under reducing conditions.

The relative clinical importance of the different mite allergens in *D. pteronyssinus* has not yet been established. Allergen *Der p I* is a major protein component of extracts of faecal-enriched culture material and was the first allergen isolated from this source. It is less concentrated in mite body extracts. Because it was available and used as the basis for a number of assays of both allergen content and for the presence of specific IgE antibodies, much of the quantitative experimental data on mite allergens has been accumulated using *Der p I* as the allergen marker. The assumption of a dominant position of this allergen has not always been supported by further research although it is generally accepted that *Der p I* is an important mite allergen. It is important therefore to define which allergens constitute the group of allergens important in a high percentage, say 95%, of mite-allergic subjects. On the basis of our results outlined here and, assuming that the major IgE-binding components are also the clinically important allergens, it should not be too difficult to pick out the vast majority (>90%) of *D. pteronyssinus*-allergic subjects by the diagnostic employment of a small number (perhaps only 2-4) of allergens.

At present, the surest and most practical way of examining the specificity of the allergic response of many individuals to a complex allergen source has been to utilize the powerful and versatile analytical tool of protein blotting. Such analyses, however, are still subject to uncertainties of the type addressed in this communication. In addition, protein blotting for allergen detection is an analytical and not a preparative method. Hence, it is of little value for providing material for immunotherapy. One possible solution under investigation is to produce, by recombinant methods, the IgE-binding determinants of allergens (Baldo and Donovan, 1988). This will require the application, and perhaps development of, approaches for both the detection and synthesis of allergenic determinants and may, in time, contribute greatly to both the understanding and the therapeutic control of allergic diseases.

ACKNOWLEDGEMENTS

This work was supported by the National Health and Medical Research Council of Australia.

REFERENCES

Baldo, B.A. (1983). Standardization of allergens. Examination of existing procedures and the likely impact of new techniques on the quality control of extracts. Allergy, 38, 535.

Baldo, B.A. and Donovan, G.R. (1988). The structural basis of allergenicity: Recombinant DNA-based strategies for the study of allergens. In press.

Baldo, B.A. and Tovey, E.R. (1988). Protein blotting for the detection and standardization of allergens. In: Protein Blotting: Methodology, Research and Diagnostic Applications (ed. B.A. Baldo and E.R. Tovey). Karger, Basel, in press.

Bronswijk, J.E.van and Sinha, R.N. (1971). Pyroglyphid mites (*Acari*) and house dust allergy. J. Allergy, 47, 31.

Chapman, M.D. and Platts-Mills, T.A.E. (1980). Purification and characterization of the major allergen from *Dermatophagoides pteronyssinus* - antigen P1. J. Immunol., 125, 587.

Chua, K.Y., Stewart, G.A., Thomas, W.R., Simpson, R.J., Dilworth, R.J., Plozza, T.M. and Turner, K.J. (1988). Sequence analysis of cDNA coding for a major house dust mite allergen *Der p I*. Homology with cysteine proteases. J. Exp. Med., 167, 175.

Ford, S.A. and Baldo, B.A. (1986). A re-examination of ryegrass (*Lolium perenne*) pollen allergens. Int. Archs. Allergy appl. Immun., 81, 193.

Ford, S.A. and Baldo, B.A. (1987), Identification of Bermuda grass (*Cynodon dactylon*)-pollen allergens by electroblotting. J. Allergy Clin. Immunol. 79, 711.

Ford, S.A., Baldo, B.A., Geraci, D. and Bass, D. (1986). Identification of *Parietaria judaica* pollen allergens. Int. Archs. Allergy appl. Immun., 79, 120.

Ford, A.W., Rawle, F.C., Lind, P., Spieksma, F.Th.M., Lowenstein, H. and Platts-Mills, T.A.E. (1985a). Standardization of *Dermatophagoides pteronyssinus*: assessment of potency and allergen content in ten coded extracts. Int. Archs. Allergy appl. Immun., 76, 58.

Ford, A.W., Seagroatt, V., Platts-Mills, T.A.E. and Lowenstein, H. (1985b). A collaborative study on the first international standard of *Dermatophagoides pteronyssinus* extract. J. Allergy Clin. Immunol., 75, 676.

Ford, S.A., Tovey, E.R. and Baldo, B.A. (1985). Identification of orchard grass (*Dactylis glomerata*) pollen allergens following electrophoretic transfer to nitrocellulose. Int. Archs. Allergy appl. Immun., 78, 15.

Kabat, E.A. (1976). Structural Concepts in Immunology and Immunochemistry. 2nd edition. Holt, Rinehart and Winston, New York.

Krilis, S., Baldo, B.A., Sutton, R. and Basten, A. (1984a). Antigens and allergens from the common house dust mite *Dermatophagoides pteronyssinus*. Part I. Demonstration of multiple allergens by immunochemical and biologic analyses. J. Allergy Clin. Immunol., 74, 132.

Krilis, S., Baldo, B.A. and Basten, A. (1984b). Antigens and allergens from the common house dust mite *Dermatophagoides pteronyssinus*. Part II. Identification of the major IgE-binding antigens by crossed radioimmunoelectrophoresis. J. Allergy Clin. Immunol., 74, 142.

Le Mao, J., Dandeu, J.P., Rabillon, J., Lux, M. and David, B. (1983). Comparison of antigenic and allergenic composition of two partially purified extracts from *Dermatophagoides farinae* and *Dermatophagoides pteronyssinus* mite cultures. J. Allergy Clin. Immunol., 71, 588.

Lind, P. (1981). Characterization of a partially purified allergen preparation of *Dermatophagoides pteronyssinus* and of 2 isolated major allergens (Abstract). 14th Nordic Congr. of Allergology, Gothenburg, 1981, p.24.

Lind, P. (1985). Purification and partial characterization of two major allergens from the house dust mite *Dermatophgoides pteronyssinus*. J. Allergy Clin. Immunol., 76, 753.

Lind, P. (1986). Demonstration of close physicochemical similarity and partial immunochemical identity between the major allergen, Dp 42, of the house dust mite *D. pteronyssinus* and corresponding antigens of *D. farinae* (Df6) and *D. microceras* (Dm6). Int. Archs. Allergy appl. Immun., 79, 60.

Lind, P. and Lowenstein, H. (1983). Identification of allergens in *Dermatophagoides pteronyssinus* mite body extract by crossed radioimmunoelectrophoresis with two different rabbit antibody pools. Scand. J. Immunol., 17, 263.

Lind, P., Weeke, B. and Lowenstein, H. (1984). A reference allergen preparation of the house dust mite *D. pteronyssinus* produced from whole mite culture - a part of the DAS 76 study. Allergy, 39, 259.

Maasch, H.J., Wahl, R. and Fuchs, T. (1987). Application of the first international standard of *Dermatophagoides pteronyssinus* (house dust mite) in the evaluation of allergen extracts produced from two different source materials. Int. Archs. Allergy appl. Immun., 84, 363.

Maunsell, K., Wraith, D.G. and Cunnington, A.M. (1968). Mites and house dust allergy in bronchial asthma. Lancet, 2, 1267.

Platts-Mills, T.A.E. and Chapman, M.D. (1987). Dust mites: Immunology, allergic disease, and environmental control. J. Allergy Clin. Immunol., 80, 755.

Sarsfield, J.K. (1974). Role of house-dust mites in childhood asthma. Arch. Dis. Child., 49, 711.

Stewart, G.A. (1982). Isolation and characterization of the allergen Dpt 12 from *Dermatophagoides pteronyssinus* by chromatofocusing. Int. Archs. Allergy appl. Immun., 69, 224.

Sutton, R., Wrigley, C.W. and Baldo, B.A. (1982). Detection of IgE- and IgG-binding proteins after electrophoretic transfer from polyacrylamide gels. J. Immunol. Methods, 52, 183.

Tovey, E.R. and Baldo, B.A. (1984). Standardization of allergens. Qualitative definition of house dust mite extracts following electroblotting and detection of components with antibody and lectin probes. Int. Archs. Allergy appl. Immun., 75, 322.

Tovey, E.R. and Baldo, B.A. (1987a). Comparison by electroblotting of IgE-binding components in extracts of house dust mite bodies and spent mite culture. J. Allergy Clin. Immunol., 79, 93.

Tovey, E.R. and Baldo, B.A. (1987b). Characterization of allergens by protein blotting. Electrophoresis, 8, 452.

Tovey, E.R., Ford, S.A. and Baldo, B.A. (1987). Protein blotting on nitrocellulose: some important aspects of the resolution and detection of antigens in complex extracts. J. Biochem. Biophys. Methods, 14, 1.

Towbin, H., Staehelin, T. and Gordon, J. (1979). Electrophoretic transfer of proteins from polyacrylamide gels to nitrocellulose sheets: procedure and some applications. Proc. Natl. Acad. Sci. USA, 76, 4350.

Voorhorst, R., Spieksma-Boezeman, M.I.A. and Spieksma, F.T.M. (1964). Is a mite (*Dermatophagoides* sp.) the producer of the house dust allergen? Allergie und Asthma, 10, 329.

Yman, L. (1985). Selection of reference preparations. In: Regulatory Control and Standardization of Allergenic Extracts. Third International Paul-Ehrlich-Seminar, 1983. (Eds. H.D. Brede and E. Stevens) Gustav Fischer Verlag, Stuttgart, pp.189-192.

The Cloning of a Mite (*Dermatophagoides Pteronyssinus*) Recombinant Protein that Shares IgE-binding Determinants with a 14—15 kD Component of Mite Extracts

Euan R. Tovey[†], Gary S. Cobon[*], M. Clare Johnson[*], and Brian A. Baldo[‡]

[†]Kolling Institute of Medical Research, Royal North Shore Hospital of Sydney, St. Leonards, NSW 2065;
[*]Biotechnology Australia Pty. Ltd., P.O. Box 20, Roseville, NSW 2069;
[‡]Department of Medicine, University of Sydney, NSW 2006, Australia

ABSTRACT

An IgE-binding recombinant protein has been cloned from mites (*Dermatophagoides pteronyssinus*) using *E. coli* infected with recombinant lambda gt11 containing mite cDNA. Plaques were initially screened with a serum pool from mite-allergic subjects and positive plaques selected. IgE eluted from the recombinant protein on blots rebound to a 14-15 kD molecular weight component on blots of the mite extract. Conversely, IgE eluted from this component on blots of mites, and not from other mite components, rebound to the blotted recombinant protein. We suggest that protein blotting may have a useful role in indicating which IgE-binding determinants can be cloned, in identifying the native origin of the recombinant proteins which have been cloned and in selecting serum from allergic subjects to be used for screening.

KEYWORDS

Allergens; IgE; cloning; blotting; mites; determinants; antigens; recombinant.

INTRODUCTION

The cloning of recombinant proteins from allergenic sources offers the possibility of obtaining pure allergens that would be extremely difficult, if not prohibitive, to produce by conventional biochemical fractionation methods. In addition, the recombinant approach allows the production of large quantities of purified allergenic components - an objective which previously has not been practical for most allergen sources. Such products could have widespread application for the diagnosis and specific immunotherapy of allergic disease, function as standards for allergens extracts and aid investigations of the cellular and humoral mechanisms involved in the development, maintenance and symptomatology of hypersensitivity diseases. Further, and perhaps most important of all, the molecular technologies employed would allow new approaches to the study of allergenic structure [1].

Printed in Great Britain

As part of our interest in allergen structure and the identification of allergenic determinants, we are undertaking a programme to produce some IgE-binding recombinant proteins. The initial candidate chosen for study was the house dust mite *(Dermatophagoides pteronyssinus)* where we already have extensive experience identifying the IgE-binding components using protein blotting methods [2]. There are reasons to suggest that the IgE-binding determinants identified by protein blotting may have much in common with those produced by cloning in procaryotic hosts. When blotted, such determinants either survive the denaturing (and sometimes reducing) conditions of electrophoresis or readily refold into a suitable IgE-binding conformation on the blotting membrane [3,4]. This stability exhibited during protein blotting suggests they are probably composed in linear sequences of amino acids. Similarly, the solubilised recombinant proteins from procaryotic hosts generally lack the tertiary conformation of the native molecule although the antigenic determinants formed by the linear sequence of the molecule may be detected under suitable conditions. IgE-binding determinants identified on blots may therefore be suitable candidates for production by cloning and may, in addition, serve as markers for the whole allergen molecule.

The commonly used approach to identify a desired recombinant antigen is to screen the cDNA expression libraries with specific heterologous antibodies, usually obtained by immunisation with the specific native antigen [5] or synthetic peptides [6]. However, this approach is not applicable for the identification of recombinant proteins expressing IgE-binding determinants since the purification of sufficient quantities of specific IgE to the many possible mite IgE-binding determinants is not feasible. We have, however, used a procedure that should be applicable to the identification of such determinants from any allergen source. For the initial screen of the cDNA expression library, IgE-binding determinants on the recombinant proteins were identified using a pool of sera from mite-allergic subjects. The identity of the native protein sharing determinants with the recombinant protein was then established by using IgE eluted from the blotted cloned protein to bind to the corresponding component on blots of the native allergen source. Identification was confirmed by the converse experiment of using the limited amounts of specific IgE obtained by elution from the native blotted allergen to identify the cloned determinant and comparing the results obtained with IgE eluted from other components of the blotted native allergen source. Application of this method to the identification of a 14-15 kD component of mite body extracts (MBE) which shares IgE-binding determinants with a recombinant protein is described, and possible future strategies using human atopic sera for identifying IgE-binding determinants are outlined.

METHODS AND RESULTS

Selection of Plated Colonies with IgE/Anti-IgE

The general method used is outlined in Fig. 1. Briefly, mRNA was prepared from live mites and poly A^+ mRNA was used to prepare a cDNA library in lambda gt11 (approximately 2×10^5 recombinants). Bacteria (RY1090, 10^9 per 9 cm plate) were plated on TSB-agar plates and infected with recombinant lambda gt11 (6.5×10^5 pfu/plate)[10]. After growth for 1h at 42°C, a nitrocellulose filter soaked in 10 mM IPTG was overlaid for 2 h at 37°C followed by a second filter placed on the plates for a further 2 h.

Nitrocellulose filters were blocked for 1 h at room temperature in PTS (PBS,

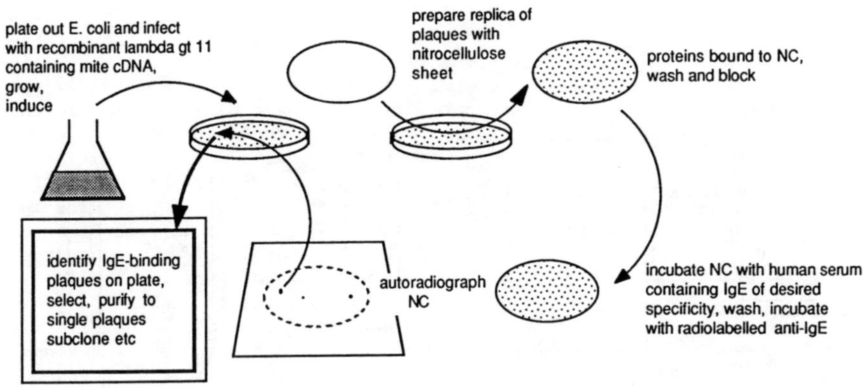

Fig. 1. A simplified schematic diagram of the general method used for the identification of lambda gt11 plaques expressing IgE-binding recombinant proteins. For a more detailed description see the methods and results section.

0.1% Tween 20, 0.1% sodium azide) and then incubated overnight at room temperature in PTS containing a 1/10 dilution of serum pooled from 20 allergic patients that had been pre-incubated with a lysate of *E. coli* strain RY1089 lysogenised with lambda gt11.

The nitrocellulose filters were then washed three times for 20 minutes in PTS and incubated for 5 h at room temperature in PTS containing ^{125}I-labelled goat anti-human IgE antibodies. Following four washes in PTS, the filters were exposed to X-ray film for 2-3 days and plaques reacting on the filters were purified to single plaques by repeated rounds of re-screening.

Following purification, individual plaques were used to infect *E. coli* strain RY1090 in liquid culture, induced with IPTG and incubated until fusion proteins could be visualised on SDS gels. These gels were blotted onto nitrocellulose paper and the blots probed with human IgE followed by radio-labelled anti-IgE. Plaques showing a positive reaction were used in subsequent experiments and the results from one of these will be described below.

Identification of the Native Allergen that Shares IgE-binding Determinants with the Cloned Protein

The general method employed is shown schematically in Fig. 2 and some of the autoradiographs are shown in Fig. 3. Elution of IgE from blots is after the method for eluting rabbit antibodies from DPT paper developed by Olmsted [7]. The optimum conditions for the elution of IgE were investigated by experiment (see Table 1). Two points are important. One is the use of acid conditions (glycine-HCl, pH 2.5) rather than other dissociating methods such as chaotropic agents, and the other is the addition of 1% BSA to the solution to act as a stabilising agent. The efficiency of dissociation of IgE from

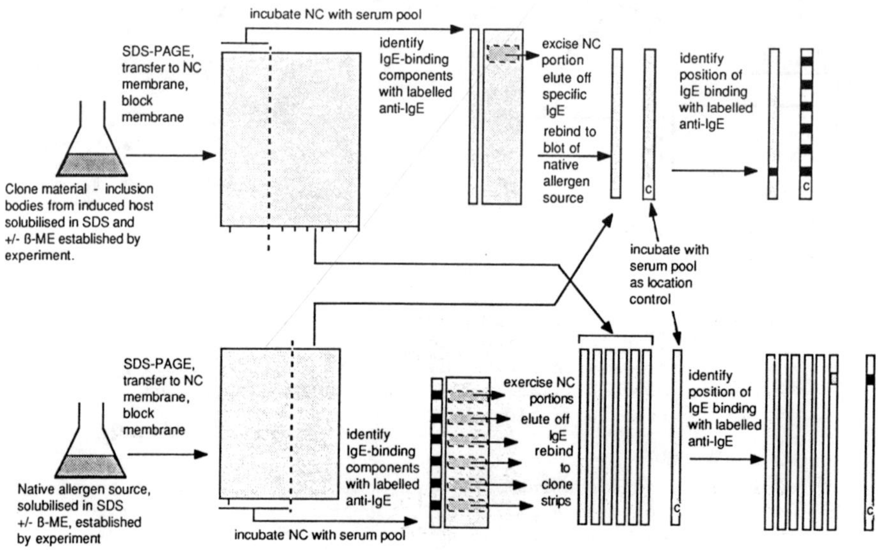

Fig. 2. A schematic diagram of the methods used to identify the mite protein component whose cDNA has been cloned into lambda gt11 and expressed in *E. coli*. Briefly, the harvested inclusion bodies from *E. coli* containing the cloned protein were separated using SDS-PAGE and electrotransfer to nitrocellulose membranes and then incubated with a pool of serum from mite-allergic subjects. The IgE specific for the cloned protein was located on the blot and eluted with glycine-HCl, 1% BSA, pH 2.5 and rebound to blots of the mite extract. The location of the rebound IgE was detected with enzyme labelled anti-IgE and its position compared with similar NC strips of blotted mite components which have been incubated with a pool of serum from mite-allergic subjects. The converse experiment is also shown where specific IgE from each of the IgE-binding mite components was separately eluted and incubated with separate strips of blots of the inclusion bodies containing the recombinant protein, followed by its localisation by staining with enzyme-labelled anti-IgE. Only IgE eluted from the mite component which has been cloned showed binding to the recombinant protein on blots of the inclusion bodies.

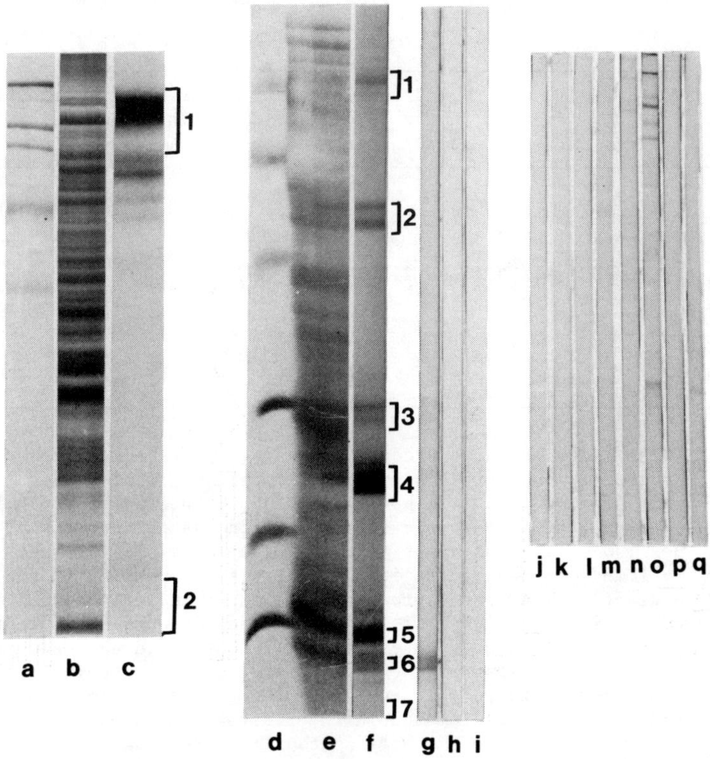

Fig. 3. Some results from experiments to identify and confirm which IgE-binding mite (*Dermatophagoides pteronyssinus*) component had been cloned. This was shown in schematic form in Fig. 2. Strip (a) and (b) shows protein staining of a SDS-PAGE 18% gel loaded with (a) (Bio-Rad®) high molecular weight markers (200, 116, 92.5, 66.2 and 45 kD) and strip (b) shows the reduced inclusion bodies from *E. coli* transformed with recombinant lambda gt11 containing mite cDNA. Strip (c) shows an autoradiograph of the detection with ^{125}I-anti-IgE of IgE binding to a NC blot of the gel shown in (b). Strips (d) and (e) show the protein staining with amido black of transfer to NC from a gradient SDS-PAGE separation of molecular weight markers (92.5, 67, 43, 31, 21 and 14.4 kD) and a mite body extract (unreduced). Strip (f) shows the position of IgE-binding components detected on an NC strip as in (e) which has been probed with a pool of mite allergic sera followed by staining using alkaline phosphatase-labelled anti-IgE. The detection with enzyme staining of IgE eluted from a blot of the cloned material at positions (1) and (2) shown on (c) and rebinding to a blot of the mite extract, is shown in strips (g) and (h) respectively. Strip (i) is a control, only incubated with the enzyme-labelled anti-IgE. The detection with enzyme staining of IgE eluted from the positions on a mite blot shown by the numbers (1) to (7) on strip (f) are shown binding to strips of the clone material in strips (j-q). Note that only IgE eluted from position (6) on the blot of mite extracts rebinds to the blot of the clone material. Detection of IgE binding on blots with enzyme-labelled anti-IgE shown in (o) is not as quantitative as the detection with radio-labelled anti-IgE shown in (c), but it generally has a lower background.

Table 1. Elution of IgE from solid phase-bound mite antigen

Eluting agent	IgE remaining on the 1st Ag-solid phase after elution	Initial IgE recovered from eluant using 2nd Ag-solid phase
	(%)	(%)
0.1 M glycine-HCl, pH 2.5	35	12
0.1 M glycine-HCl, 1% BSA, pH 2.5	29	84
3.5 M KCNS, pH 7.0	72	12
3 M $MgCl_2$	67	8
5 M Kl	28	6

These experiments were conducted using a mite body extract covalently bound to 5.5 mm diam. Immobilon discs. Discs were blocked with milk powder and then incubated with 50 µl of a pool of mite atopic sera. Elution was performed with 3 x 100 µl eluting agent over a 4 minute period, followed by immediate neutralisation of dialysis, incubation of the elutant with a second group of mite discs and determination of IgE binding with ^{125}I-anti-IgE. The percentages of IgE eluted and rebound were estimated from standard curves using dilutions of serum of the appropriate volume. Other experiments demonstrated that 80-90% of the IgE eluted in a 15 minute period was removed in the first 4-6 minutes when the excised portions of NC bearing the IgE were shaken with successive small volumes (approx. 100-300 µl/min/cm² membrane) of eluting agent.

the solid phase-antigen did not correlate with the amount that could be recovered by rebinding to fresh antigen-solid phase. This suggests that in the presence of some eluting agents IgE loses at least some of its capacity to rebind to antigen.

The results in Fig. 3 demonstrate that IgE eluted from the IgE-binding portion of the blotted extract of the transformed *E. coli* rebound to blots of the mite body extract (MBE) at the position of a known IgE-binding component of MW 14-15 kD. The specificity of the eluted IgE was confirmed by rebinding it to blots or cloned material after its incubation with MBE blots (results not shown). On Fig. 3, bands on strip (f) corresponding to recognised allergens are bands (3) *Der p III*; (4) *Der p I* and (5) *Der p II*. The component which has been cloned corresponds to band (6) and not yet been named in the standardized nomenclature.

In protein blotting studies from our laboratory presented in another paper at this meeting, the frequency of binding of IgE from mite-allergic sera (n=96) to the 14-15 kD MBE component was 48%. Separate (unpublished) studies of the frequency of IgE-binding using blotted cloned material with a different group of mite-allergic sera (n=44) showed a binding frequency of 52%.

There was a strong correlation between the intensity of IgE-binding to the band on the clone and the binding to the 14-15 kD region on mite blots. Occasional sera were identified that bound a 14-15 kD component on MBE blots but did not bind to the clone. This is still under investigation. On close examination of some of these autoradiographs, the 14-15 kD component of MBE which binds IgE can be resolved into a pair of very closely spaced bands, whereas those sera binding a component of this MW but not binding the clone, appear to bind a single fine band of this MW.

Confirmation of Identity of Cloned Component from MBE by Binding of IgE Eluted from This, and Not Other, Mite Components

The procedure employed is shown schematically in the lower part of Fig. 2 and autoradiographs of the strips are shown in Fig. 3. The sharing of antigenic identity between the cloned determinants and the mite component of 14-15 kD was confirmed by the binding of IgE eluted from the 14-15 kD component, and the absence of binding of IgE eluted from other mite components, to blots of the cloned allergen. The specificity of the eluted IgE was confirmed by rebinding it to MBE blots after its incubation with blots of the cloned protein (results not shown).

DISCUSSION

Strategies developed to produce IgE-binding determinants from an allergen source have allowed the cloning and identification of a 14-15 kD component of MBE which binds IgE. This component is about 1 kD less in molecular weight than the important IgE-binding component *Der p II* on these SDS-PAGE gradient gels run under non-reducing conditions. The frequency and relative intensity of IgE-binding to the recombinant protein on blots was similar to the binding of the MBE component on blots. The most likely explanation for the recognition of a blotted 14-15 kD MBE component by occasional sera which did not bind to the clone is that MBE contains two IgE-binding components of this molecular weight. This is supported by the different widths of the bands on the autoradiographs. It is also possible that the cloned portion of the component does not contain all of the IgE-binding determinants of the 14-15 kD allergen. Irrespective of the true explanation, it does suggest that where they are available, individual cloned IgE-binding determinants may offer a more precise method for analysing the specificities of the human IgE response at the determinant level than protein blotting which, until now, is the most precise and appropriate technique available for this purpose.

Studies with protein blotting of allergen sources suggest that there may be many IgE-binding determinants which could be candidates for cloning and the methods of identification outlined here should have wide application. Over the past 7 years, protein blotting has been used to identify the component allergens from over 30 different sources, including pollens, grains, insect venoms, fungi, mites etc.; see references 3 and 4 for reviews. Indeed, no allergen sources are known from which some allergenic components cannot be identified by blotting. All of the previously identified, and so-called "major" allergens so far investigated by blotting have been detected by this technique and, in addition, blotting studies have indicated the existence of several important allergens previously unrecognised by other methods.

The application of recombinant DNA technology to the investigation of allergens will allow the routine derivation of amino acid sequence data without the prior tedious and difficult purification that must otherwise often be employed [8]. In addition, the manipulation of the cloned genes by deletion and insertion procedures to investigate the relationships between structure and antigenicity will allow new insights into our understanding of the structure of IgE-binding determinants and regions of the molecules responsible for biological activity detected in cellular studies. Amino acid sequence data can also be the basis of investigation utilising synthetic peptides [9].

The development of further strategies for the selection and isolation of clones expressing IgE-binding determinants is required. These could include the use of a "cascade" approach to screening, where a pool of atopic serum can be used to identify an IgE-binding determinant in the initial screening of plaques. The isolated determinant can then be used to prepare an adsorption column for the preparation of a serum pool free of IgE antibodies of this specificity. The serum pool may then be used to identify clones bearing other determinants. By this process of successive screening and affinity absorptions it should be possible to continue to identify and isolate different IgE-binding cloned recombinant proteins (Fig. 4).

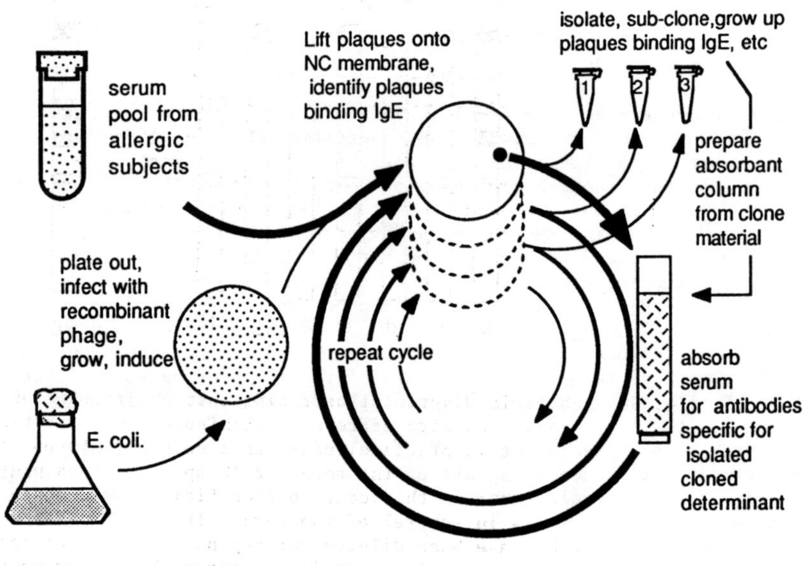

Fig. 4. Schematic diagram showing the "cascade" approach to successively selecting plaques expressing different IgE-binding determinants. The pooled serum is initially used to select plaques expressing IgE-binding determinants, these expressed recombinant proteins are then used to prepare an absorption column, and the sera absorbed for specificity for that determinant. This absorbed sera is again used to select new determinants and the cycle of absorption and reselection repeated.

In addition, our experience with blots has shown that many components which are detected easily by only occasional individual sera will not be recognised by a pool because the level of specific IgE is diluted out below the detection threshold. This is shown diagrammatically in Fig. 5. It is suggested, therefore, that for selection of specific components, a large group of individual sera should be screened by blotting and only sera showing strong binding to the specified components on blots should be chosen for use. Indeed, a series of serum pools could be created with known high specificity for each IgE-binding component. Such sera chosen by blotting probably recognise determinants of a linear form similar to those which will be produced by cloning. Allergen-reactive IgE antibodies eluted from individual blotted bands of the native allergen source could also be employed.

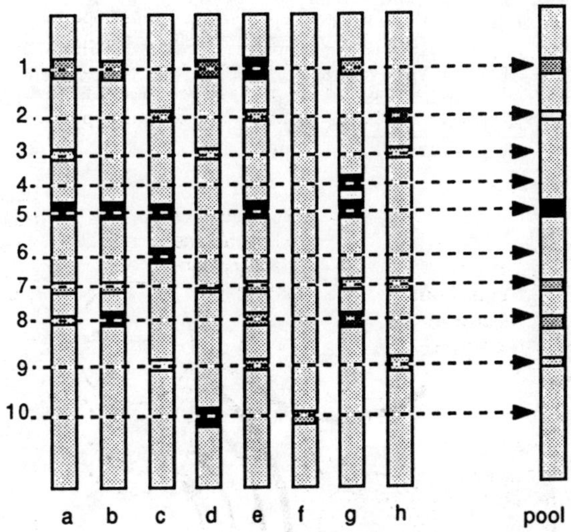

Fig. 5. Schematic diagram illustrating that the pooling of 8 sera each with different IgE-binding specificities does not in practice result in a pool capable of readily detecting all of the total of 10 specificities present, but rather in this case, 6 specificities which are each present in several of the sera. The other IgE specificities have been diluted out beyond the limit of detection of the system. It is suggested that prior to the use of sera to select for cloned proteins, a large number of individual sera be screened by blotting using the native antigen and individual pools be created for the identification of each desired component. This diagram is based on general experience rather than a single experiment.

ACKNOWLEDGEMENTS

E. Tovey and B. Baldo are both supported by grants from the National Health and Medical Research Council of Australia.

REFERENCES

1. Baldo, B.A. and Donovan, G.R. (1988). Allergy (in press).
2. Tovey, E.R. and Baldo, B.A. (1987). J. Allergy Clin. Immunol., 79, 93.
3. Tovey, E.R. and Baldo, B.A. (1987). Electrophoresis, 8, 452.
4. Baldo, B.A. and Tovey, E.R. (1988). Protein blotting for the detection and standardization of allergens. In: Monographs in Allergy series; Protein Blotting: Methodology, Research and Diagnostic Applications (eds. Baldo and Tovey), Karger, Basel (in press).
5. Coppel, R.L., Cowman, A.T., Lingelbach, K.R., Brown, G.V., Saint, R.B., Kemp, D.J. and Anders, R.F. (1983). Nature, 306, 751.
6. Oesch, B., Westaway, D., Walchli, M., McKinley, M.P., Kent, S.B.A., Aebersold, R., Barry, R.A., Tempst, P., Teplow, D.B., Hood, C.E., Prusiner, S.B. and Weissman, C. (1985). Cell, 40, 735.
7. Olmsted, J.B. (1981). J. Biol. Chem., 256, 11955.
8. Chua, K.Y., Stewart, G.A., Thomas, W.R., Simpson, R.J., Dilworth, R.J., Plozza, T.M. and Turner, K.J. (1988). J. Exp. Med., 167, 175.
9. Geysen, H.M., Rodda, S.J., Mason, T.J., Tribbick, G. and Schoofs, P.G. (1987). J. Immunol. Methods, 102, 259.
10. Kaplan, D.A., Naumouski, L. and Collier, R.J. (1981). Gene, 13, 211.

Fungal Allergens: Their Characterization with Special Reference to *Aspergillus Fumigatus*

Joan L. Longbottom

Paediatric Department, Cardiothoracic Institute, Brompton Hospital, London, UK

ABSTRACT

Fungi, because of their inherent genetic and morphological variability, are possibly the most difficult sources of allergenic components to study. Fungal spores of many different types are almost certainly the sensitizing elements and yet most allergenic extracts are prepared from sporulating mycelia or culture filtrates which may contain many of the same allergenic molecules or determinants. In this presentation some of the problems encountered in preparation, the cross reactivity and allergenicity of fungal allergens will be discussed. Characterization of two major *Aspergillus fumigatus* allergens is described to illustrate methods used for fungal allergen identification and purification.

KEYWORDS

Fungi; allergen preparation; cross-reactivity; XRIE; *Aspergillus fumigatus*.

INTRODUCTION

Fungi are eukaryotic organisms, basically formed of chains of cells (hyphae) with rigid (cellulose or chitin) cell walls, in which reproduction may be by simple fragmentation of the mycelium or by highly specialized cells (spores/ conidia) produced on elaborate fruiting structures.

Four major groups contain the 61 or more fungal genera that have been reported as potentially allergenic: Phycomycetes (Zygomycetes), eg. *Mucor*, *Rhizopus*; Ascomycetes, eg. yeasts, mildews, cup fungi; Basidiomycetes, eg. rusts, smuts, mushrooms; and Deuteromycetes (Fungi Imperfecti), eg. *Aspergillus*, *Penicillium*, *Alternaria*. This last group which contains most recognized fungal aeroallergens is an artificial grouping of asexual stages and since many are now known to be the imperfect stages of ascomycete fungi they may have two distinct generic names (Burge, 1985). Thus many fungi produce morphologically different sexual and asexual spores that become airborne, for example both *Stemphylium* (asexual) and *Pleospora* (sexual) can be

Advances in the Biosciences Vol. 74

identified in samples.

Although undoubtedly inhaled fungal spores are responsible for production of upper and lower respiratory tract symptoms their actual importance *per se* in any given patient is more difficult to ascertain, since sensitivity, as demonstrated by skin test reactivity, mostly occurs in highly atopic individuals with reactivity to multiple environmental aeroallergens, eg. pollens, animal danders, house dust mite etc. Sensitivity may also be demonstrated to several species belonging to different genera and this has suggested allergenic cross-reactivity between them although more likely this represents parallel but independent fungal allergies.

Circumstantial evidence to incriminate airborne fungi in allergic disease also comes from symptoms being related to recovery of vast numbers of spores of certain fungal species by various air sampling 'Spore-trap' techniques.

Preparation of Fungal Allergens

One of the major problems associated with fungal antigen/allergen preparations stems from the ability of fungi to adapt to a variety of environmental conditions under which natural 'spontaneous' variations may occur causing morphological changes even to one form or spore producing phase metamorphosing into another. Thus, largely depending on growth conditions (inoculation procedure, type culture (surface/shake), medium, temperature and period of growth) morphological appearance as well as biochemical and antigenic composition have been found to vary considerably.

As yet there are no recognised standardized methods for production of fungal allergens. Generally where the fungus can be cultured, optimal conditions for growth on an appropriate fully-synthetic dialysable medium (to avoid the possibility of media-derived allergenic components) are chosen, and after a suitable period, the fungal mycelial mat is harvested from the culture medium. Both have been used as sources of allergenic material.

Most commercially available allergen preparations are extracts of freeze-dried mixed mycelial/spore material, although in our experience, for *A. fumigatus* in particular, the culture filtrates obtained from long-term sporulating cultures, considered to contain a full spectrum of metabolic, cytoplasmic (due to lysis) and spore derived components, have proved to be highly allergenic. Comparison of spore and mycelial components of *Alternaria alternata* (Hoffman *et al.*, 1981) demonstrated that although many were similar or identical, the spore extracts contained 8 antigens (some allergenic) that were not found in the extracts of mycelium.

Where culture of the fungus in the laboratory is difficult, eg. most Basidomycetes and Ascomycetes, collections of spores from appropriate fruiting bodies have been used (Santilli *et al.*, 1985; Hasnain *et al.*, 1985; Lehrer *et al.*, 1986) and these studies have now indicated the potential allergenic importance of these genera. In other instances use has been made of non-sporing cultures derived from airborne ascospores, for example, in the investigation of responses to *Didymella exitialis*. Such cultures probably represent the imperfect (asexual) states which, in the case of *Didymella*, may belong to several different forms, eg. genera *Phoma* and *Ascochyta*, and since these may be closely related to imperfect states of *Alternaria* spp. cross-reactivity with *A. alternata*, as demonstrated, would be expected (Harries *et al.*, 1985).

Allergenicity of Fungal Extracts

In the main, the allergenic nature of extracts has been evaluated by their skin test and/or direct RAST reactivity, whilst for allergenic identity or cross-reactivity studies, RAST inhibition has mainly been used. Comparisons of antigenic (precipitating) reactivity with hyperimmunized rabbit antisera have also been extensively employed, particularly with recently developed quantitative immunoelectrophoretic techniques.

There have been reports of biochemical and allergenic/antigenic variability in different strains of a fungus, eg. *Cladosporium herbarum* (Aukrust, 1979), *Aspergillus fumigatus* (Wallenbeck *et al*., 1984) and *Alternaria alternata* (Schumacher and Jeffery, 1976), but possibly more important were those between different batches of the same fungal strain grown and processed under identical conditions. For example, for the 10 strains of *A. fumigatus* studied (Wallenbeck *et al*., 1984) total protein and carbohydrate content varied considerably and of 27-41 antigenic components identified, 7-13 were recognized as allergens. However, similar findings were reported for 6 extracts prepared from a single strain in that antigens varied between 32-42 with 10-14 associated as allergenic (Longbottom and Austwick, 1986) although differences here were shown to be quantitative rather then qualitative in nature.

Comparable findings have been reported for other fungi, eg. for *Cladosporium herbarum* 57-62 antigenic components were identified of which >20 were allergenic by XRIE (Lowenstein *et al*., 1977; Aukrust, 1979) and for *Alternaria alternata (tenuis)* again much antigenic variability was found between strains and batch-to-batch preparations from the same strain (Schumacher and Jeffery, 1976) and 3 (major) of 22 antigens were shown to be allergenic (Bush and Voss, 1983).

The presence of allergenic cross-reactivity between fungi has been suggested by simultaneous sensitivity to multiple fungal allergens. This is particularly where different genera and/or species have been named. In general, however, little or no cross-reactivity has been found between genera, for example *Aspergillus*, *Alternaria* or *Cladosporium*, with the exception of those genera which are morphologically closely related such as *Alternaria*, *Stemphylium*, *Curvularia* and *Ulocladium* and genera with both sexual/asexual states, *Didymella/Phoma* and *Leptospheria/Alternaria*. Different species of a genus may show some similarities but also significant differences, as many also occur even between different strains of the same species. These differences/ similarities, however, may be difficult to determine when based on single batches since because of the great variability in growth, as mentioned above, even batch to batch differences in a single fungal strain have been noted.

Characterized Major Allergens

The most commonly found airborne fungi of both indoor and outdoor environments include *Cladosporium*, *Penicillium*, *Alternaria*, *Stemphylium*, *Curvularia* and *Aspergillus* species. Most attention has centred on three fungi in particular namely *Alternaria alternata*, *Aspergillus fumigatus* and *Cladosporium herbarum* as being among the most important of the allergenic fungi and major allergens have been characterized (Table 1).

Partially purified preparations of some of these major allergens have followed their initial identification by immunoelectrophoretic techniques (as described later for *A. fumigatus* allergens). Two groups independently

Table 1. Characterized fungal allergens

Fungus/ New Nomenclature	Old Nomenclature	M.W. kD gel filtration/ SDS PAGE	pI	References
Alternaria alternata				
Alt a I	Alt-I	25-50/	4.2	Yunginger *et al.*, 1980.
	(Ag 1) (Ag 8)	60/	4.0	Nyholm *et al.*, 1983.
Alt a II	'basic peptide'	6/	9.5-9.8	Budd *et al.*, 1983.
Aspergillus fumigatus				
Asp f I	Ag 3 (?Ag 10)	24/18		Longbottom, 1983 & 1986. Wallenbeck *et al.* 1984.
Asp f II	*Ag 7	150-200/36	4.8	Harvey & Longbottom, 1986.
Cladosporium herbarum				
Cla h I	Ag 32	13/	3.4-4.4	Aukrust & Borch, 1979.
Cla h II (intermediate)	Ag 54	22/19	5	Swärd-Nordmo *et al.*, 1985.

*Also of major importance as antigen for precipitating (IgG) antibody in ABPA patients.

isolated a major allergen from *Alternaria alternata*, ie. Alt-I (Yunginger *et al.*, 1980) and Ag 1 (Nyholm *et al.*, 1983). These were shown to share antigenic identity, although different variants sharing partial identity may exist in different strains (Nyholm *et al.*, 1983). A basic peptide allergen has also been isolated and partially characterized (Budd *et al.*, 1983).

For *Cladosporium herbarum* the major allergen Ag 32 proved to be very heterogeneous, occurring over a wide MW range on gel filtration and in 5 isomeric forms with pI's between 3.4 and 4.4. It also appeared to be poorly antigenic in rabbits, with only 1 of 12 rabbits responding (Aukrust, 1979). However, it is possible that this apparent lack of precipitating antibodies was due to insensitivity of protein staining similar to both Ag 3 of *A. fumigatus* (Longbottom, 1983) and Ag 1 of *Alternaria alternata* (Nyholm *et al.*, 1983) which produced almost invisible immunoprecipitates but were visualized by their strong binding of IgE and ^{125}I-anti IgE. This apparent poor immunogenicity in rabbits may be related to the relatively low molecular weights of these components, ie. 13-24 kD.

C. herbarum Ag 54 which has been classed an 'intermediate' allergen on XRIE classification (see below) was of glycoprotein nature, containing 19.5% protein and 80.2% carbohydrate (mannose : galactose : glucose in ratio 1.0 : 0.6 : 1.3) with MW approximately 20 kD.

Identification of Two Major Allergens of *Aspergillus fumigatus*

Since immediate skin reactivity to *A. fumigatus* is found in a proportion (16-28%) of patients with asthma and is a criterion for diagnosis of allergic bronchopulmonary aspergillus (ABPA), the characterization of two major allergenic components Ag 3 and Ag 7 of this fungus will be described to illustrate the various methods that have been used for identification and partial purification of fungal allergens in general.

With the introduction of crossed radioimmunoelectrophoresis (XRIE) (Lowenstein and Weeke, 1973) and related methods, it has been possible to identify individual allergenic components in complex antigenic extracts. This method presupposes that in the immunization of rabbits precipitating antibodies have been produced to all the allergenic components within the extract and that after immunoprecipitation in the gel sufficient sites are still available on the potential allergen for the reaction with the specific IgE of the patient's serum. The individuality of patient's responses can be seen in Fig. 1 which shows the XRIE patterns for 3 ABPA patients. Although each varies, one component which was barely visualized in the protein stained immunoplate showed strong uptake of radioactivity (indicated by the arrow).

The allergenic importance of the various antigenic components may be assessed by evaluating such tests in the form of an "allergogram" which indicates the number of sera reacting (IgE) with each component. Thus 'major' allergens have been defined as those components which bind specific IgE from at least 50% of sera tested with strong uptake by the majority, 'minor' allergens, those to which less than 10% of sera show usually only weak binding, and 'intermediate' allergens, those falling between those two values. Since the majority of sera tested reacted strongly with this component it was regarded as a 'major' allergen of *A. fumigatus*.

Further characterization of *A. fumigatus* components was achieved by using a modification of XRIE in which ABPA patients serum containing both IgG and IgE antibodies replaced the polyclonal rabbit antiserum incorporated in the gel (ie. self-XRIE). When a reference ABPA 'serum pool' was used, fifteen both IgG and IgE binding components were identified and a numbered pattern of reactivity recognized (Longbottom, 1983). By comparing 32 individual ABPA serum responses to the reference pattern it was also possible to differentiate two types of components from the immunoprecipitates, those readily protein stained (antigens) and those visualized predominantly by autoradiography (allergens). Thus three components Ag 2, Ag 3 (in particular) and Ag 15 were identified as major allergens and two components Ag 7 and Ag 13 as major antigens although Ag 7 also showed considerable uptake of IgE (Fig. 2a,b).

By incorporating the rabbit antiserum in an intermediate gel (Fig. 2c,d) the Ag 3 of the ABPA reference pattern was shown to be the previously identified 'major' allergen in conventional XRIE.

Gel filtration (Sephacryl-S200) of a highly allergenic protein precipitate of the *A. fumigatus* antigen monitored by fused-rocket radioimmunoelectrophoresis (FRRIE) incorporating the same ABPA reference serum pool (Fig. 3) showed that (1) most of the visible 'precipitinogenic' components were of MW

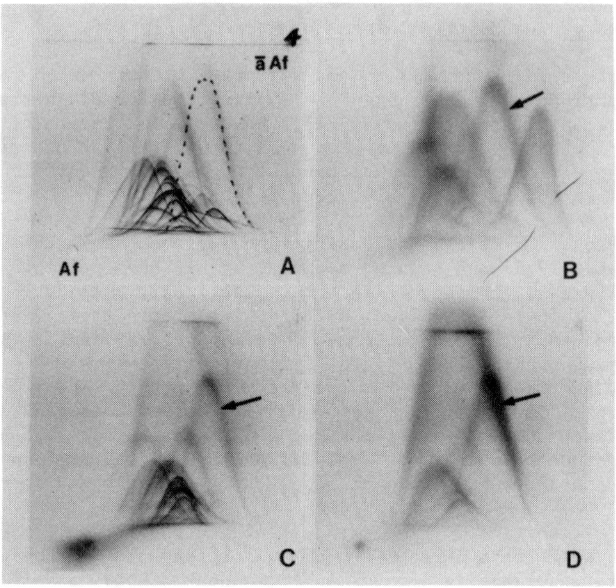

Fig. 1. Crossed radioimmunoelectrophoresis (XRIE) tests for 3 ABPA sera.

A. XIE precipitin pattern (protein stained) for *A. fumigatus* antigen (Af) electrophoresed first dimension and then into a rabbit antiserum (ā Af) containing gel in the second dimension.

B–D. XRIE autoradiographs for 3 ABPA sera.

Note: The peak showing strong uptake of radioactivity for all three sera tested is indicated by the arrow and its associated XIE peak which was barely visualized indicated by the dotted line.

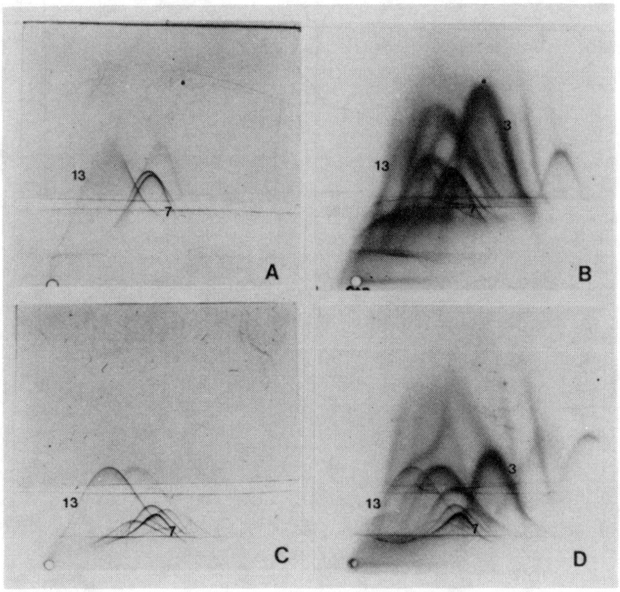

Fig. 2. A & B. Reference self-crossed radioimmunoelectrophoresis pattern using the ABPA serum pool both incorporated in the gel (precipitating IgG) and as a source of specific IgE.
C & D. As above, but with an intermediate gel incorporating the rabbit antiserum.
(A & C - protein stained plates; B & D - corresponding autoradiographs.)

Note: Absorption of the peaks by the rabbit antiserum in particular the major allergen peak, ie. Ag 3. Ag 7 the major antigen and allergen and Ag 13 a major antigen but minor allergen (Harvey and Longbottom, 1987b) are also indicated.

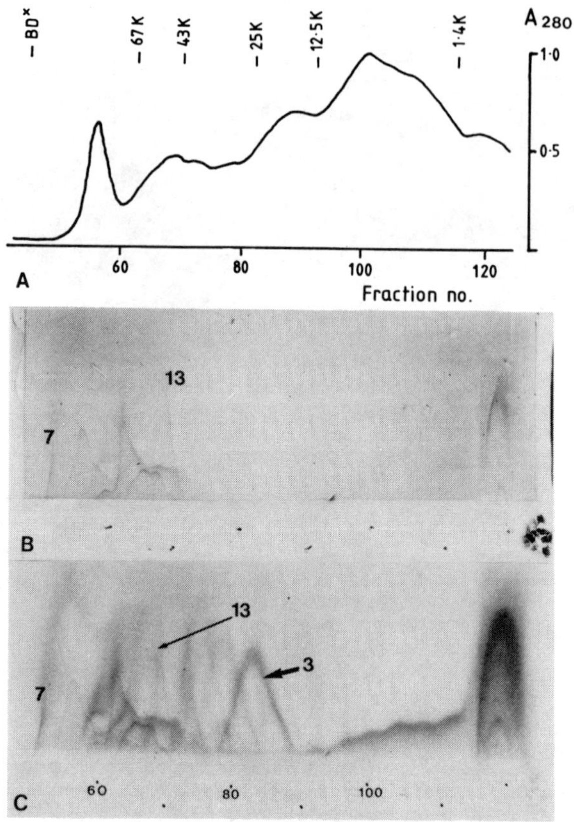

Fig. 3. Sephacryl S200 fractionation of a protein precipitated fraction of *A. fumigatus*.
A. O.D. profile with molecular weight markers.
B & C. Self-fused-rocket radioimmunoelectrophoresis profile with the ABPA serum pool incorporated in the gel.
(B - protein stained; C - autoradiograph.)

Note: Ag 3 (peak with maximum at fraction 84, MW 24 kD) is not visualized in B but shows strong uptake of radioactivity in C, whilst Ag 7 (in the initial peak eluted MW 150-200 kD) appears protein stained and with strong radioactive uptake, ie. both antigenic and allergenic.

>43 kD, (2) the first peak from the column MW approximately 200 kD showed both precipitins and uptake of IgE and (3) the presence of a component MW 24 kD (fract. 84) which was not visualized by protein staining but showed strong IgE uptake - as previously shown for Ag 3 on XRIE. By XRIE of this fraction a component with mobility of Ag 3 was identified (Fig. 4a,b) and by combining

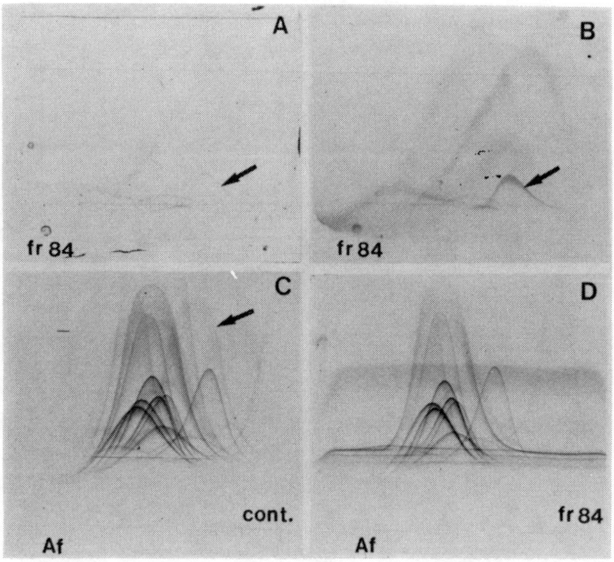

Fig. 4. A & B. XRIE of fraction 84 from the gel filtration of *A. fumigatus* protein precipitate (see Fig. 3).

Note: Ag 3 peak indicated by the arrow.

C & D. Crossed-line immunoelectrophoresis of *A. fumigatus* antigen (Af) C control and D with fraction 84 incorporated into an intermediate gel.

Note: In D "absorption", ie. removal of the Ag 3 peak indicated in C.

this fraction in an intermediate gel, the peak corresponding to Ag 3 was completely removed - absorption of antibodies *in situ*.

Further studies showed Ag 3 to be heat labile, non-Con A binding (therefore lacking α-D-glucoside or mannoside terminal residues) (Fig. 5) and on SDS-PAGE to have a MW of 18 kD (Fig. 6). On immunoblotting using a patient's serum known to have high IgE to Ag 3 this 18 kD component also showed strong uptake of IgE (Fig. 7).

From its electrophoretic mobility Ag 3 is probably identical with antigen 10, a major allergen identified in a XRIE study of various strains of *A. fumigatus* antigens (Wallenbeck *et al.*, 1984) and the component shown to have a MW of 18 kD by electroblotting (Bengtsson *et al.*, 1986).

Analysis of the initial peak from the Sephacryl S200 gel filtration (Fig. 3) showed the presence of one main and several minor antigenic components. By incorporating a monospecific antiserum produced to the main component (by

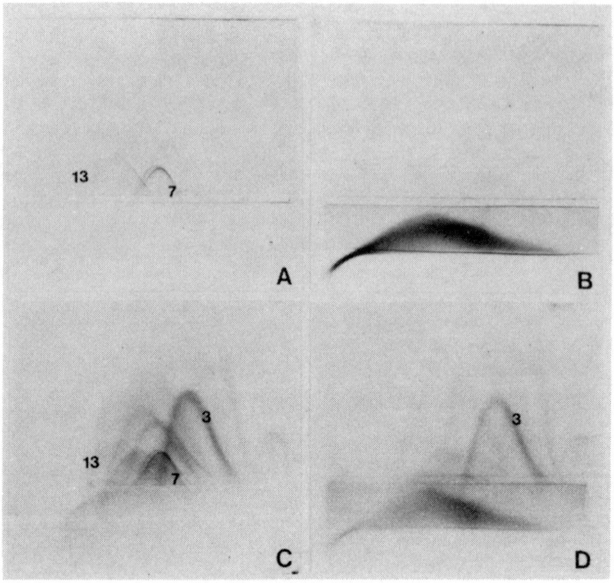

Fig. 5. A & C. Reference ABPA self-crossed immunoelectrophoresis pattern and B & D with an intermediate gel containing Con A.
(A & B - protein stained plates; C & D - corresponding autoradiographs.)

Note: Ag 3 is not affected whereas Ag 7 (and Ag 13) are completely absorbed by the Con A.

immunization of rabbits with excised immunoprecipitates) into an intermediate gel, this component was identified as Ag 7 of the reference ABPA pattern (Harvey and Longbottom, 1986), ie. not only a major antigen (precipitin) but a component also able to bind considerable amounts of IgE antibody.

Ag 7 proved to be a heat-stable acidic glycoprotein with α-D manno or α-D-glucopyrannoside terminal residues (Con A binding) (Fig. 5). It appeared to exist naturally with a MW 150-200 kD, but composed of 36 kD subunits held together by weak covalent forces, ie. dissociated by SDS (Fig. 6). Ag 7 was present in greater amounts in mycelial rather than culture filtrate extracts and is presumed to be a cytoplasmic rather than metabolic component due to its late appearance in culture filtrate extracts (Harvey and Longbottom, 1987) and the inability to demonstrate binding of the Ag 7 specific antiserum to *A. fumigatus* spores or germlings in indirect fluorescent antibody tests would also indicate it is not present as a surface component (Taylor and Longbottom, 1988). Skin prick tests with affinity purified Ag 7 at the same concentration as in the crude extract, confirmed its allergenicity for 8/10 ABPA patients so tested.

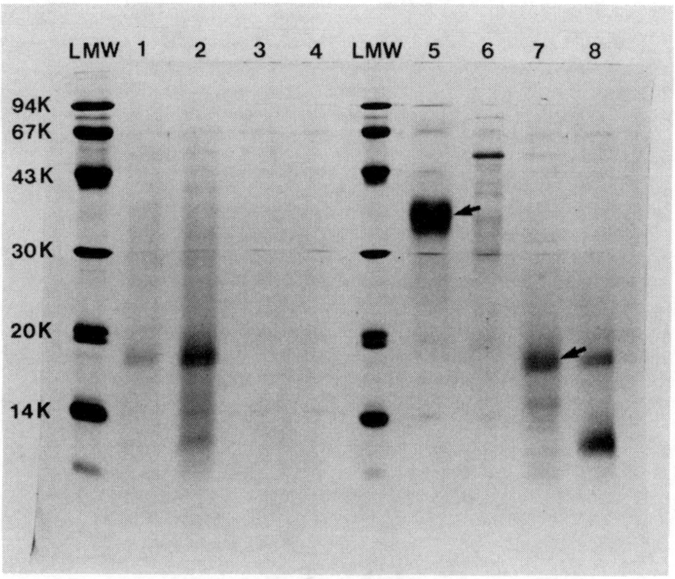

Fig. 6. SDS-PAGE (15% gel) of *A. fumigatus* antigenic fractions obtained by gel filtration (see Fig. 3): 1, crude (Af) Antigen; 2, Af salt precipitated protein; 3 & 4, Af salt supernatant; 5, fractions 54-86; 6, fractions 70-73; 7, fractions 81-87; 8, fractions 93-100; LMW, low molecular weight markers.

Note: 36 kD component, ie. Ag 7 in 5 (initial eluted peak MW 150-200 kD) (indicated by arrow). Major component with MW 18 kD in 7 (indicated by arrow).

Thus, using combinations of biochemical, immunochemical and separation techniques, two major allergenic components Ag 3 and Ag 7 (also a dominant precipitinogen) have been characterized from the mosaic of antigenic/allergenic components present in extracts of *A. fumigatus*. This study has been used to illustrate methodology applicable not only to *A. fumigatus* but to any allergenic source material for identification and characterization of 'major' allergens - such components being required before further molecular biological investigations can be carried out.

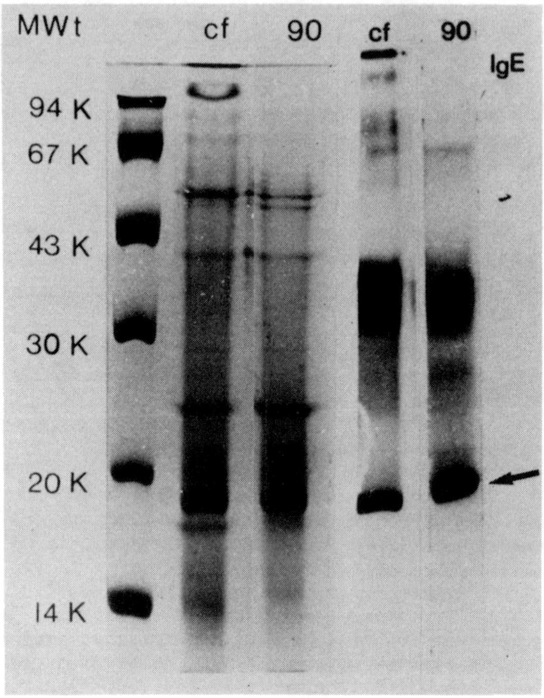

Fig. 7. IgE immunoblot analysis of *A. fumigatus* culture filtrate antigen (CF) and fraction '90' containing Ag 3. LHS protein stained SDS-PAGE separated components together with MW markers and RHS the autoradiograph obtained after 'blotting' onto nitrocellulose, and subsequently incubated with serum of ABPA patient with high IgE antibody titre to Ag 3, followed by ^{125}I-anti IgE.

Note: Intense radioactive uptake on component MW 18 kD.

REFERENCES

Aukrust, L. (1979). Crossed radioimmunoelectrophoretic studies of distinct allergens in two extracts of *Cladosporium herbarum*. Int. Archs. Allergy Appl. Immunol., 58, 375-390.

Aukrust, L. and Borch, S.M. (1979). Partial purification and characterization of two *Cladosporium herbarum* allergens. Int. Archs. Allergy Appl. Immunol., 60, 68-79.

Bengtsson, A., Karlsson, A., Rolfson, W. and Einarsson, R. (1986). Detection of allergens in mould and mite preparations by a nitrocellulose electroblotting technique. Int. Archs. Allergy Appl. Immunol., 80, 383-390.

Budd, T.W., Kuo, C.Y., Yoo, T.J., McKenna, W.R. and Cazin, J. (1983). Antigens of *Alternaria* 1. Isolation and partial characterization of a basic peptide allergen. J. Allergy Clin. Immunol., 71, 277-282.

Burge, H. (1985). Fungus Allergens. Clin. Rev. Allergy, 3, 319-329.

Bush, R.K. and Voss, M.J. (1983). Detection of *Alternaria* allergens by crossed radioimmunoelectrophoresis. J. Allergy Clin. Immunol., 71, 239-244.

Harries, M.G., Lacey, J., Tee, R.D., Cayley, G.R. and Newman-Taylor, A.J. (1985). *Didymella exitalis* and late summer asthma. Lancet, i, 1063-1066.

Harvey, C. and Longbottom, J.L. (1986). Characterization of a major antigenic component of *Aspergillus fumigatus*. Clin. Exp. Immunol., 65, 206-214.

Harvey, C. and Longbottom, J.L. (1987a). Release of antigens and allergens during shake-culture of *Aspergillus fumigatus*. Allergy, 42, 359-365.

Harvey, C. and Longbottom, J.L. (1987b). Characterization of a second major antigen Ag 13 (C antigen) of *Aspergillus fumigatus* and investigation of its immunological reactivity. Clin. Exp. Immunol., 70, 247-254.

Hoffman, D.R., Kozak, P.P., Gillman, S.A., Cummins, L.H. and Gallup, J. (1981). Isolation of spore specific allergens from *Alternaria*. Ann. Allergy, 46, 310-316.

Hasnain, S.M., Wilson, J.D., Newhook, F.J. and Segedin, B.P. (1985). Allergy to basidiospores: immunologic studies. N.Z. Med. J., 98, 393-396.

Lehrer, S.B., Lopez, M., Butcher, B.T., Olson, J., Reed, M. and Salvaggio, J.E. (1986). Basidiomycete mycelia and spore-allergen extracts: skin test reactivity in adults with symptoms of respiratory allergy. J. Allergy Clin. Immunol., 78, 478-485.

Longbottom, J.L. (1983). Allergic bronchopulmonary aspergillosis. Reactivity of IgE and IgG antibodies with antigenic components of *Aspergillus fumigatus* (IgE/IgG antigen complexes). J. Allergy Clin. Immunol., 72, 668-675.

Longbottom, J.L. (1986). Antigens and allergens of *Aspergillus fumigatus*. II. Their further identification and partial characterization of a major allergen (Ag 3). J. Allergy Clin. Immunol., 78, 18-24.

Longbottom, J.L. and Austwick, P.K.C. (1986). Antigens and allergens of *Aspergillus fumigatus*. I. Characterization by quantitative immunoelectrophoretic techniques. J. Allergy Clin. Immunol., 78, 9-17.

Lowenstein, H., Aukrust, A. and Gravensen, S. (1977). *Cladosporium herbarum* extract characterized by means of quantitative immunoelectrophoretic methods with special attention to immediate type allergy. Int. Archs. Allergy Appl. Immunol., 55, 1-12.

Nyholm, L., Lowenstein, H. and Yunginger, J.W. (1983). Immunochemical partial identity between two independently identified isolated major allergens from *Alternaria alternata* (Alt-I and Ag 1). J. Allergy Clin. Immunol., 71, 461-467.

Santilli, J., Rockwell, W.J. and Collins, R.P. (1985). The significance of the spores of the Basidiomycetes (mushrooms and their allies) in bronchial asthma and allergic rhinitis. Ann. Allergy, 55, 469-471.

Schumacher, M.J. and Jeffery, S.E. (1976). Variability of *Alternaria alternata*: Biochemical and immunological characteristics of culture filtrates from seven isolates. J. Allergy Clin. Immunol., 58, 263-277.

Swärd-Nordmo, M., Wold, J.K., Paulsen, B.S. and Aukrust, L. (1985). Purification and partial characterization of the allergen Ag 54 from *Cladosporium herbarum*. Int. Archs. Allergy Appl. Immunol., 78, 249-255.

Taylor, L.M. and Longbottom, J.L. (1988). Partial characterization of a rapidly released antigenic/allergenic component (Ag 5) of *Aspergillus fumigatus*. J. Allergy Clin. Immunol., 81, 548-557.

Wallenbeck, I., Aukrust, L. and Einarsson, R. (1984). Antigenic variability of different strains of *Aspergillus fumigatus*. Int. Archs. Allergy Appl. Immunol., 73, 166-172.

Weeke, B. and Lowenstein, H. (1973). Allergens identified in crossed radioimmunoelectrophoresis. Scand. J. Immunol., 2, 149-154.

Yunginger, J.W., Jones, R.T., Nesheim, M.E. and Geller, M. (1980). Studies on *Alternaria* allergens III. Isolation of a major allergenic fraction (Alt-I). J. Allergy Clin. Immunol., 66, 138-147.

Discussion: Session One

Chairman: Jack Pepys

Jack Pepys. Are there any questions relating to the mite allergens?

Robert Hamburger. Since the very first presentation this morning, I have been thinking about the uses of the Western blotting technique. Human anti-serum, as you indicate, may exhibit a response to eight or ten or even thirty different epitopes either on one or several distinct antigens. How many of these seemingly different sized and different antigens are in fact cross-reacting, because they have either the same or similar epitopes to the original antigen? I was going to ask if you had done any absorption studies, but then I saw those which Dr Tovey presented, and now I wonder if there is another interpretation to those absorptions. Instead of a low amount of IgE antibody to the ones that disappeared, could they not indicate the existence of cross-reacting or identical epitopes on different parts of protein-sized structures?

Brian Baldo. I think that is a very good question. It is quite likely that if, for example, we detect thirty IgE-binding proteins, some ten or twenty of those will share cross-reacting determinants. We have not really had much of an opportunity yet to look at this, although together with Dr Elsayed in Bergen we have made monoclonal antibodies to codfish allergen M, which most people would agree is the best defined allergen so far. The literature tells us that allergen M is the only codfish allergen that matters, but Western blots tell a different story: if you look at an extract of cod you can find IgE binding proteins everywhere. It is clear although allergen M has its own unique antigenic and allergenic determinants it also has cross-reacting determinants because if you elute the antibodies off allergen M, they will bind to some of the other cod allergenic proteins. Therefore some of the bands that we detect are probably reacting to the same population of IgE molecules.

Robert Hamburger. The reason I ask is that some years ago we tried fractionating some Bermuda grass pollen extracts, then refractionated the major peaks only to see them split into the peaks that we thought we had seen in the first place. I wondered how many were repeating examples from the same original protein molecule.

Brian Baldo. Again I think it is the problem that with antigens, because

Advances in the Biosciences Vol. 74

of the classic work of Heidelberger, Kabat and others, we know a lot about the size of antigens which can be accommodated in their complementary antibody sites, but when it comes to IgE binding we know very little.

Euan Tovey. We have compared the binding of individual sera to the clone and to the component on blots of the mite extract, and in general have got a good correlation. But there do seem to be odd exceptions. We're not sure why this is, on blots it's possible that two components may sit one behind the other. We think that cloned determinants may be a better way of analysing the specific IgE serum, then at least you know that you are dealing with a single defined IgE binding determinant.

Alec Sehon. I think that it is appropriate to remind everyone interested in pure allergens that twenty-nine years ago, right around the corner from here, i.e., at Caltech in Pasadena, the late Professor Dan Campbell had the initiative to establish the first N.I.H. Committee on Standardization of Allergens. I am glad to see here Dr T.P. King who was a member of that Committee. I believe that the answer to Dr Hamburger's questions is documented in the literature also with respect to other pollen allergens. Thus, for example, we found many years ago that the average sedimentation coefficient of the crude aqueous extract of ragweed pollen dropped from about 6S to 3.5S within about 30-40 minutes at room temperature. We interpreted this to be due to enzymatic degradation of the high molecular weight constituents in the pollen extract. After all, the pollen is essentially a "living" cell and contains a variety of compounds including enzymes. I suggest that the answer to the question as to whether or not the epitopes, which are detected on different bands, are identical, will be resolved once the different allergenic constituents of a given pollen are produced by recombinant DNA technology. Furthermore, the allergenic products synthesized by this new methodology are expected to be stable in the absence of added enzymes, unless the allergens happen to be enzymes with auto-catalytic activity.

Larry Schwartz. I would agree with what has just been said and in particular we need to wait for the sequencing of these cDNA molecules so that the amino acid sequence can be correlated with the antigenic determinant. Then it will be clear, without a shadow of a doubt, which epitopes are associated with which antigenic fragments and which peptide fragments are truly not only cross-reactive but cross-identical. A simpler question that I have is: 'Have any localisation studies of these numerous molecules being recognised by IgE been done? If so, what organ of the mite, or what excretory product of the mite may they be derived from, and does that give us any indication of their potency as a clinically relevant allergen?'

Brian Baldo. Euan Tovey and I have spent much time in trying to answer this very question and in fact part of his thesis covers this. The problem is in making sections of the mite: once you try and make thin sections it is difficult to preserve any structural integrity. However, you can make frozen sections with reasonable clarity, and we have some quite good pictures of probing frozen sections with what is said to be anti-*Der p I* which clearly show localisation to salivary parts, the gut, and faecal material. But in some other cases it seems that the cuticle binds and we are not really sure of the significance of that. We tried to make good thin sections embedded in methacrylate, JB4 and those sorts of embedding media, but the definition wasn't anywhere near as good as we would have liked. There's much that could be still done to clarify this question.

Sheldon Spector. I'm wondering about the clinical implication of this epitope heterogeneity. The other day Jack Pepys and I were discussing the number of deaths in England associated with mite immunotherapy. Could the explanation for some of the problems that patients have run into be related to differences in the mixtures and the heterogeneity of the mixtures that have been given?

Brian Baldo. I think it is most important to emphasise the heterogeneity in the immune response. Until we started our blotting studies to define allergens, we were unaware just how important the unique sensitivity spectrum was. From some of the slides we showed it's clear that while some people react quite strongly to maybe eight or nine components, others react to only one or two. If the patient who reacts to eight or nine components is being diagnosed or treated with an extract which lacks one or more of these components, then it seems to be that there is only a remote possibility that the outcome will be successful. Put another way, if you wanted to place any confidence in the treatment you would have to know what is in the extract and what the unique allergen sensitivity or recognition spectrum of that patient is. There can be very few clinicians who practise this way, and perhaps that is why in some cases it has been reported to work and in others it hasn't.

Martin Chapman. It seems to me that you have essentially confirmed what you were criticising in your opening remarks. You spend a lot of time talking about *Der p I* and *Der p II*, yet the only way you are able to do this is because these allergens have been purified by several different people and also that specific monoclonal antibodies have been made to them. That's just by way of a comment; of course, there may well be other mite allergens, but it is really incumbent on people to actually purify those allergens and demonstrate their activity in a variety of biological and IgE-binding systems. The other comment I'd like to make concerns the use of electroblotting for standardisation: you say there are thirty or forty different allergens of varying degress of importance, so how are you going to use that technique at a practical level, when Bob Esch outlined some of his company's problems with standardisation. Suppose for instance a mite extract doesn't contain all of your bands, are you going to tell the manufacturer you cannot use this extract because it doesn't contain antigens x, y and z; antigen bands defined by an *in vitro* system whereas the extract may have very potent skin test reactivitiy. The other aspect of the technique is that it is not quantitative, and so you cannot actually convey to the manufacturer methods by which he can then go ahead and improve his product. I think really we need answers to these questions before we can really look to blotting as a serious alternative to some of the other techniques that are available at present.

Brian Baldo. Don't get me wrong, I think that the groups that have purified components and made monoclonal antibodies to them deserve nothing but praise and of course the more of that we do the better. The quicker we identify the allergens, purify them, and prepare reagents that can detect them quantitatively, the more I am pleased, and I don't think I said anything to disagree with that. As far as using blotting to define extracts, it seems to me already clear that some of the major houses that prepare extracts (e.g. Pharmacia) have decided that that's what they want to use. I would be the last person to claim blotting is quantitative, but that does not detract from the ability of the method to provide a qualitative definition - this is what I have been talking about here. What we have to do is to define what is there at a qualitative level, then develop reagents which can then quantify those allergens which we have decided are important. I don't see any inconsistency.

Martin Chapman. I'm just commenting on how you are going to use this technique on a practical basis.

Brian Baldo. On the contrary, I think it's a very practical method, more practical than those already advocated. I mentioned CRIE, the problem with CRIE is that its a damned hard method to standardise; we have pored over crossed IEP's prepared in different labs to the same extract, and to identify with confidence the same band prepared with different antisera is a very, very difficult task. It should be possible with a technique like blotting to eventually arrive at a definition which we all agree on, and I don't see that as a problem. I don't think the companies that have been using it see it as a problem either.

Sola Alaba. I'd like to address the question of using the Western blot that the last speaker commented on. We use the Western blot routinely to qualitatively assess allergenic extracts, and we have done a lot of work to quantify the system. Although it is very difficult, it is possible to standardise each part of the procedure, the blotting time, immunoreaction time etc. We have got all this down to the point where we can be certain that an extract contains similar amounts from preparation to preparation, and use densitometric measurements to quantify to nanogram levels of particular molecules in each extract.

Jack Pepys. Are there any questions for Dr Longbottom?

Alec Sehon. I should like to bring out some limitations of *in vitro* techniques as they apply to the diagnosis of allergy of patients. There is no question that such procedures as immunoblotting identify the main allergens. However, the *in vitro* results may sometimes differ from the clinician's observations, because *in vitro* techniques require intensive washing in order to eliminate, or at least minimize, non-specific binding. Consequently, IgE of low affinity will be washed away and, therefore, will go undetected. By contrast, *in vivo* there is no "washing" and IgE antibodies attached to mast cells and basophiles bind the allergen which is in equilibrium with the allergen in tissue fluids or in blood; consequently these cells may be triggered even by low affinity IgE antibodies. To illustrate my point I shall refer to some of our experience with one mouse monoclonal antibody of the IgE isotype directed against ovalbumin. This monoclonal antibody would not have been detected by *in vitro* tests because of its extremely low affinity for ovalbumin. For example, it did not remain attached to an ovalbumin immunosorbent during the washing step. However, after sensitizing the skin of mouse or rat with this antibody, on challenge of the animal with ovalbumin we detected a strong PCA reaction; of course, in this particular case, since we used monoclonal IgE antibodies we had to challenge the animals with a polymerized form of ovalbumin. In other words, had we not resorted to the *in vivo* PCA technique, we would have never detected this IgE antibody. Clearly, this example emphasizes the difference of the reactivity of some IgE antibodies *in vitro* and *in vivo*, and that for low affinity IgE antibodies the clinician's skin test would be diagnostically more meaningful than *in vitro* tests which would give a false negative.

Joan Longbottom. I take your point, but I can only show you the antibodies that have bound, and add the proviso that the binding of IgE and mouse monoclonals may be quite different.

Alec Sehon. I made the statement so as to underscore the fact that clinicians who do not have ready access to a laboratory should not take as gospel truth every result obtained by an *in vitro* test. For this reason I believe that a dramatic result seen by the prick test may occasionally be missed by an *in vitro* test, particularly if the IgE antibody has a low affinity for the allergen.

John Warner. Moulds have always seemed to me to be an enigma because, as you showed from the measurements made in your own bedroom, we are exposed to fairly high concentrations of these respirable particles, and yet clinically they seem to affect a minority of patients by comparison with housemite or pollen. As you have shown that some of the principal components are coming from the mycelia, does this mean perhaps, that in cases of allergic bronchopulmonary aspergillosis, this must be growing in the airways?

Joan Longbottom. That may be so, but I really couldn't say with certainty that all components are in the mycelia; certainly there is limited growth in the airways of ABPA patients. However, we have, I must admit, also found component antigen 5, which is released from the spores within two to four hours, so early release components may also spell trouble.

Robert Hamburger. Thank you for your elegant presentation. Putting on my clinicans hat I must say that, fortunately for us, we don't have to have every single allergen or antigen to which the patient responds to get some positive results from immunotherapy. If we did we would have had zero success over the last fifty years.

Session Two

MOLECULAR BIOLOGY

Chairman: J. Watson

Toward a Total Human Immune Response Fingerprint: the Allergy Model

David G. Marsh, Patty Zwollo and Aftab A. Ansari

Johns Hopkins University School of Medicine, Good Samaritan Hospital, Baltimore, MD 21239, USA

ABSTRACT

Atopic allergy provides a good model for studying the genetics of human immune responsiveness and has the potential for "fingerprinting" responsiveness to a wide variety of antigens. Immune response fingerprints may then be related to appropriate genetic markers involved in the immune response, including those of the *HLA* complex. Approaches toward defining such fingerprints are described using two model systems, the *Lol p* I, II and III allergens where responsiveness is associated with DR3, and the *Amb* V allergens where responsiveness is associated with DR2/Dw2. Emphasis is placed on the use of a new technology for gene amplification, the polymerase chain reaction, which has been used to amplify *HLA-D* region gene sequences of *Amb a* V responders and non-responders. Specific hypervariable *HLA-D* gene sequences implicated in encoding the Ag-binding sites (for the Ia/T-cell epitopes of allergens) are then identified either by dot-blotting or dideoxy sequencing. Future studies will be directed toward the T-cell receptor (TcR) and relating particular TcR sequences involved in the recognition of specific Ia/T-cell epitopes on the Ag (allergen) molecules. These approaches will be highly relevant in elucidating the genetics of human immune responsiveness and susceptibility to immunologic diseases.

KEYWORDS

Allergen; HLA; *Ir* gene; immune response; PCR; gene cloning; DNA sequencing.

ABBREVIATIONS

Ab, antibody; Ag, antigen; bp, base pair; DARIA, double-antibody radioimmunoassay; Ia, immune-associated; Ir, immune response; HLA, human leukocyte antigen; MHC, major histocompatibility complex; MLR, mixed lymphocyte reaction; PCR, polymerase chain reaction; RFLP, restriction fragment length polymorphism; Rx, immunotherapy; SSO, sequence-specific oligonucleotide; TcR, T-cell receptor.

Advances in the Biosciences Vol. 74

THE ALLERGY MODEL

We postulated that atopic allergy would provide a good model for studying the genetics of human immune responsiveness (Marsh *et al.*, 1973) and have elaborated on this model in subsequent publications (e.g. Marsh 1975, 1986, 1988b). The important features of the model are: 1. Allergy is a common disease which afflicts about 20% of Caucasoid populations and usually somewhat greater proportions of non-Caucasoids (Friedhoff, 1988). 2. It has long been known that allergy exhibits marked familial aggregation (Wyman, 1872; Cooke and Vander Veer, 1916; Wiener *et al.*, 1936; Schwartz, 1952; Marsh, 1975; Black and Marsh, 1976; Marsh, 1988b). It has been presumed, although not definitively proved, that the primary basis for this aggregation is genetic. 3. Many allergens have now been isolated in highly purified form (Marsh *et al.*, 1986; Marsh, 1988a), which facilitates studies of specific immune responses. 4. Because there are so many more allergens that could relatively easily be purified, the allergy model offers a way of "fingerprinting" human immune responsiveness to a wide variety of macromolecules and relating patterns of specific responsiveness to the presence of particular patterns of defined genetic markers, especially those of the HLA system. Unlike workers in the field of allergen standardization, researchers studying the genetics of human immune responsiveness welcome the complexity and the number of different allergens implicated in atopic allergy. 5. A most important feature of the model is that natural exposure is usually toward exquisitely limiting low Ag doses. This is especially the case for most inhaled allergens where, characteristically, the annual doses are less than 1 μg and hourly doses are under 1 ng (Marsh, 1975). 6. Beside measuring IgE antibodies (Abs) and IgE-mediated sensitivities, one can study IgG Abs and other classes of humoral antibodies, as well as specific T-cell immunity. 7. Finally, one can study gene interaction, particularly between *HLA*-linked[1] immune response (*Ir*) genes and non-*HLA*-linked, IgE-regulating genes (Marsh, 1975; Willcox and Marsh, 1978).

Table 1 summarizes some of the known properties of highly purified pollen allergens. From this information and sequence data on these and other allergens, one finds that there is no unique property which distinguishes allergens as a unique subset of Ags (cf, Marsh, 1975). Allergens cover a wide range of relative molecular masses (from 4400 for *Amb t* V to 57,000 for *Lol p* IV in the cited examples) and isoelectric points (pI 3.9 to >11 in the examples). From the known properties of these and other allergens, no particular type of amino acid sequence or prosthetic group has been found that separates allergens from other Ags. Thus, it is appropriate to use atopic allergens as a representative array of Ags having many different molecular structures. We will illustrate our approach in using allergens for "human Ir fingerprinting" by two model systems, rye grass *Lol p* I, II and III and the *Amb* V homologues.

HLA-IR ASSOCIATIONS — STATISTICAL STUDIES

Table 2 summarizes the major, statistically significant associations that we have previously reported, among *atopic* individuals, between particular HLA types and the ability of individuals possessing these phenotypes to make specific immune responses to highly purified allergens (see Marsh and Bias, 1988 and Marsh, 1988b for further discussion). Most of the cited data

---

[1] Italics are being used for the designation of genes and normal type for the designation of polypeptide gene products.

Table 1. Properties of some highly purified pollen allergens

Source	Systematic name	Original name(s)	Rel. molr. mass, M_r	No. amino acid residues	pI
Poaceae (Gramineae; grasses)					
Lolium perenne (perennial rye)	*Lol p* I	Group I (Rye I)	B: 27,000[a]	233	5.9[a]
	Lol p II	Group II (Rye II)	B: 11,000[a]	---	5.1[a]
	Lol p III	Group III (Rye III)	B: 11,000[a]	---	9.0[a]
	Lol p IV	HMBA (Gp IV?)	56,800	508	9.9
	Lol p X	Cytochrome c	12,000	112	~10
Phleum pratense (timothy)	*Phl p* V	Ag25	15,000	---	4.5
	Phl p VI	Ag19	15,000	---	4.5
	Phl p VII	Ag30	34,000	300	9.4
	Phl p VIII	Ag3	~8,000	82	3.9
Dactylis glomerata (orchard)	*Dac g* I	Dg1	33,000	---	5.9
Poa pratensis (Kentucky blue/june)	*Poa p* X	Cytochrome c	12,000	112	9.9
Ambrosiinae (ragweeds)					
Ambrosia artemisiifolia (elatior) (short ragweed)	*Amb a* I	AgE	B: 37,800[a]	337	5.1[a]
	Amb a II	AgK	A: 38,200[a]	331	5.9[a]
	Amb a III	Ra3I	A: 12,300[a,b]	101[b]	8.6[a]
	Amb a IV	Ra4 (BPA-R)	22,800	189	8.0
	Amb a V	Ra5	A: 4,990[a,b,c]	45[b]	9.6[a]
	Amb a VI	Ra6	A: 11,500[a,c]	108	>11[a]
Ambrosia trifida (giant ragweed)	*Amb t* V	Ra5G	4,390[a,b]	40[b]	8.3[a]
Chenopodiaceae (goosefoots)					
Salsola pestifer (Russian thistle)	*Sal p* I	RT_1	A: 39,000	359[d]	6.7
Betulaceae (birches)					
Alnus incana (alder)	*Aln i* I	AI4	22,500	138[e]	4.8
Betula verrucosa (silver birch)	*Bet v* I	Ag23 (BV45?)	17,000	---	5.25
Corylus avellana (hazel)	*Cor a* I	HIa	13,500	124[e]	5.4
Taxodiaceae (bald cypresses)					
Cryptomeria japonica (Japanese cedar)	*Cry j* I	$SBPa_1$	40,000	350[e]	9.2

[a]M_r and pI of 'A' or 'B' isoallergenic variant indicated. Other forms have similar M_r and pI values.

[b]From sequence data (allowing for any carbohydrate present in the case of the M_r value). Other M_r values and numbers of amino acid residues were computed from physicochemical analysis or amino acid compositions.

[c]Each of these components may exist in two isoallergenic forms having slightly different sequences. The apparent M_r values cited were determined by gel filtration; lower apparent M_r values of 8000 were found by SDS–PAGE.

[d]Data exclude analyses of half-cystine and tryptophan residues.

[e]Data exclude analysis of tryptophan residues.

Adapted from Marsh (1988a).

Table 2. Non-parametric statistical analysis of HLA associations with specific antibody responsiveness toward some highly purified pollen allergens in Caucasoid subjects

Systematic Name*	Old Name	Rel. molr. mass, M_r	Primary Association	p values† 1st study	2nd study	Overall
***Ambrosia artemisiifolia* and *Ambrosia trifida* (short and giant ragweeds):**						
Amb a I‡	AgE	37,800	None	–	ND	–
Amb a III	Ra3	12,300	A2	4×10^{-2}	NS	–
Amb a V	Ra5	5,000	DR2/Dw2	2×10^{-5}¶	4×10^{-10}¶	$<10^{-9}$¶
Amb a VI	Ra6	11,500	DR5	7×10^{-7}	1×10^{-3}	$<10^{-6}$
Amb t V	Ra5G	4,400	DR2/Dw2	1×10^{-2}¶	3×10^{-2}¶	$<10^{-3}$¶
***Lolium perenne* (perennial rye):**						
Lol p I‡	Rye I	27,000	DR3	1×10^{-2}	6×10^{-3}	$<10^{-3}$
Lol p II	Rye II	11,000	DR3	4×10^{-2}	7×10^{-3}	10^{-3}
Lol p III	Rye III	11,000	DR3	7×10^{-3}	1×10^{-5}	$<10^{-5}$
Lol p X	Cytochrome c	12,000	None	–	–	–

* WHO Allergen Nomenclature (Marsh *et al.*, 1986).

† p values are for studies of Ab responses resulting primarily from natural allergenic exposure. IgE Ab data were used in most cases; IgE-mediated skin-test data were used for *Amb a* I and *Amb a* III and IgG Ab data were used where there were an insufficient number of IgE Ab+ subjects (*Amb t* V and *Lol p* X). Two-tail Fisher's Exact Tests have been used for analysis of the first study and one-tail tests for the second study.

‡ Highly purified allergens (≥97% pure); all other allergen preparations estimated to be ≥99.5% pure.

¶ Data from Marsh *et al.* (1982a) and Roebber *et al.* (1986). The stronger associations (for which p values are cited) were found with the Dw2 MLR-defined specificity found in *ca* 90% of DR2+ Caucasoid individuals.

relate to specific serum IgE Ab positivity, but in all cases IgG Abs have also been found to be similarly associated with the particular HLA type. (This is not altogether surprising since specific serum log [IgE Ab] and log [IgG Ab] levels are significantly correlated in allergic subjects; Marsh and Bias, 1988.) For measurement of specific responsiveness to allergens, we prefer to use quantitative double-Ab radioimmunoassay (DARIA) of serum Abs (Marsh *et al.*, 1982a). Measurement of IgE-mediated sensitivity *in vivo* by skin testing or leukocyte histamine release assay, or semi-quantitative measurement of serum Abs by crossed radioimmunoelectrophoersis (CRIE), Western blotting or IEF immunoprinting (Roebber, 1988) has proved to be less satisfactory. Before considering the HLA-association data further, it is necessary to discuss both the genetics of the *HLA* complex and the molecular basis of Ag presentation to provide a background for the molecular biological studies.

THE *HLA* COMPLEX

Figure 1A shows a diagram of the *D* region of *HLA*, the human major histocompatibility complex (MHC), which is located on the short arm of chromosome

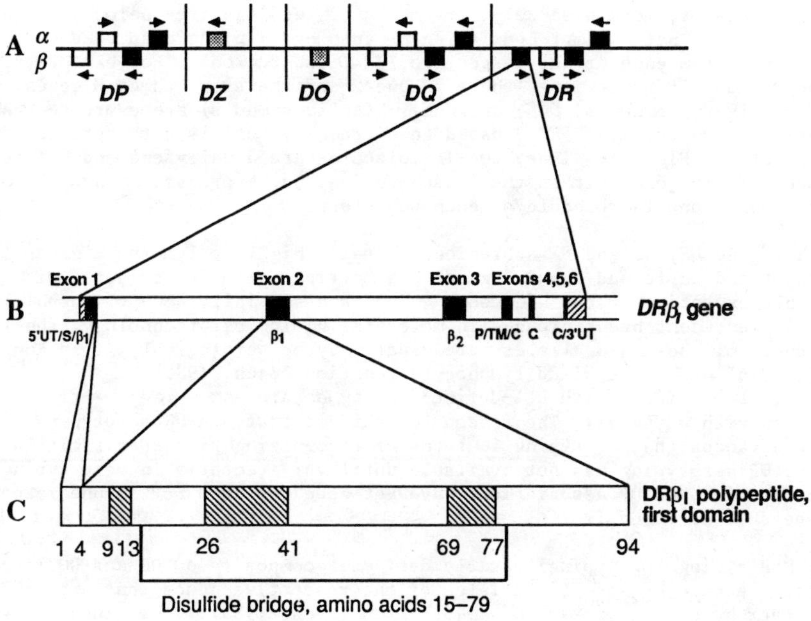

Fig. 1. *A*, The human *HLA-D* region and its sub-regions. The genes shown by solid boxes are expressed as protein products; those shown by open boxes are not expressed. The genes shown by stippled boxes are weakly expressed and/or produce inactive products. Arrows indicate the direction of gene transcription from the 5' to the 3' end of one or other anti-parallel DNA strands. The class II *Ir* genes include both α genes (upper boxes) and β genes (lower boxes). The number of unexpressed β "pseudo-genes" within the *DR* sub-region varies somewhat among the different *DR* subtypes, but most have two expressed β genes, $DR\beta_I$ and $DR\beta_{III}$.
B, the arrangement of exons (which produce RNA transcripts) and introns (non-expressed regions) of the $DR\beta_I$ gene. Untranslated regions and encoded peptides of the $DR\beta_I$ chain are indicated as follows: 5' UT, the untranslated region located at the 5' end of the gene; S, signal (or leader) peptide (which is removed upon formation of the mature $DR\beta_I$ polypeptide); β1 and β2, the two extracellular domains of the $DR\beta_I$, polypeptide; P, the connecting peptide between the extracellular β2 domain and the trans-membrane (TM) peptide; C, the cytoplasmic peptide; 3' UT, the untranslated region located at the 3' end of the gene. The other *HLA-D* region genes have exon arrangements quite similar to $DR\beta_I$.
C, a schematic representation of the first domain of the $DR\beta_I$ polypeptide, emphasizing the hypervariable regions (shaded), which are involved in binding to the Ag fragment and the TcR (*cf* Fig. 2). Other expressed *HLA-D* genes have similar hypervariable regions, except *DR*α which is essentially invariant.
[Data from Trowsdale *et al.* (1985), Hardy *et al.*(1986), Andersson *et al.* (1987) and Jonsson *et al.* (1987); diagram adapted from Marsh (1988b)]

6. We will focus on the *D* region, which encodes the MHC class II (or Ia) specificities. *HLA-D* contains several subregions, of which *DR*, *DQ* and *DP* encode the Ia molecules primarily involved in the immune recognition of soluble Ags or, more precisely, fragments of such Ags (see below). Each Ia molecule is a heterodimer consisting of an α and a β chain encoded by the α-β genes from each of the respective *HLA-D* sub-regions. For example, in the case of most *DR* subregions, there is one α and there are three β genes, of which only two β genes, $DR\beta_{I}$ and $DR\beta_{III}$ (solid symbols) are expressed as peptide products; the $DR\beta_{II}$ pseudogene (open symbol) is not expressed (footnote to Fig. 1). Thus, two Ia molecules are usually encoded by the expressed α-β genes within the *DR* sub-region, but expressed *DQ* and *DP* genes encode only one Ia molecule of each sub-class.

Each of the *DR*, *DQ* and *DP* subregions contains highly polymorphic genes (i.e. each of the expressed α and β genes has multiple allelic variants, with the notable exception of the *DR*α gene, which is essentially non-polymorphic). Until the recent breakthroughs in molecular biological technology, the Ia products of these α-β alleles were usually typed serologically (van Rood *et al.*, 1976) or by the mixed lymphocyte reaction (Bach, 1985). The Ir associations with HLA-defined *serotypes* are stronger with DR than with DQ types. The reason for this is that HLA-DR serotyping is more advanced than DQ typing in terms of recognizing fine specificities. Since DP serotyping was not available until very recently in most typing laboratories, no DP associations have yet been reported for immune responsiveness to allergens.

The MLR typing (Dw typing) usually defines a composite of DR and DQ specificities encoded by specific alleles of the respective genes that are held together by strong *linkage disequilibrium* (Bach, 1985). The linkage disequilibrium between particular α and β alleles of the *DR* and *DQ* subregions means that one finds a particular *DR* sub-region associated with a particular *DQ* subregion on the same chromosome much more frequently than expected by chance. For example, the Caucasoid MLR type, Dw2 is defined by both DR2 and DQw1 proteins or, more specifically, by the sub-specificities DR2.2 and DQ1.2, that are encoded by alleles in strong linkage disequilibrium. This precludes one being able to evaluate, definitively, which of the DR or DQ Ia molecules on the haplotype is involved in the specific immune response. Conversely, *HLA-DR* and *DP* alleles are *not* in tight linkage disequilibrium (because of a recombination 'hot-spot' between them). Thus, the finding of a significant DR-Ir (or Dw-Ir) association would tend to preclude the involvement of DP.

Thus, we have focused our molecular biological studies on the specific responses to allergens which are strongly associated with DR specificities, bearing in mind that the association may be with $DR\alpha\text{-}\beta_{I}$, $DR\alpha\text{-}\beta_{III}$, or DQα-β Ia molecules. Our two model systems, rye grass *Lol p* I, II and III on the one hand and the *Amb* V allergens on the other, exhibit significant associations with DR3 and DR2/Dw2, respectively.

ANTIGEN PRESENTATION

Figure 2 summarizes the key elements concerning MHC-restricted presentation of Ag fragments to the T-cell receptor (TcR), derived from the mouse studies of Unanue, Grey, Gefter and their collaborators (Babbitt *et al.*, 1985; Unanue and Allen, 1987; Buus *et al.*, 1987; Sette *et al.*, 1987; Guillet *et al.* 1987). Data from these experiments strongly suggest that the foreign Ag is processed by an Ag-presenting cell (APC), which primarily involves proteo-

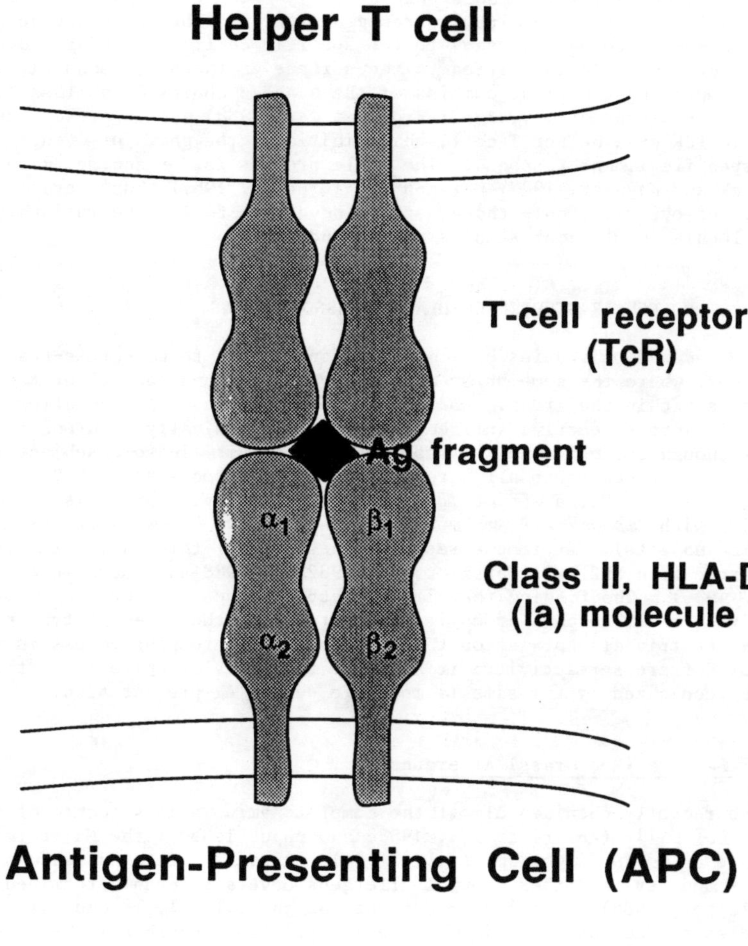

Fig. 2. The Ia —— Ag —— TcR complex: a simplified model. From Marsh (1988b).

lytic degradation. Class II- (HLA-D-) restricted Ag presentation (shown here) is the more important in the case of soluble Ags (like most allergens), whereas class I- (HLA-A, B or C-) restricted presentation is more important for particulate Ags such as viruses (Schwartz, 1985). Because of the ultra-limiting Ag doses involved in atopic allergy, the APC is likely to be a B cell, which has an immunoglobulin receptor specific for the undegraded Ag, allowing the Ag to be "focused" to the specific cell. Following processing, a fragment of the Ag is carried to the surface of the APC, bound within a cleft comprising the first domains of the α and β chains of a class II Ia molecule (Bjorkman *et al.*, 1987; Brown *et al.*, 1988) and is presented to a specific TcR on a helper T cell, which initiates the chain of events leading to a specific immune response. The whole process is, of course, more complex (Marrack and Kappler, 1987; Oettgen and Terhorst, 1987) than presented in this brief overview; only those aspects needed to follow the rationale of our molecular biological studies are included.

MOLECULAR STUDIES OF HLA—IR ASSOCIATIONS

We have focused on studies of immune responsiveness to non-cross-reacting allergens, where the same DR/Dw associations are found for all of the allergens within the group. *Lolium perenne* (rye grass) pollen allergen *Lol p* I is not cross-reactive antigenically or allergenically with *Lol p* II and III (although *Lol p* II and III show cross-reactivity in some subjects); yet immune responsiveness to all three allergens is associated with DR3 (Ansari *et al.*, 1988a,b; Friedhoff *et al.*, 1988). Similarly, *Amb t* V is non-cross-reactive with *Amb a* V and *Amb p* V[2] (*Amb a* V and *Amb p* V show cross-reactivity in some subjects); yet immune responsiveness to all three allergens is associated with DR2/Dw2 (Marsh *et al.*, 1982a,b, 1988 and unpublished; Marsh, 1986). The finding that immune responsiveness to the non-cross-reactive Ags within each group is associated with the same HLA type precludes the trivial explanation that concordance of responsiveness is merely a result of cross-reactivity; it suggests that similar epitopes on these Ags may be recognized by the same Ia molecule during Ag presentation.

Lolium Perenne (Rye Grass) Allergens

We have recently obtained almost the complete amino acid sequence of *Lol p* II and *Lol p* III (Ansari *et al.*, 1988c and unpublished); the NH_2-terminal sequence of *Lol p* I (Cottam *et al.*, 1986) and partial internal sequences of *Lol p* I and several other Group I allergens have also been determined (Esch and Klapper, 1988). Homologous segments of the *Lol p* I, II and III sequences are compared in Fig. 3. We investigated possible Ia/T-cell sites on these molecules using an analytical method devised by Berzofsky and associates (Margalit *et al.*, 1987). These authors found that 18/23 (78%) of the known T-cell/Ia sites on protein Ags that are recognized by mouse T-cell clones are amphipathic. (An amphipathic structure contains distinct hydrophobic and hydrophilic regions which, in peptides, are often alternate faces of α-helical structures.) Using the Berzofsky algorithm to analyze all of the known sequence information for *Lol p* III and II, we found that

---

[2]Trace cross-reactivity was seen in an occasional individual between *Amb a* V and *Amb t* V (Roebber *et al.*, 1985).

	56				60					65					70	
***Lol p* III:**	**F**	**L**	**S**	K	**G**	**G**	**M**	**K**	**N**	**V**	**F**	**D**	**E**	**V**	**I**	**P**
***Lol p* II:**	**F**	**V**	**S**	E	K	**G**	**M**	**R**	**N**	**V**	**F**	**D**	**D**	**V**	**V**	**P**
***Lol p* I:**	Y	T	T	E	**G**	**G**	T	**K**	S	E	**F**	**E**	**D**	**V**	**I**	**P**

Fig. 3. Sequences of homologous regions in the amino acid sequences of *Lol p* III, II and I (Ansari *et al*., 1988c; Ansari, Shenbagmurthi and Marsh, unpublished; Esch and Klapper, 1988). Residue numbering is according to a provisional *Lol p* III sequence (Ansari, Shenbagmurthi and Marshm unpublished). The *Lol p* I sequence has a variant with valine at position 66. Residues that are either identical or closely homologous to the *Lol p* III sequence are shown in bold type. For the purpose of comparison, the following residues have been considered to be equivalent: acidic residues, aspartic (D) and glutamic (E) acids, basic residues lysine (K) and arginine (R) and hydrophobic residues valine (V), leucine (L) and isoleucine (I).

the regions around residues 61-67[3] are amphipathic in both molecules. There is a similar sequence in *Lol p* I around this region (Fig. 3). With the limited amount of sequence data available on *Lol p* I, we were not able to show whether this region was amphipathic. Based on these data, we postulate that DR3 may be associated with the recognition of similar sites in or around these homologous regions in all three *Lol p* molecules. We emphasize that the amphipathicity model is not absolutely predictive of Ia/T-cell sites. Obviously, we have to perform the cellular immunological experiments in order to test these intriguing predictions.

Beside the strong DR3 associations in two groups of allergic subjects (Table 2), we have noted secondary, significant associations between responsiveness to *Lol p* III and DR5 in both study groups, as well as evidence for a weak association with DRw6 in one of these groups (Ansari *et al*., 1988a,b). There was no evidence for any DQ association in either group. We examined all the known sequences of DRβ_I, DRβ_{III}, DQα and DQβ polypeptides, particularly the highly polymorphic regions of the first domains that are encoded by the respective second exons (Figs. 1B and 1C). We found that the β_I chains of DR3, DR5 and the two known DRw6 subtypes all have the same sequence, EYSTS, at amino acid residues Nos. 9-13 (Ansari *et al*., 1988a,b). No other HLA-D polypeptide contains this sequence, which may be important in the interaction with the major Ia/T-cell site on *Lol p* III (and perhaps the other *Lol p* molecules). The relationship between these findings and the sequence homologies of the three allergens (Fig. 3) remains to be evaluated experimentally.

---

[3]Tentative residue numbering based on a provisional *Lol p* III sequence.

		1				5					10					15
*Amb a*V:		L	V	P	C	A	W	A	G	N	V	C	G	E	K	R
*Amb t*V:	D	D	G	L	C	-	Y	E	G	T	N	C	G	K	V	G

	16				20					25					30
*Amb a*V:	A	Y	C	C	S	D	P	G	R	Y	C	P	W	Q	V
*Amb t*V:	K	Y	C	C	S	P	I	G	K	Y	C	-	-	-	-

	31				35					40					45
*Amb a*V:	V	C	Y	E	S	S	E	I	C	S	K	K	C	G	(K)
*Amb t*V:	V	C	Y	D	S	K	A	I	C	N	K	N	C	T	

Fig. 4. Comparison of the amino acid sequences of the homologues, *Amb a* V and *Amb t* V. The sequences have been aligned to optimize homology, particularly with respect to the alignment of the half-cystine (C) residues. Homologous residues are shown in bold type (see also footnote to Fig. 3). The boxes enclose regions 17-26 and 31-35 which have closely similar sequences in the two *Amb* V homologues and are believed to be disulfide-bonded. See also Mole *et al.* (1975), Goodfriend *et al.* (1985) and Marsh (1986).

Amb V Allergens

The amino acid sequences of *Amb a* V from *Ambrosia artemisiifolia* (short ragweed) and *Amb t* V from *A. trifida* (giant ragweed) are known (Mole *et al.*, 1975; Goodfriend *et al.*, 1985). There is a region of 10 amino acids, residues Nos. 17-26, which shows homology in the two *Amb* V molecules (Fig. 4). Our analyses using the Berzofsky algorithm (Margalit *et al.*, 1987) shows that this region is amphipathic in both molecules, suggesting that it is a candidate for an Ia/T-cell site.[4] This is an attractive hypothesis, since *Amb a* V and *Amb t* V are virtually non-cross-reactive, but responsiveness to both molecules is significantly associated with DR2/Dw2. [See Marsh (1986) for further discussion of *Amb* V structures.] In our early studies of the *Amb* V allergens, we found that 36/38 (95%) of individuals having IgE Ab to *Amb a* V after natural exposure to short ragweed pollen had HLA-DR2$^+$/Dw2$^+$ phenotypes (Marsh *et al.*, 1982a). Coulter *et al.* (1988) and Blumenthal *et al.* (1985) subsequently confirmed these findings. Further experiments using sera from people who had received ragweed immunotherapy (Rx) with higher *Amb a* V doses than found by natural pollen exposure suggested that all Dw2$^+$ subjects are potentially *Amb a* V responders (Marsh *et al.*, 1982b). Subsequent studies with *Amb t* V showed that immune responsiveness is also DR2/Dw2-associated (Roebber *et al.*, 1985); furthermore, 13/13 of *concordant*

[4]The other (smaller) homologous region, residues Nos. 31-35, is not amphipathic in either of the *Amb* V molecules.

responders to both *Amb* V molecules have DR2/Dw2 (Goodfriend *et al*., 1985; Roebber *et al*., 1985).

We have recently carried out the first prospective use of allergen immunotherapy to study the genetics of human immune response (Marsh *et al*., 1987 and unpublished), focusing on the *Amb* V model. We immunized two matched groups of 12 patients ($DR2^+$ and $DR2^-$) sequentially with crude ragweed extracts [*A. trifida* (giant), *A. psilostachya* (western) and *A. artemisiifolia* (short)] and studied the IgG Ab responses to each of the component *Amb* V molecules. We found significant positive associations between the presence of DR2, and especially Dw2, and the quantity of IgG Ab produced toward each of the *Amb* V molecules, which confirms and extends previous findings.

One interesting subject in this study, who responded after natural, ultra-low dose exposure to *Amb a* V, typed DR2 serologically and by MLR typing he was a questionable Dw2. By RFLP mapping of his genomic DNA, he had an *Eco*RI 6.5 kb restriction fragment detectable with a *DR*β probe (which is characteristic of DR2), but he did not have the *Eco*RI 2.3 kb *DQ*β restriction fragment characteristic of Dw2 (Zwollo *et al*., 1988a). Usually, we found individuals with the $DR2^+/Dw2^-$ phenotype (and who do not have the *Eco*RI 2.3 kb *DQ*β restriction fragment) to be *non*-responders; yet here is an individual who is responding after only natural pollen exposure.

MOLECULAR BIOLOGICAL STUDIES OF *HLA-D* REGION GENES

In order to explore the DR2—*Amb a* V association further, especially in atypical subjects like the one just mentioned, we examined the $DR\beta_I$, $DR\beta_{III}$ and $DQ\beta_I$ second-exon gene sequences of our patients. According to data from many laboratories (summarized in Todd *et al*., 1987 and Horn *et al*., 1988), the second exons encode most of the hypervariability within these genes. (The *DR*α gene has not been studied because it is essentially non-polymorphic; the polymorphic *DQ*α gene is currently under study.) X-ray crystallographic studies of class I HLA molecules (Bjorkman *et al*., 1987) and computer simulations of class II structures (Brown *et al*., 1988), strongly suggest that the hypervariable regions of the class II molecules (*cf*. shaded boxes in Fig. 1C) are the key to exploring the molecular basis of Ia—Ag fragment interactions.

We have used the polymerase chain reaction (PCR) to amplify the second exons of *DR*β and *DQ*β genes (Saiki *et al*., 1985; Erlich *et al*., 1986; Fig. 5). As is the case with many revolutionary developments in science, the PCR is elegant yet simple, in concept as well as in practice — you need just three heating blocks, or you can buy an automated temperature-cycling machine costing much more if you wish. Following an initial heating step to ensure complete DNA denaturation and deactivation of any enzymes present in the sample, three basic steps — denaturation, primer annealing and DNA chain extension (using the heat-stable *Taq* DNA polymerase) — are repeated, usually through 25-30 cycles. The oligonucleotide primers are complementary to the flanking regions of the *alternate* strands of the second exons of *DR*β or *DQ*β. Since these regions are almost completely conserved throughout the $DR\beta_I$ and $DR\beta_{III}$ or the *DQ*β alleles, one can use essentially the same primer sets for each of the expansions of *DR*β genes on the one hand or *DQ*β on the other. (A similar approach can also be used for *DQ*α.) As noted in Fig. 5, one obtains an exponential expansion of the segment of interest (i.e. the second exon containing the hypervariable regions), but only a linear expansion of the longer chromosomal segments that are synthesized in the first cycle. Following the PCR, the amplified segments can easily be visualized

1. To ~1μg genomic DNA add large excess of:
   a. Primer oligonucleotides corresponding to the sequences of ~18 – 22 bases at each 5' end of the opposite DNA strands flanking the region of interest.
   b. All 4 deoxynucleoside triphosphates.
2. Denature/inactivate sample (95° for 10 min)
3. Cool and add *Taq* DNA polymerase (from the thermophilic bacterium *Thermus aquaticus).*
4. Heat sample at 95° for 2 min to ensure that DNA is single-stranded.
5. Anneal the primers (2 min at 37°– 55°).
6. Extend DNA strands (1–3 min at 72°).
7. Repeat steps #4 – 6 approx. 24 times.

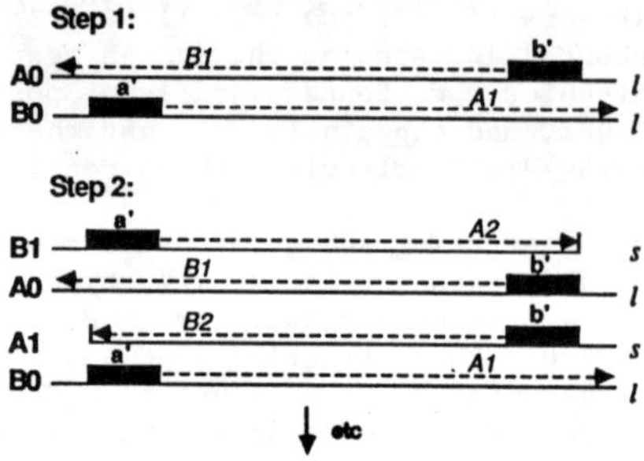

Theoretical number of short *('s')* strands produced at step 'n' = $2^n - 2$.
25 steps should produce ~3 x 10^7 short strands. (Yield ~ 10^6 strands.)

Fig. 5. The polymerase chain reaction (PCR). The temperature selected for the annealing Step 5 is optimized for the particular primers being used.

1. Polymerase Chain Reaction- (PCR-) amplification of polymorphic second-exon gene segments in *HLA-DRβ* or *HLA-DQβ* genes from genomic DNA samples of responders and non-responders to *Amb a* V.
2. Dot-blot analysis of the amplified DNA from responders and non-responders using sequence-specific oligonucleotides (SSOs) for hypervariable regions (HVRs) in *HLA-DRβ* or *HLA-DQβ.*
3. M13 cloning and dideoxy sequencing of PCR-amplified *HLA-DRβ* and *HLA-DQβ* second exons of exceptional patients with atypical phenotypes.
4. Analysis of the DNA sequence information using the BIONET computer-analysis system to search for sequences that are specific for *Amb a* V responders and non-responders.
5. *Future:* Design of SSOs specific for responder (and certain non-responder) phenotypes, followed by testing of these SSOs on PCR-amplified DNA of a large number of ragweed-allergic, *Amb a* V responders and non-responders.

Fig. 6. The approach used in defining *HLA-D* gene sequences involved in specific immune responsiveness to *Amb a* V.

after gel electrophoresis, where they are by far the most prominent bands. Thus, one can effectively "extract" the approximately 3 *hundred* bps of an *HLA* gene from the 3 *billion* bps of the human genome in just a few hours!

Our approach is summarized in Fig. 6. We have screened the amplified second exons by dot-blot analysis (Angelini *et al.*, 1986; Erlich *et al.*, 1986) using sequence-specific oligonucleotides (SSOs) specific for the known hypervariable regions of $DR\beta_I$, $DR\beta_{III}$ and $DQ\beta_I$. We have also cloned the second-exon segments into appropriate M13 vectors and sequenced the PCR-amplified second exons by the dideoxy method (*cf.* Erlich *et al.*, 1986; Todd *et al.*, 1987; Horn *et al.*, 1988). For the sequencing project, we have concentrated especially on the exceptional individuals who do not fit the typical *Amb a* V—DR2/Dw2 association (i.e. $DR2^-/Dw2^-$/*Amb a* V^+ and, also, $DR2^+/Dw2^+$/*Amb a* V^- subjects). Once we have developed a sufficient body of data from analyzing this sequence information, we expect to be able to design SSOs for *Amb a* V responder (and certain related non-responder) phenotypes and test these SSOs on amplified DNA from a large number of ragweed-allergic, *Amb* V responders and non-responders. This could well be done on thousands of people, because it is easy to do the PCR dot-blotting.

Figure 7 shows an example of a set of SSO dot-blotting experiments: a "framework" (F.W.) oligonucleotide from a completely conserved region of $DR\beta_I$ and $DR\beta_{III}$ serves as a positive control to quantitate the $DR\beta$ gene segments expanded in the PCR, and a $DR\beta_I2.2$-specific oligonucleotide probe from the second hypervariable region has been used to test for the presence of a DNA sequence normally found in $Dw2^+$ individuals (Zwollo *et al.*, 1988b). The first three samples, from HLA-D homozygous cells, demonstrate the specificity of the probe for the sequence found in $Dw2^+$ cell lines, but not in cell lines from $Dw21(AZH)^+$ or $Dw12^+$ subjects, both of which are DR2 variants. The next four samples are from *Amb a* V^+ subjects, one $Dw12^+$ Oriental individual, two typical $Dw2^+$ subjects and the unusual Dw2? subject discussed earlier. The next three subjects are typical $DR2^+/Dw2^-$ *Amb a* V *non*-responders, or relatively weak responders, who do not react with the probe. Finally, the last two subjects, who are both $Dw2^-$ *Amb a* V *responders*, are *negative* for the probe.

As indicated in the footnotes to Fig. 7, we have been sequencing the $DR\beta$ and $DQ\beta$ gene segments of several individuals. This work is not yet sufficiently complete to be able to formulate any definite hypotheses concerning the molecular basis of *Amb a* V recognition by Ia molecules on APCs. However, both the SSO and dideoxy sequencing data show that most *Amb a* V responders have $DR\beta_I$, $DR\beta_{III}$ and $DQ\beta_I$ sequences typically found in $Dw2^+$ subjects (Zwollo *et al.*, 1988b and unpublished). The $DR2^+/Dw2^-$ non-responders all have sequences typical of $Dw21^+$ subjects. The few atypical individuals are particularly important in understanding the molecular basis of *Amb a* V recognition. The Dw2?/*Amb a* V^+ subject already discussed has $DR\beta_I2.2$ and $DR\beta_{III}2.2$ sequences (like $Dw2^+$ responders), but has a $DQ\beta_I1.21$ sequence (like $Dw21^+$ non-responders). This person appears to be a recombinant, an unusual individual, of a type not previously reported in the literature. Results from this individual, along with other data, *suggest* that a DR Ia molecule *may* be involved in the recognition of *Amb a* V, but we still have to resolve some conflicting results with other atypical subjects for whom DQ is indicated. Experiments using cloned *Amb a* V-specific T-cell clones will be required to test these and other hypotheses.

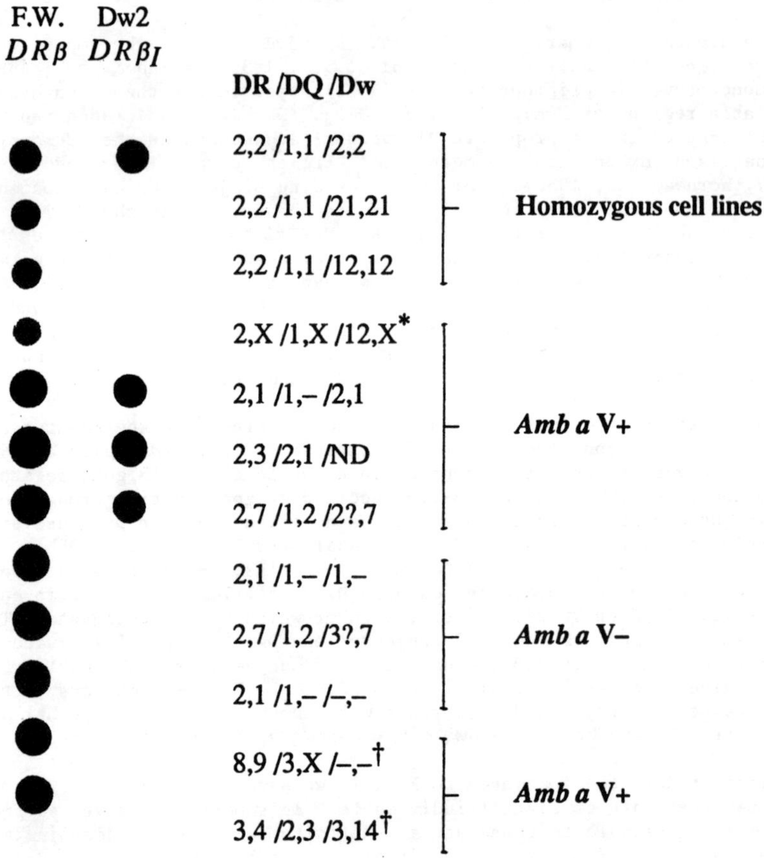

Fig. 7. Photograph of an autoradiogram of a set of two SSO dot-blots. A control framework (F.W.) SSO, which hybridizes with all $DR\beta_I$ and $DR\beta_{III}$ second exons, was used in the left-hand experiment. Following stripping this probe from the filter, an SSO specific for the second hypervariable region of $DR2.2\beta_I$ (found in $Dw2^+$ subjects) was used in the right-hand experiment.

*From additional dot-blot experiments using other SSO probes.

†From sequence information and further dot-blot experiments.

CONCLUDING REMARKS

The foregoing examples illustrate how molecular biological techniques may be used in the field of allergy for studying the molecular basis of specific human immune responsiveness to defined Ag molecules. As noted earlier, there is an almost limitless variety of allergens that can be isolated from common allergenic particulates. So far, only a few allergens have been isolated from pollens and other materials (e.g. Table 1), but the number should rapidly increase with the availability of many new isolation techniques. In the future, allergen sequencing is more likely to be performed on "*allergenes*" by DNA technology rather than by amino acid sequencing of the allergenic proteins themselves. Cloning and sequencing of cDNAs is usually much faster and the cDNA clones are useful for many further studies. Definition of Ia/T-cell epitopes still requires a great deal of work, but we believe that simpler procedures will be developed in that area and the prediction of such sites by computer analysis will be improved. The sera of atopic individuals can readily be studied by DARIA (and other methods) for the presence of specific Abs to many well-defined allergens; this will allow the definition of an "immune response fingerprint" for each subject to a great variety of different macromolecules. Using a combination of PCR and dot-blot technology, supplemented by dideoxy sequencing where necessary, a considerable amount of information on the hypervariable regions of *HLA-D* genes can rapidly be obtained. Thus, it should be possible to relate the person's specific *HLA-D* gene sequences, and the polypeptides they encode, to the sequences of the Ia/T-cell epitopes on the array of allergens to which he responds.

The PCR truly represents a major step in the development of recombinant DNA technology. We anticipate using this methodology to investigate the molecular structure of the hypervariable regions of the TcRs of allergen-specific cloned T-cells. Both peptide chemistry and recombinant DNA technology, using site-directed mutagenesis of allergen cDNAs, will certainly prove useful in the investigation of Ia/T-cell sites on allergen molecules. We believe that, beside the study of *HLA-D* genes, a further key to fingerprinting human immune responsiveness is relating particular TcRs to the recognition of particular Ia/T-cell epitopes on Ag (allergen) molecules. Then, we will have to put everything together and, hopefully, we can really learn something about the genetics of human immune response and susceptibility to immunologic diseases.

REFERENCES

Andersson, G., Larhammar, D., Widmark, E., Servenius, B., Petersen, P.A. (1987). Class II genes of the human major histocompatibility complex: organization and evolutionary relationship of the DRβ genes. J. Biol. Chem., 262, 8748-8758.

Angelini, G., de Preval, C., Gorski, J., Mach, B. (1986). High-resolution analysis of the human HLA-DR polymorphism by hybridization with sequence-specific oligonucleotide probes. Proc. Natl. Acad. Sci., 83, 4489-4493.

Ansari, A.A., Kihara, T.K. and Marsh, D.G. (1987). Immunochemical studies of *Lolium perenne* (rye grass) allergens *Lol p* I, II and III. J. Immunol., 139, 4034-4041.

Ansari, A.A., Freidhoff, L.R. and Marsh, D.G. (1988a). Molecular genetics of human immune response to *Lolium perenne* (rye) allergens, *Lol p* I and *Lol p* II. In: Proc. 10th Collegium Internationale Allergologicum Meeting, Matinique (Sehon, A.H. and Capron, A., eds.). In press.

Ansari, A.A., Freidhoff, L.R., Meyers, D.A., Bias, W.B. and Marsh, D.G.

(1988b). Human immune responsiveness to *Lolium perenne* (rye) grass pollen allergen *Lol p* III (Rye III) is associated with HLA-DR3 and DR5 and possibly with a unique DRβ_I sequence. Submitted for publication.

Ansari, A.A., Shenbagamurthi, P. and Marsh, D.G. (1988c). Comparison of amino acid sequences of *Lolium perenne* allergens *Lol p* II and III. J. Allergy Clin. Immunol., 81, 307.

Babbitt, B.P., Allen, P.M., Matsueda, G., Haber, E. and Unanue, E.R. (1985). Binding of immunogenic peptides to Ia histocompatibility molecules. Nature, 317, 359-361.

Bach, F.H. (1985). The *HLA* class II genes and products: the *HLA-D* region. Immunol. Today, 6, 89-94.

Bjorkman, P.J., Saper, M.A., Samraoui, B., *et al.* (1987). The foreign antigen binding site and T cell recognition regions of class I histocompatibility antigens. Nature, 329, 512-518.

Black, P.L. and Marsh, D.G. (1976). The genetic basis for atopic allergy in man. In: Bronchial Asthma: Mechanisms and Therapeutics (M.S. Segal and E.B.Weiss, eds.). Little, Brown and Co., Boston, pp.53-64.

Blumenthal, M., Awdeh, Z., Alper, C. and Yunis, E. (1985). Ra5 immune responses, HLA antigens and complotypes. J. Allergy Clin. Immunol., 75, 155 (Abstract).

Brown, J.H., Jardetsky, T., Saper, M.A., *et al.* (1988). A hypothetical model of the foreign antigen binding site of Class II histocompatibility molecules. Nature, 332, 845-850.

Buus, S., Sette, A., Colon, S.M., *et al.* (1987). The relation between major histocompatibility complex (MHC) restriction and the capacity of Ia to bind immunogenic peptides. Science, 235, 1353-1358.

Cooke, R.A. and VanderVeer, A. (1916). Human sensitization. J. Immunol., 1, 201-305.

Cottan, G.P., Moran, D.M. and Standring, R. (1986). Physicochemical and immunochemical characterization of allergenic proteins from rye grass (*Lolium perenne*) pollen prepared by a rapid and efficient purification method. Biochem. J., 234, 305-310.

Coulter, K., Dorval, G. and Goodfriend, L. (1988). Genetic control of IgE antibody responses in man: the *Amb* V (Ra5) model. In: Genetic and Environmental Factors in Clinical Allergy (D.G. Marsh and M.N. Blumenthal, eds.). University of Minnesota Press. In press.

Erlich, H.A., Sheldon, E.L. and Horn, G. (1986). HLA typing using DNA probes. Biotechnology, 4, 975-981.

Esch, R.E. and Kalpper, D.G. (1988). Isolation and characterization of a major cross-reactive grass Group I allergenic determinant. In press.

Freidhoff, L.R. (1988). Epidemiology of atopic allergy. In: Genetic and Environmental Factors in Clinical Allergy (D.G. Marsh and M.N. Blumenthal, eds.). University of Minnesota Press. In press.

Freidhoff, L.R., Kautzky, E.E., Meyers, D.A., Hsu, S.H., Bias, W.B. and Marsh, D.G. (1988). Association of HLA-DR3 and total serum immunoglobulin E level with human immune response to *Lol p* I and *Lol p* II allergens in allergic subjects. Tissue Antigens. In press.

Goodfriend, L., Choudhury, A.M., Klapper, D.G., *et al.* (1985). Ra5G, a homologue of Ra5 in giant ragweed pollen: isolation, HLA-DR-associated activity and amino acid sequence. Mol. Immunol., 22, 899-906.

Guillet, J.G., Lai, M.Z., Briner, T.J., *et al.* (1987). Immunological self, nonself discrimination. Science, 235, 865-870.

Hardy, D.A., Bell, J.I., Long, E.O., Lindsten, T. and McDevitt, H.O. (1986). Mapping of the class II region of the human major histocompatibility complex by pulsed-field gel electrophoresis. Nature, 323, 453-455.

Horn, G.T., Bugawan, T.L., Long, C.M. and Erlich, H.A. (1988). Allelic sequence variation of the HLA-DQ loci: relationship to serology and to insulin-dependent diabetes susceptibility. Proc. Natl. Acad. Sci. (USA),

In press.

Jonsson, A.K., Hyldig, Nielsen, J.J., Servenius, B., *et al.* (1987). Class II genes of the human major histocompatibility complex: comparisons of the *DQ* and *DX* α and β genes. J. Biol. Chem., 262, 8767-8777.

Margalit, H., Spouge, J.L., Cornette, J.L., *et al.* (1987). Prediction of immunodominant helper T-cell antigenic sites from the primary sequence. J. Immunol., 138, 2213-2229.

Marrack, P. and Kappler, J. (1987). The T-cell receptor. Science, 238, 1073-1078.

Marsh, D.G. (1975). Allergens and the genetics of allergy. In: The Antigens, Vol.III (M. Sela, ed.). Academic Press, NY, pp.271-359.

Marsh, D.G. (1986). Defining human immune response fingerprints toward ultra-pure allergens: immunochemical and genetic aspects of responsiveness toward the *Amb* V (Ra5) homologues. In: Proc. XII Internatl. Congress Allergol. Clin. Immunol., Washington, D.C. (C.E. Reed, ed.). C.V. Mosby Co., St. Louis, Mo. J. Allergy Clin. Immunol., 78 (Suppl.): 242-248.

Marsh, D.G. (1988a). Allergens and allergen nomenclature. In: Genetic and Environmental Factors in Clinical Allergy (D.G. Marsh and M.N. Blumenthal, eds.). University of Minnesota Press. In press.

Marsh, D.G. (1988b). Immunogenetic and immunochemical factors determining immune responsiveness to allergens: studies in unrelated subjects. In: Genetic and Environmental Factors in Clinical Allergy (D.G. Marsh and M.N. Blumenthal, eds.). University of Minnesota Press. In press.

Marsh, D.G. and Bias, W.B. (1988). The genetics of atopic allergy. In: Immunological Diseases, Fourth Edition (M. Samter, D.W. Talmage, M.M. Frank, K.F. Austen and H.N. Claman, eds.). Little, Borwn and Co., Boston, Ma, pp.1027-1045.

Marsh, D.G., Bias, W.B., Hsu, S.H. and Goodfriend, L. (1973). Associations between major histocompatibility (HL-A) antigens and specific reaginic antibody responses in allergic man. In: Mechanisms in Allergy: Reagin-mediated Hypersensitivity. (L. Goodfriend, A.H. Sehon and R.P. Orange, eds.). Marcel Dekker, Inc., New York, pp.113-129.

Marsh, D.G., Hsu, S.H., Roebber, M., Kautzky, E.E., Freidhoff, L.R., Meyers, D.A., Pollard, M.K. and Bias, W.B. (1982a). HLA-Dw2: a genetic marker for human immune response to short ragweed pollen allergen Ra5. I. Response resulting primarily from natural antigenic exposure. J. Exp. Med., 155, 1439-1451.

Marsh, D.G., Meyers, D.A., Freidhoff, L.R., Kautzky, E.E., Roebber, M., Norman, P.S., Hsu, S.H. and Bias, W.B. (1982b). HLA-Dw2: a genetic marker for human immune response to short ragweed pollen allergen Ra5. II. Response after ragweed immunotherapy. J. Exp. Med., 155, 1452-1463.

Marsh, D.G., Goodfriend, L., King, T.P., Løwenstein, H. and Platts-Mills, T.A.E. (1986). Allergen nomenclature. Bull. WHO, 64, 767-770. Reprinted in: J. Allergy Clin. Immunol., 80, 639-645, 1987; Mol. Immunol., 24, 1383-1388, 1987.

Marsh, D.G., Freidhoff, L.R., Golden, D.B.K., *et al.* (1987). Genetic studies of immune response to the *Amb* V homologues. Fed. Proc., 46, 1047 (Abstract).

Mole, L.E., Goodfriend, L., Lapkoff, C.B., *et al.* (1975). The amino acid sequence of allergen Ra5. Biochem., 14, 1216-1220.

Oettgen, H.C. and Terhorst, C. (1987). The T-cell receptor-T3 complex and T-lymphocyte activation. Hum. Immunol., 18, 187-204.

Roebber, M. (1988). The relevance of allergen standardization and purity to epidemiologic and genetic studies of immune responses to allergens. In: Genetic and Environmental Factors in Clinical Allergy (D.G. Marsh and M.N. Blumenthal, eds.). University of Minnesota Press. In press.

Roebber, M., Klapper, D.G., Goodfriend, L., Bias, W.B., Hsu, S.H. and Marsh, D.G. (1985). Immunochemical and genetic studies of *Amb t* V (Ra5G), an

Ra5 homologue from giant ragweed pollen. J. Immunol., 134, 3062-3069.
Saiki, R.K., Scharf, S., Faloona, F., *et al.* (1985). Enzymatic amplification of β-globin genomic sequence and restriction site analysis for diagnosis of sickle cell anemia. Science, 230, 1350-1354.
Schwartz, M. (1952). Heredity in bronchial asthma. Acta Allergol., 5 (Suppl. 2), 1-288.
Schwartz, R.H. (1985). T-lymphocyte recognition of antigen in association with gene products of the major histocompatibility complex. Ann. Rev. Immunol., 3, 237-261.
Sette, A., Buus, S., Colon, S., *et al.* (1987). Structural characteristics of an antigen required for its interaction with Ia and recognition by T cells. Nature, 328, 395-399.
Todd, J.A., Bell, J.I. and McDevitt, H.O. (1987). HLA-DQβ gene contributes to susceptibility and resistance to insulin-dependent diabetes mellitus. Nature, 329, 599-604.
Trowsdale, J., Young, J.A.T., Kelly, A.P., *et al.* (1985). Structure, sequence and polymorphism in the *HLA-D* region. Immunol. Rev., 85, 5-43.
Unanue, E.R. and Allen, P.M. (1987). The basis for the immunoregulatory role of macrophages and other accessory cells. Science, 236, 551-557.
van Rood, J.J., van Leenvan, A., Keuning, J.J. and Blusse van oud Alblas, A. (1975). The serological recognition of the human MLC determinants using a modified cytotoxicity technique. Tissue Antigens, 5, 73-80.
Wiener, A.S., Zieve, I. and Fries, J.H. (1936). The inheritance of allergic disease. Ann. Eugen., 7, 141-162.
Willcox, H.N.A. and Marsh, D.G. (1978). Genetic regulation of antibody heterogeneity: its possible significance in human allergy. Immunogenetics, 6, 209-225.
Wyman, M. (1872). Autumnal Catarrh (Hay Fever). Cambridge, Mass, Hurd and Houghton, p.82.

ACKNOWLEDGEMENT

This work was supported by NIH Grants Nos. AI-19727, AI-20059 and AI-25372.

The Interaction of Human IgE with Class-specific Fc-Receptors

Birgit A. Helm

Department of Biophysics, Cell and Molecular Biology,
King's College, University of London, 26–29 Drury Lane,
London WC2B 5RL, UK

ABSTRACT

A series of overlapping recombinant human epsilon gene fragments (rFc_ε) was expressed in *E. coli*. In order to establish site(s) on IgE involved in the binding to high affinity receptors ($Fc_\varepsilon R1$), the capacity of rFc_ε peptides to inhibit passive sensitisation of human mast cells was assayed *in vivo* (Prausnitz-Kustner = P-K test) and *in vitro* (sensitisation of peripheral blood basophils). To map structures on IgE necessary for the interaction with its low affinity receptor ($Fc_\varepsilon R2$), the ability of rFc_ε fragments to bind to the human lymphoblastoid cell-line RPMI 8866 was assayed by indirect immunofluorescence. Furthermore, the domain specificity of monoclonal antibodies (mAb) was determined, using the same spectrum of peptides, in an attempt to correlate their binding site with their capacity to interfere with the binding of IgE to $Fc_\varepsilon R2$. The outcome of these studies demonstrates that glycosylation is not vital for the interaction of IgE with its Fc receptors and shows that different structural requirements exist for the binding to $Fc_\varepsilon R1$ and $Fc_\varepsilon R2$, although there may be a common site on the CH3 domain.

KEYWORDS

Human Fc_ε gene; cDNA cloning; bacterial expression; affinity for IgE-Fc receptors; monoclonal anti-IgE antibodies.

INTRODUCTION

IgE is best known for its role in allergy, an aberrant immune response arising from an imbalance of helper and suppressor lymphocytes in atopic individuals [1]. The physiological requirement for IgE is less widely appreciated, but compelling evidence now exists to show it is important in the immune defense against e.g. parasites [2]. Serum concentrations of IgE rarely exceed 1 µg/L (IgG=5-15 g/L), and the complex regulation of its production by a network of T- and B-lymphocytes [1] may be a reflection of the cytotoxic processes mediated by this Ig isotype. IgE is a cytotropic antibody, and at present we are aware of two distinct classes of Fc receptors for

IgE ($Fc_{\varepsilon}R$). Following the discovery of IgE as the mediator of the allergic response (class I hypersensitivity) by the Ishizakas in 1967 [3], the interaction of IgE with receptors on mast cells and basophils was extensively studied. It could soon be demonstrated that IgE binds to high affinity receptors ($Fc_{\varepsilon}R1$, ka in the nmolar range) via the Fc portion of the molecule [4]. Prepared by papain digestion from human IgE, this fragment encompasses the paired CH2-4 domain of the Ig (aa 227-547). The range of fragments that could be produced by proteolysis was rather limited and the papain-Fc_{ε} fragment was the smallest peptide known to bind $Fc_{\varepsilon}R1$. Due to the absence of a human cell line expressing $Fc_{\varepsilon}R1$, extensive studies on the high affinity receptor were facilitated by the existence of the rat basophilic leukaemia cell line (RBL-III) [5], which expresses functional $Fc_{\varepsilon}R1$. Recent investigations have given rise to a model [5] suggesting that the receptor consists of four associated polypeptide chains. The single alpha chain, which has been cloned [6,37] contains the binding site for IgE. The alpha chain is associated with a single beta, and two disulphide-linked gamma chains. The interaction of IgE-Fc with the alpha chain sensitises, but does not trigger target cell exocytosis unless the ligand is aggregated [8]. Under natural conditions, the binding of specific antigen to IgE occupied $Fc_{\varepsilon}R1$ is thought to crosslink these receptors and elicit, via a series of yet unidentified steps, the release of preformed mediators such as histamine, bradykinin, and ECF-A, as well as the synthesis of new ones such as leukotrienes [5].

For several years the interaction of IgE molecules with class specific receptors was thought to involve only receptors on mast cells and basophils. In 1975, however, Lawrence *et al.* and Capron *et al.* [9,10] described the interaction of IgE with receptors on a diverse population of white blood cells such as neutrophils, eosinophils, monocytes, macrophages, and platelets, collectively known as inflammatory cells. Such cell populations carry specific IgE receptors ($Fc_{\varepsilon}R2$) distinct from those present on mast cells and basophils, and they bind IgE with low affinity (Ka in the μmolar range). Subsequent investigations demonstrated $Fc_{\varepsilon}R2$ occupancy under conditions where serum levels of IgE were raised pathologically, i.e. in allergy or parasitic infestations, and this was found to be associated with an increase in the number of cells bearing $Fc_{\varepsilon}R2$ receptors [2]. Furthermore, there is evidence to link the selective activation of inflammatory cell sub-populations with mast-cell exocytosis in that e.g. eosinophil-mediated cytotoxicity requires accessory mast cells or their mediators [11], and that the mast-cell derived chemotactic factor of anaphylaxis (ECF-A) plays a vital part in enhancing the expression of $Fc_{\varepsilon}R2$ on both rat and human eosinophils. In turn, eosinophils generate predominantly platelet activating factor (PAF) and eosinophil peroxidase, whereas IgE-stimulated macrophages secrete a number of lysosomal enzymes, leukotrienes, PAF, and interleukin 1. IgE activates platelets to release a variety of still unidentified cytocidal mediators and oxygen metabolites. The co-operative interaction of this network of inflammatory cells will lead to the selective IgE-dependent killing of parasites, and in atopic patients, seems to be responsible for the higher levels and faster release of inflammatory mediators from the activated cell population [2,11].

Indirect evidence suggests that a similar mechanism may operate in the killing of tumour cells [12,13,14]. The destruction of tumour cells by oxygen radicals liberated by eosinophils and macrophages is well established, and the co-operative action of macrophages and platelets in antibody-mediated suppression of tumour growth has been demonstrated, although the antibody isotype participating has not been identified. The mechanism involved in tumour cell killing appear indistinguishable from those activated by the IgE-mediated triggering of effector cells bearing low affinity receptors, when these kill parasites [2,11,12,13,14].

Human $Fc_{\epsilon}R2$ has been cloned by three independent groups [15,16,17]. The nucleotide sequence of the cloned cDNA suggests that $Fc_{\epsilon}R2$ consists of 321 AA (MW 36,281 daltons) and is oriented with its N-terminus on the cytoplasmic side, while the C-terminus, comprising the IgE binding site, shows a striking homology with the chicken asialoglycoprotein receptor. Furthermore, this receptor shows identity with the B-cell restricted, activation-association CD23 differentiation antigen found on mature and EBV transformed B cells, but not T cells (excepting HIV-transformed T cell lines) [16]. Evidence suggests that IgE, B-cell stimulating factor, and interleukin 4 (IL4) are responsible for the induction of this receptor which is rapidly shed as a proteolytic fragment upon isotype switching to gamma, alpha, or epsilon bearing cells [18]. An increase in soluble $Fc_{\epsilon}R2$, complexed to IgE, has been observed in the serum of atopic patients [16,18]. Furthermore, preliminary evidence suggests that this fragment possesses growth-promoting activity similar to autocrine B-cell growth factor [21]. In essence, these observations indicate a fundamental role for $Fc_{\epsilon}R2$ not only in the regulation of IgE production but also in B-cell differentiation.

It is our aim to identify site(s) on the IgE molecule involved in the interaction with both high- and low-affinity receptors and proteolytic fragments derived from the latter in order to understand the structural requirements involved in the binding. This may in time help us to design a rational inhibitor that interferes with the pathological manifestations of the IgE isotype (allergy) without inhibiting its potentially important physiological functions. Furthermore, it might enable us to exploit the cytotoxic potential of IgE-mediated processes involving its low-affinity receptor in the killing of tumour cells and parasites through the construction of chimaeric IgE molecules of pre-selected specificity without precipitating an anaphylactic response.

RESULTS

Structure-activity studies on reaginic antibodies were hampered in the past by both the dearth of material and the difficulty in introducing specific changes into the protein for functional studies. In order to overcome these problems, the gene coding for the Fc portion of human IgE was cloned and expressed in *E. coli*, since eukaryotic cells do not support the expression of Ig-Fc fragments [19,20]. The demonstration that recombinant IgE fragments, despite their lack of glycosylation possess effector functions indistinguishable from those of human myeloma IgE [19,20,21,22,23] allowed us to investigate the biological activity of a series of overlapping rFc_{ϵ} gene products. In order to identify site(s) on IgE interacting with high- and low-affinity receptors, these fragments were tested for their ability to sensitise, or inhibit passive sensitisation of cells bearing $Fc_{\epsilon}R1$ [24], and to bind to $Fc_{\epsilon}R2$ [25]. Furthermore, these peptides were employed to determine the domain or epitope specificity of monoclonal antibodies (mAb) that inhibit the binding of IgE to $Fc_{\epsilon}R2$ [26].

The spectrum of fragments used in these studies is shown in Fig. 1; and their capacity to bind to high- and low-affinity receptors is summarised in Table 1. These data show that the minimum structural requirements for binding to $Fc_{\epsilon}R1$ is associated with a fragment spanning aa 301-367, while the IgE binding site for $Fc_{\epsilon}R2$ on B-cells is located in the CH3 and/or CH4 domain. In contrast, mAbs that inhibit the binding of IgE to $Fc_{\epsilon}R2$ map to the C-terminal CH2 and the N-terminal CH3 domain (Table 2).

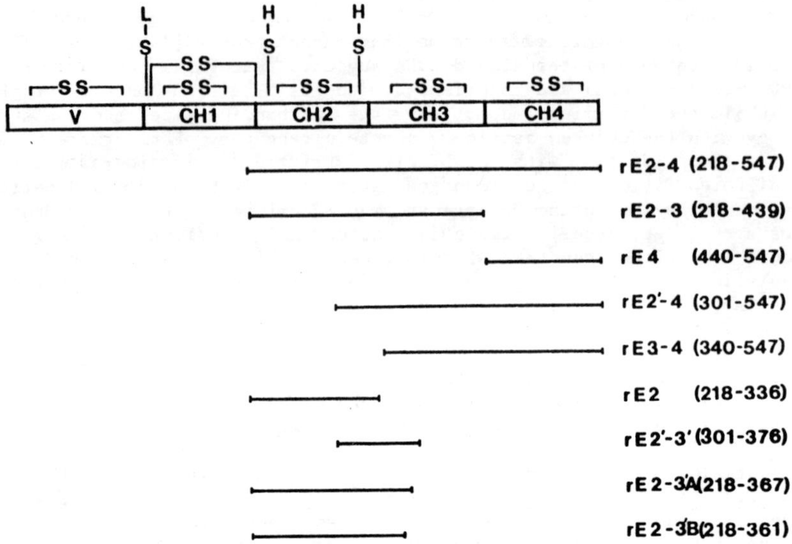

Fig. 1. Recombinant IgE fragments employed in the delineation of high- and low-affinity receptor-binding sites on the human epsilon chain. The peptides were generated by sub-cloning restriction fragments into suitable vectors [19,24,33,38]. The boundaries were determined by sequencing, and the cloned gene products were purified by established methods [19,33].

Table 1. Summary of binding of rFcε peptides to the human high-affinity (FcεR1) and low-affinity (FcεR2) receptors. The precise boundaries of these peptides are shown in Fig. 1 (for experimental details see ref. 24).

	rE2-4	rE2-3	rE4	rE2'-4	rE3-4	rE2	rE2'-3'	rE2-3'A	rE2-3'B
FcεR1	+	+	−	+	−	−	+	+	−
FcεR2	+	−	−	+	+	−	−	−	−

Table 2. Monoclonal antibodies to human IgE as structural and functional probes. (Methods employed for the determination of domain specificity are described in ref. 26.)

mAb	Domain Specificity	Inhibition of binding of IgE to $Fc_{\varepsilon}R2$
D.C.	CH2 (aa 218-301)	-
A.S. 7.12	CH2 (aa 218-301)	-
R.P. 3	CH2 (aa 301-336)	+
B.S. 17	CH2 (aa 301-336)	+
I.C. 272	CH2 (aa 301-336)	-
L.E. 27	CH3 (aa 340-361)	-
R.P. 1	CH3 (aa 340-361)	-
I.C. 27	CH3 (aa 367-376)	+
I.C. 173	CH4 (aa 440-547)	-

DISCUSSION

The demonstration that rFc_{ε} (aa 218-547) is indistinguishable in its physiological properties from natural IgE justifies the utilisation of recombinant DNA technology for the cloning of an overlapping family of epsilon gene fragments, in order to determine the minimal structural requirements necessary for the binding of human IgE to its high- and low-affinity receptors [19,24].

The mast cell binding site on human IgE was determined in an *in vivo* assay by measuring the capacity of these recombinant peptides to inhibit the binding of antigen-specific IgE (ragweed) in the Prausnitz-Küstner (P-K) test [35]. Results showed that recombinant peptides containing the sequence aa 301-376 were capable of inhibiting the wheal and flare reaction. The efficacy of rE2'-3' compared favourably with human myeloma IgE (PS), on a molar basis, it is about 30% as effective, and skin sites injected with 1 μmol of rE2'-3' or IgE (PS) remained refractory to passive sensitisation for 14 and 19 days respectively [24]. In addition, this sequence retains the target-cell triggering activity characteristic of cross-linked IgE [8]. This rules out earlier claims that a decapeptide sequence in CH4 is essential for IgE-mediated target cell exocytosis [27]. Although rE2-4 formed epsilon chain dimers efficiently [19,21], some of the smaller biologically active fragments failed to do so, suggesting that a monomeric fragment can form the active site [24].

At present, the shortest sequence on human IgE known to inhibit in the P-K test spans aa 301-367. $Fc_{\varepsilon}R1$ interaction requires structures from the carboxy-terminal CH2 domain (aa 301-336) and the amino-terminal CH3 domain (aa 340-367), since peptides containing the separate sequences are inactive. This confirms earlier observations which showed that cleavage by pepsin at position 338/9 releases two inactive fragments [32]. Furthermore, the protection against proteolytic digestion of this region of rat IgE when complexed to $Fc_{\varepsilon}R1$, suggested that this part of the Ig molecule was involved in the binding to $Fc_{\varepsilon}R1$ [34].

These fragments were subsequently employed to determine the domain-specificity of a panel of anti-IgE mAbs. The outcome of such studies is shown in Table 2, and a correlation of their domain specificity with their ability to inhibit the binding of human IgE to $Fc_{\varepsilon}R2$ implies that when certain epitopes in either the carboxy-terminal CH2 domain or the amino-terminal CH3 domain are occupied by either anti-IgE mAb R.P.3 or I.C.27, or their Fab fragments, IgE can no longer make contact with $Fc_{\varepsilon}R2$. Furthermore, these studies demonstrated that when IgE is first pre-incubated with RPMI 8866 cells, mAb I.C.27 and mAb R.C.3 can still engage receptor-bound IgE. These observations indicate that IgE binds to a single $Fc_{\varepsilon}R2$ molecule involving epitope(s) from only one of the two epsilon chains, leaving the other epitope(s) localised on the second epsilon chain free to interact with anti-IgE mAb. Alternatively, the site recognised may be close, but not identical with the site on IgE that interacts with $Fc_{\varepsilon}R2$.

The domain specificity of anti-IgE mAbs that inhibit the binding of IgE to $Fc_{\varepsilon}R2$ suggested that sites on IgE involved in the binding to this receptor are close to, or identical with sites on IgE required for $Fc_{\varepsilon}R1$ interaction. However, an investigation of the bindability of rFc_{ε} fragments to the RPMI 8866 cell-line by indirect immunofluorescence, showed that only rE2-4 (aa 218-547), rE2'-4 (aa 301-547), and rE3-4 (aa 340-547) engage the B-cell receptor. All other fragments are inactive. Since human $Fc_{\varepsilon}R1$ and $Fc_{\varepsilon}R2$ are highly dissimilar, it is not surprising that a different spectrum of recombinant peptides binds to the low-affinity receptor. In view of the sequence homology of $Fc_{\varepsilon}R2$ with the asialoglycoprotein receptor [16,17,18] it was somewhat unexpected to observe activity with non-glycosylated recombinant peptides, indicating that carbohydrate is not crucial for the binding of IgE to its Fc receptors.

These data indicate that site(s) on IgE interacting with $Fc_{\varepsilon}R2$/CD23 are found on structures contributed by the CH3 and/or CH4 domain, although mAbs that block the binding to $Fc_{\varepsilon}R2$ map to the carboxy-terminal CH2 and amino-terminal CH3 domain.

When these apparently contradictory observations are analysed in the context of a high resolution model of human IgE-Fc [28,29], an attractive interpretation presents itself. In Fig. 2 the amino acid sequence common to all peptides capable of inhibiting the P-K reaction (aa 301-367) is highlighted. The model predicts that only two of the beta strands in CH2 (aa 307-314, and aa 324-334) and the short amino-terminal strand from CH3 (aa 362-370) are exposed to solvent and hence available to the receptor. The portions of the

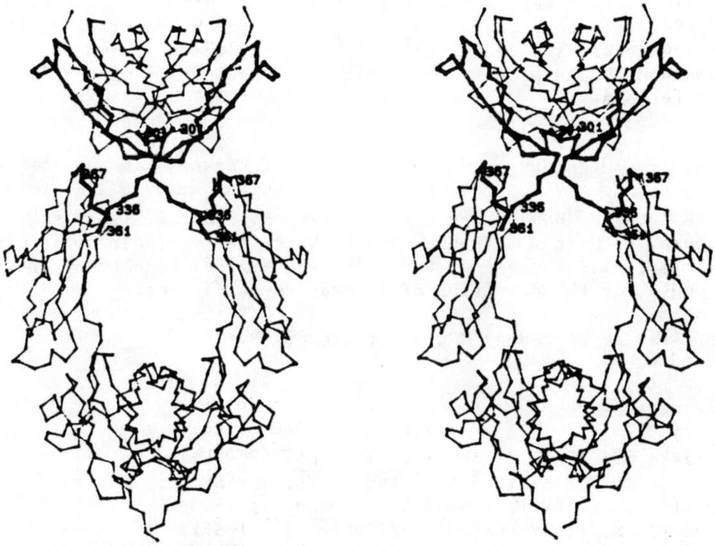

Fig. 2. High-resolution stereoscopic model of human IgE-Fc. (Courtesy of E. Padlan, see ref. 28). The amino acid sequence common to all peptides capable of inhibiting the P-K reaction is highlighted. Note the close proximity of the predicted antigenic site in the short N-terminal strand on CH3 (aa 362-370) to one of the B strands on CH2 (aa 307-314).

sequence that are not buried in the interior of IgE form a cleft between the CH2 and CH3 domains. It is tempting to speculate that the alpha chain of $Fc_{\epsilon}R1$, which protrudes from the mast cell [36], will engage at least one site from each domain, and the observed high affinity of IgE for $Fc_{\epsilon}R1$ is consistent with the involvement of more than one binding site. It is interesting to note in this context that the interaction of protein A with IgG requires structural elements from two Ig domains [30,31].

When one considers possible site(s) on IgE interacting with $Fc_{\epsilon}R2$ in the context of the 3D model structure, the close proximity of the sequence defined by mAb I.C.27 (aa 367-376) to one of the beta strands in CH2 (aa 307-314) becomes apparent, especially when one keeps in mind that the model predicts an antigenic site involving aa 367-369. Since the binding of Fab fragments of mAb R.P.3 and I.C.27 to IgE inhibit its interaction with $Fc_{\epsilon}R2$, it appears likely that residues on IgE near or involving sites defined by these antibodies contribute to $Fc_{\epsilon}R2$ interaction. However, rFc_{ϵ} peptides devoid of CH2 will still engage $Fc_{\epsilon}R2$, and this implies that mAb R.P.3 inhibits because it binds to an epitope in CH2 that is very close to the site where IgE occupies $Fc_{\epsilon}R2$, i.e. the epitope defined by mAb I.C.27.

Although these observations indicate that the third beta strand in CH3 may contribute residues for the binding to both high- and low-affinity receptors, it is important to point out that the rE2'-3' peptide containing the $Fc_{\epsilon}R1$ binding site does not bind to the B-cell receptor, in contrast, rE3-4, which possesses the $Fc_{\epsilon}R2$ binding site, has no affinity for the mast cell receptor.

The availability of peptides that differentially bind to high- as opposed to low-affinity receptors for human IgE is vital to our understanding of the interplay of various effector cells in cellular immunity and B-cell differentiation and should provide the basis for a rational manipulation of IgE-mediated processes.

ACKNOWLEDGMENTS

I wish to thank J. Banchereau, I. Crétien, R. Geha, K. Ishizaka, T. Ishizaka, P. Marsh, E. Padlan and D. Vercelli for their collaboration, R. Pumphries, A. Saxon and D. Capra for the gift of monoclonal antibodies, and H. Gould for the provision of laboratory facilities.

This work was supported by MRC grant G860674CB.

REFERENCES

1. Ishizaka, K.(1984). Ann. Rev. Immunol., 2, 159.
2. Capron, A., Dessaint, J.P., Capron, M., Joseph, M., Amiesen, J.C. and Tonnel, A.B. (1986). Immunology Today, 7, 15.
3. Ishizaka, K., Ishizaka, T. and Terry, W. (1967). J. Immunol., 99, 849.
4. Stanworth, D.R., Humphrey, J.H., Bennich, H. and Johansson, S.G.O. (1968) Lancet, 2, 17.
5. Metzger, H., Alcaraz, G., Hohman, R., Kinet, J.P., Pribluda, V. and Quarto, R. (1986). Ann. Rev. Immunol., 4, 419.
6. Kinet, J.P., Metzger, H., Hakimi, J. and Kochan, J. (1987). E. J. Biochemistry, 26, 4605.
7. Hempstead, B., Parker, C., Kulzcycki Jr., A. (1983). Proc. Natl. Acad. Sci. USA, 80, 3050.
8. Segal, D.M. and Hurwitz, E. (1977). J. Immunol., 118, 1328.
9. Lawrence, D.A., Weigle, W.O. and Spiegelberg, H.L. (1975). J. Clin. Invest., 55, 368.
10. Capron, A., Dessaint, J.P., Capron, M. and Bazin (1975). Nature, 254, 474.
11. Tonnel, A.B. (1986). Ann. Inst. Pasteur/Immunol., 137C, 358.
12. Jong, E.C. and Klebanoff, S.J. (1980). J. Immunol., 124, 1949.
13. Nathan, C.F. and Klebanoff, S.J. (1980). J. Exp. Med., 155, 198.
14. Nathan, C.F. and Klebanoff, S.J. (1982). J. Exp. Med., 192, 1291.
15. Ikuta, K., Takami, M., Kim, C.W., Honjo, T., Myoshi, T., Tagaya, Y., Kwabe, T. and Yodoi, J. (1987). Proc. Natl. Acad. Sci., 84, 819.
16. Kikutani, H., Inui, S., Sato, R., Barsumian, E.L., Owaki, H., Yamasaki, K., Kaisho, T., Uchibayashi, N., Hardy, R.R., Hirano, T., Tsunasawa, S., Sakiyama, F., Suemura, M. and Kishimoto, T. (1986). Cell, 47, 657.
17. Ludin, C., Hofstetter, H., Sarfarti, M., Levy, C.A., Suter, U., Alaimo, D., Kirchherr, E., Frost, H. and Delespesse, G. (1987). EMBO J., 6, 109.
18. Gordon, J., Webb, A.J., Walker, L., Guy, G.R. and Rowe, M. (1986). Eur. J. Immunol., 16, 1627.
19. Kenten, J.H., Helm, B.A., Ishizaka, T., Cattini, P. and Gould, H.J. (1984). Proc. Natl. Acad. Sci. USA, 81, 2955.
20. Lui, F.T., Albrandt, K.A., Bry, C.G. and Ishizaka, T. (1984). Proc. Natl. Acad. Sci. USA, 81, 5369.
21. Coleman, J., Helm, B.A., Stanworth, D.R. and Gould, H.J. (1985). Eur. J. Immunol., 15, 966.
22. Geha, R.S., Helm, B.A. and Gould, H.J. (1985). Nature, 315, 577.
23. Ishizaka, T., Helm, B.A., Hakimi, J., Niebl, J., Ishizaka, K. and Gould, H.J. (1986). Proc. Natl. Acad. Sci. USA, 83, 8323.

24. Helm, B., Marsh, P., Vercelli, D., Padlan, E., Gould, H. and Geha, R. (1988). Nature, 331, 181.
25. Vercelli, D., Helm, B., Marsh, P., Hakimi, J., Geha, R. and Gould, H. (1988). FASEB Journal, 2, A1441.
26. Cretien, I., Helm, B.A., Marsh, P.J., Padlan, E.A., Wijdenes, J. and Banchereau, J. (1988), J. Immunol., in press.
27. Stanworth, D.R., Kings, M., Roy, P.D., Moran, J.M. and Moran, D.M. (1979). Biochem. J., 180, 665.
28. Padlan, E.A. and Davies, D.R. (1986). Mol. Immunol., 23, 1063.
29. Pumphries, R. (1986). Immunol. Today, 7, 174.
30. Deisenhofer, J. (1981). Biochemistry, 20, 2361.
31. Kronvall, G. and Frommel, D. Immunochemistry, 7, 124.
32. Dorrington, K.J. and Bennich, H. (1978). Immunol. Rev., 41, 3.
33. Gould, H.J., Helm, B.A., Marsh, P. and Geha, R. (1987). Int. Archs. Allergy appl. Immunol., 82, 392.
34. Perez-Montfort, R. and Metzger, H. (1982). Mol. Immunol., 19, 1113.
35. Prausnitz, C. and Kustner, H. (1921). Zbt. Bkt. Orig., 86, 166.
36. Holowska, D., Conrad, D.H. and Baird, B. (1985). Biochemistry, 24, 6260.
37. Shimizu, A., Tepler, I., Benefy, P.N., Berenstein, E.H., Siraganian, R.P. and Leder, P. (1988). Proc. Natl. Acad. Sci. USA, 85, 1907.
38. Marsh, P., Helm, B. and Gould, H.J. (1988). Allergy, in press.

Chimaeric Antibodies

Marianne Brüggemann

Division of Immunology, Department of Pathology,
University of Cambridge, Addenbrooke's Hospital,
Laboratories Block, Hills Road, Cambridge CB2 200, UK

ABSTRACT

Immunoglobulin DNA can be manipulated *in vitro* and then reintroduced into mammalian cells. This allows the production of novel chimaeric antibodies which can be used in therapeutic procedures.

KEYWORDS

Chimaeric antibodies; DNA manipulation; transfection.

INTRODUCTION

The production of monoclonal antibodies (Köhler and Milstein, 1975) involves the immortalisation of specific lymphocytes expressing particular antibody molecules. Since the first production of such monoclonal antibodies, it has been desirable to produce large quantities of human monoclonal antibodies as therapeutically useful reagents. Monoclonal antibodies have already had considerable impact on biology and medicine and work has concentrated on isolating therapeutic monoclonal antibodies not only of the desired specificity but also exhibiting useful effector functions. In particular, a substantial effort has been made to establish immortalized human B-cell lines; major problems are encountered both with efficient *in vitro* immunization and by the establishment of lines secreting large quantities of human antibody. However, the application of recombinant DNA technology to immunoglobulins now allows one to design novel antibodies with desired characteristics rather than being limited by the repertoire of naturally occurring antibody producing cells. The production of engineered antibodies involves *in vitro* manipulation of DNA to construct the desired immunoglobulin genes followed by expression of the engineered genes by transfection into a suitable host cell (reviewed by Brüggemann and Neuberger, 1988).

Antibody polypeptides consist of heavy (H) chains and light (L) chains which are held together by disulphide linkage to form, for example, an H_2L_2 dimer. The variable regions of heavy and light chains (V_H and V_L) are implicated in

Advances in the Biosciences Vol. 74

antigen binding. There are five different heavy chain constant regions (C_μ, C_δ, C_γ, C_ε, C_α) - corresponding to the five immunoglobulin classes - which can exhibit a variety of different effector activities.

Antibody genes consist of a number of segments. The heavy chain variable region is composed of a variable gene, a joining segment and a diversity segment which are juxtaposed to a heavy chain constant region in a DNA re-arrangement process (V_H-D-J_H joining). Similarly, a light chain is composed of a variable gene, a joining segment (V_L-J_L joining) and a constant region. After successful DNA rearrangement of heavy and light chain genes functional antibodies are produced in lymphoid cells.

It is possible to isolate antibody variable or constant regions and to manipulate these genes *in vitro* prior to their re-expression in mammalian cells. This allows the production of novel antibodies which can be used in therapeutic procedures. The *in vitro* manipulation of immunoglobulin genes essentially involves two types of procedures: either site directed mutagenesis or the re-arrangements of gene segments by use of restriction endonuclease cleavage.

PROKARYOTIC AND YEAST ANTIBODY EXPRESSION SYSTEMS

Escherichia coli is certainly the most extensively used host for the expression of transfected DNA. The advantages of producing antibodies in lower eukaryotes or in *E. coli* are the simple culture facilities and high yields. Immunoglobulin genes are, of course, eukaryotic genes and contain introns. Thus, the expression of mammalian genes in *E. coli* or yeast essentially requests the use of cDNAs. Subsequently plasmids have been constructed in which expression of immunoglobulin gene cDNAs is driven by *E. coli* or yeast promoters (Boss *et al.*, 1984 ; Cabilly *et al.*, 1984; Wood *et al.*, 1985). However, the production of intact immunoglobulin in lower organisms appears to be difficult as there are problems with efficient chain assembly, disulphide linkage formation and secretion as well as appropriate glycosylation.

To give some examples: Wood *et al.* (1985) report synthesis, processing and secretion of functional antibodies in yeast. However, efficiencies are extremely poor. Expression of heavy and light chain cDNAs in *E. coli* has also been described (Boss *et al.*, 1984; Wood *et al.*, 1984). Both chains are insoluble in bacteria, synthesized in inclusion bodies, but could be purified and *in vitro* reconstituted with very low efficiency. Although it has essentially been shown that functional antibody can be obtained in low yield from *E. coli*, greater success has been achieved in the production of immunoglobulin fragments. Cabilly *et al.* (1984) have shown the expression of truncated heavy and light chain (Fd' fragment) obtained from a cDNA of a hybridoma antibody with specificity for carcinoma embryonic antigen. The antibody consists of V_H and V_L whilst the constant regions of heavy and light chain are incomplete. Production of heavy chain fragments in *E. coli* has also been shown for human ε domains (Kurokawa *et al.*, 1983). The expression resulted in monomeric ε chain fragments of the correct molecular weight and ε domain aggregates, possibly promoted by disulphide linkage. The yield of ε chain peptides can be up to 32 μg/ml culture supernatants. Thus, production of ε peptides is very encouraging and might be extended for further competition studies with natural IgE for binding of membrane receptors. Functional activity has also been demonstrated for larger ε chain fragments encoding C_H2-C_H3-C_H4 and Kenten *et al.* (1984) report the purification of dimers on antigen columns and inhibition of binding to human Fc_ε receptors.

Thus, there is some application of these lower organisms for production of antibody fragments.

EUKARYOTIC ANTIBODY EXPRESSION SYSTEMS

Stable introduction of foreign DNA into suitable host cells requires selective pressure applied to an expressed marker gene. Initially, techniques developed for selecting transfectants made use of the thymidine kinase (tk) gene from *herpes simplex* virus (Pellicer *et al.*, 1978). Mouse tk^- L cells were transfected with vectors containing the viral tk gene (Graham and van der Eb, 1973) and tk^+ transfectants could be selected in HAT (hypoxanthine/aminopterin/thymidine) medium. The disadvantage of this powerful selection method is that only host cells deficient in thymidine kinase could be used. Subsequently vectors have been constructed with dominant selectable markers such as gpt which confers resistance to mycophenolic acid (Mulligan and Berg, 1981), neo which confers resistance to the antibiotics G418 (Davies and Jiminez, 1980) and hyg which confers resistance to hygromycin (Blochlinger and Diggelman, 1984). These markers allow the use of effectively any recipient cell type. In addition to driving expression of these bacterial drug resistance genes in mammalian cells, eukaryotic expression vectors (for example the pSV series described by Mulligan and Berg (1981)) also contain SV40 sequences which provide promoter, intron and polyadenylation signals, as well as the ampicillin-resistance gene (amp) and an origin of replication to allow selection and propagation in *E. coli*.

A pSV-based vector was constructed by Neuberger *et al.* (1985). This contains a rearranged variable region (V_HDJ_H) linked to the IgH enhancer with a downstream cloning site for heavy chain genes. Similar vector constructs contain immunoglobulin light or heavy chains (Rice and Baltimore, 1982; Oi *et al.*, 1983; Ochi *et al.*, 1983a; Neuberger, 1983; Potter *et al.*, 1984; Deans *et al.*, 1984). Using both neo and gpt selection, light and heavy chain genes can be co-transfected. Alternatively, plasmids have been constructed in which both heavy and light chain genes are incorporated into the same plasmid (Ochi *et al.*, 1983).

Expression of productively rearranged immunoglobulin genes introduced into the germline of transgenic mice has also been obtained (Grosschedl *et al.*, 1984; Storb *et al.*, 1984; Yamamura *et al.*, 1986; reviewed by Storb, 1987). It is possible that expression of transgenic immunoglobulin genes in the body fluids (eg. colostrum, milk) of transgenic mammals may prove a useful way of producing large quantities of monoclonal antibodies as well as facilitating the provision of passive immunity. It has also been shown that rearrangement of the introduced light chain locus occurs in the lymphoid lineage of mice transgenic for part of the chicken C_λ or rabbit C_κ loci injected in germ line configuration (Bucchini *et al.*, 1987; Goodhardt *et al.*, 1987). Thus there may be some scope for applications of this technology when it becomes possible to inject large DNA fragments or segments containing multiple immunoglobulin V and C genes.

TRANSFECTING DNA INTO CELLS

Various methods have been described for the introduction of exogenous DNA into mammalian cells; calcium phosphate co-precipitation (Graham and van der Eb, 1973; Oi *et al.*, 1983), DEAE dextran treatment (McCutchan and Pagano, 1968; Neuberger, 1983; Banerji *et al.*, 1983), protoplast or spheroplast fusion (Oi *et al.*, 1983; Neuberger and Williams, 1985) and electroporation

(Potter *et al.*, 1984). The efficiency of each technique is dependent upon the cells being used for transfection. Electroporation, a method developed by Zimmermann *et al.* (1982) to introduce linearized DNA into cells by high-voltage electric discharge, has been used very successfully with lymphoid cell-lines. The frequencies for stable transfectants obtained by Potter *et al.* (1984) for various cell types are in the order of 10^{-4}. Electroporation is widely applicable and is the easiest method for introducing DNA into a variety of cell types.

DETECTION OF ANTIBODY PRODUCED

Essentially, detection techniques to identify chimaeric antibodies are similar to those described for assaying monoclonal antibodies. For example, using an ELISA (enzyme-linked-immuno-sorbent-assay), RIA (radioimmunoassay) or haemagglutination assay, antibody in the culture supernatant can be detected from stably transfected cells. In transient expression, cytoplasmic staining with fluorescein-conjugated antisera can be used to detect antibody production. Other methods of detection include plaque assays, biosynthetic labelling of secreted proteins or analysis of cytoplasmic RNA.

CHIMAERIC ANTIBODIES ASSEMBLED FROM MOUSE GENES

A large number of reports describe the expression of transfected antibody heavy and light chain genes using plasmids which have been assembled from mouse immunoglobulin gene segments (reviewed by Morrison and Oi, 1984). For immunoglobulin gene expression, myeloma and hybridoma cell lines are used, as cells of the lymphoid lineage are capable of sufficient antibody secretion/processing and, in the case of making IgM or IgA these cells provide the immunoglobulin J chain. It has been shown that in lymphoid cell variants, defective for synthesis of particular heavy or light chains, specific antibody production can be restored by introduction of light (Ochi *et al.*, 1983; Pepe *et al.*, 1986) or heavy chain genes (Neuberger, 1983). A similar approach, but using a mutant heavy chain transfected into a light chain producing cell line, allowed to determine the involvement of particular heavy and light chain residues in antigen-binding (Brüggemann *et al.*, 1986a). However, high yield antibody production can not only be achieved in lymphoid tissues. Secretion of single heavy and light chains as well as complete antibodies has been shown for COS cells and Chinese hamster ovary cells (Weidle *et al.*, 1987; Whittle *et al.*, 1987). In addition, Cattaneo and Neuberger (1987) showed the secretion of a functional IgM antibody from glioma cells in the absence of J chain.

CHIMAERIC ANTIBODIES WITH HUMAN EFFECTOR FUNCTIONS

Some of the desired features of therapeutically useful antibodies might include diminished immunogenicity and the possession of human effector functions. Expression of mouse V regions coupled to human C regions has been described (Morrison *et al.*, 1984; Boulianne *et al.*, 1984; Neuberger *et al.*, 1985). Several groups have reported the construction of chimaeric human IgG1 or IgG3 antibodies specific for human carcinoma cells (Sahagan *et al.*, 1986; Sun *et al.*, 1986; Liu *et al.*, 1987). After coexpression of light and heavy chain containing vectors tumor specific antibodies have been obtained. Indeed it is clear that such chimaeric antibodies retain the antigen specificity determined by the mouse variable region and they also exhibit the effector functions determined by the human C region. It has been shown that

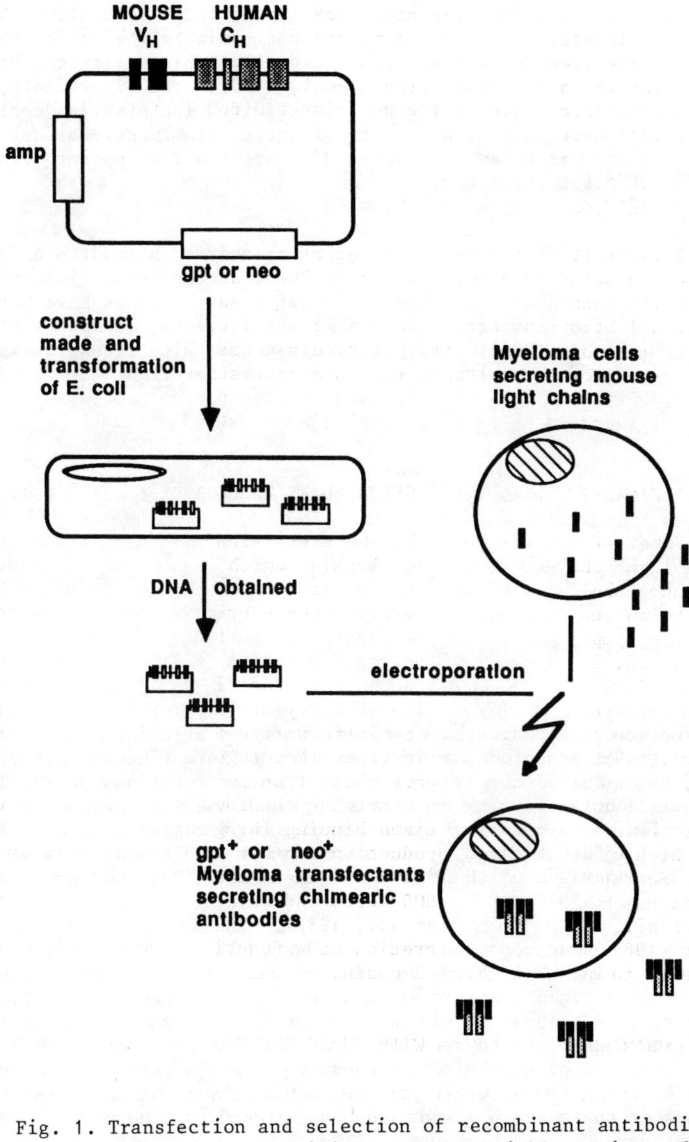

Fig. 1. Transfection and selection of recombinant antibodies. A construct containing a mouse variable region and a human constant region is amplified in *E. coli*. The DNA is being transfected into cells of a light chain secreting myeloma line by electroporation. Transfectants producing antibodies can be selected in the presence of mycophenolic acid or the drug G418.

chimaeric antibodies are glycosylated although the carbohydrate content may differ between the chimaera and the cell originally producing the antibody. A set of chimaeric human antibodies (IgM, IgG3, IgG1, IgG2, IgG4, IgE, IgA2) has been constructed that use the same mouse variable region for the heavy chain (V_{NP}, anti-4-hydroxy-3-nitro-phenacetyl) combined with the different human constant regions (Brüggemann *et al.*, 1987). The light chain, also of mouse origin, is provided by the cell line into which plasmids containing the chimaeric heavy chain gene are transfected. In functional assays it was found that chimaeric human IgG1 antibodies are the most potent regarding a variety of effector functions.

Table 1. Effector functions of chimaeric antibodies of the same specificity. The effector activities of five human antibodies of different isotypes have been compared (Brüggemann *et al.*, 1987; Bindon *et al.*, 1988). (+, positive reaction with strength indicated by multiple +; -, negative reaction)

	Human IgM	IgG1	IgG3	IgG2	IgG4
Lysis	+++	+++	+	-	-
C1q binding	+	++	+++	-	-
ADCC	-	+++	+	-	-
Protein A binding	-	+++	-	+++	+++

Thus, chimaeric antibodies have proved of use in allowing the production of a matched set of monoclonal antibodies with different human C_H regions, thus allowing a controlled comparison of the effector functions of the different classes and subclasses. An analogous approach has also proved invaluable in other species. The construction and analysis of rat chimaeric antibodies has allowed different rat C_H genes to be typed serologically as well as allowing a comparison of their effector functions (Brüggemann *et al.*, 1986; Bindon *et al.*, 1988).

As there is no human monoclonal IgE antibody of known specificity available it is worth noting that chimaeric antibody technology has allowed the production of an antigen specific monoclonal chimaeric human IgE antibody (Neuberger *et al.*, 1985). This has proved to be a useful reagent in analysis and purification of the Fc receptor for IgE. The monoclonal IgE is able to block the release of histamine from human basophils which can be triggered *in vitro* by sera from allergic patients. Clearly it is important to establish whether analogous blocking can be achieved *in vivo* using skin sensitisation assays.

ANTIBODY VARIABLE REGIONS ALTERED BY SITE DIRECTED MUTAGENESIS

The domain structure of the antibody polypeptide chains makes it possible to exchange whole domains between various immunoglobulins and still retain a functional antibody. A successful approach to substitute the complementarity -determining regions (CDRs) of a human antibody with those from a mouse antibody and simultaneously transfer the antigen specificity with the mouse CDRs has been described by Jones *et al.* (1986). A synthetic V_H region was constructed that has the framework residues of a human antibody and the CDRs of a mouse antibody. If such an approach is generally valid it may mean a substantial simplification to the application of antibody engineering technology to new monoclonal antibodies. Indeed, recently a human antibody has been constructed by introduction of the hypervariable regions of a rodent antibody already used for therapy in humans (Riechmann *et al.*, 1988). This reshaped, or fully humanized, antibody is most effective in lysis of human target cells.

Clearly site directed mutagenesis will allow a detailed study of antibody-antigen interactions. The exchange of particular variable region residues to study antibody-antigen interaction has been shown (Sharon *et al.*, 1986; Roberts *et al.*, 1987) as well as Fc effector function interaction (Baker *et al.*, 1986; Helm *et al.*, 1988; Duncan *et al.*, 1988; Duncan and Winter, 1988).

ENZYME-ANTIBODIES

Antibodies are widely used in immunoassays in which, for example, an enzyme is conjugated to the antibody. Transfectants have been established that secrete $F(ab')_2$ or Fab like molecules, thus confirming that the Fc portion of an antibody is not necessary for secretion (Sharon *et al.*, 1984; Neuberger *et al.*, 1985). Subsequently it has been shown that the C_H2 and C_H3 exons can be replaced by enzymatic moities. Recombinant "enzyme"-bodies have been made in which the nuclease of *Staphylococcus aureus* (Neuberger *et al.*, 1984) or the Klenow fragment of *E. coli* DNA polymerase I (polIK) (Williams and Neuberger, 1986) are piggy-backed onto an antibody Fab. Whereas the Fab-nuclease illustrates the potential of such hybrid proteins for immunoassays, the Fab-PolIK (an enzyme used in DNA sequencing) illustrates how tagging polIK with a hapten specific antibody Fab moiety allows easy preparation and purification. It still remains unclear how widely the fusion protein approach will be used. It is likely that almost any gene can be backpacked on an immunoglobulin molecule although problems might be encountered with the activity, toxicity or secretion of the resultant recombinant. Difficulties might also be encountered with enzymes not active as monomers, or enzymes that are sensitive to fusion to the NH_2-terminus.

OUTLOOK

Novel chimaeric antibodies can be produced by gene manipulation. Successful antibody production has been achieved in eukaryotic expression systems and to a lesser extent in prokaryotic and yeast systems. A number of chimaeric antibodies have been produced although, so far, no chimaeric antibody has been tested in therapy of human diseases. Rodent monoclonal antibodies are currently being used for tumor imaging and for depletion of alloreactive lymphocytes responsible for graft-versus-host disease in transplant patients (Ritz and Schlossman, 1982; Hale *et al.*, 1985). A current aim is therefore to alter the specific variable region of mouse or rat anti-T cell antibodies to make them "human"-like and combine them with a human constant region of desired effector function. Such an antibody should be less immunogenic and might escape the surveillance of the human immune system. In addition, site directed mutagenesis might be used for detailed mapping of residues involved in effector functions like binding of components of the complement system or Fc receptor interactions. The construction of recombinant antibodies with novel effector functions such as enzymes might not only be used in diagnostic procedures but one might conjugate toxic reagents, like the rich A chain or diphtheria toxin, onto such anti-tumor antibodies. Thus, the antibody engineering technology allows one to design 'magic bullets' based upon the antibody combining site.

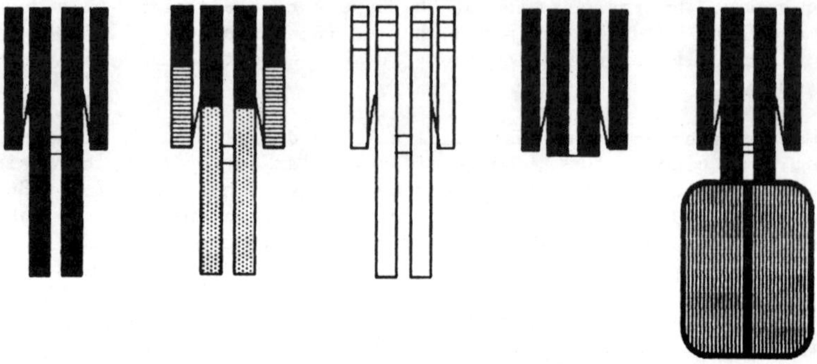

Fig. 2. Schematic view of various recombinant antibodies.

ACKNOWLEDGEMENTS

I am a recipient of a Leukemia Society of America Special Fellowship.

REFERENCES

Baker, M.D., Wu, G.E., Toone, W.M., Murialdo, H., Davis, A.C. and Shulman, M.J. (1986). A region of the immunoglobulin-μ heavy chain necessary for forming pentameric IgM. J. Immunol., 137, 1724-1728.

Banerji, J., Olson, L. and Schaffner, W. (1983). A lymphocyte-specific cellular enhancer is located downstream of the joining region in immunoglobulin heavy chain genes. Cell, 33, 729-740.

Bindon, C.I., Hale, G., Brüggemann, M. and Waldmann, H. (1988). Human monoclonal IgG isotypes differ in complement activating function at the level of C4 as well as C1q. J. Exptl. Med., in press.

Blochlinger, K. and Diggelmann, H. (1984). Hygromycin B phosphotransferase as a selectable marker for DNA transfer experiments with higher eukaryotic cells. Mol. Cell. Biol., 4, 2929-2931.

Boss, M.A., Kenten, J.H., Wood, C.R. and Emtage, J.S. (1984). Assembly of functional antibodies from immunoglobulin heavy and light chains synthesised in *Escherichia coli*. Nucl. Acids Res., 12, 3791-3806.

Boulianne, G.L., Hozumi, N. and Shulman, M.J. (1984). Production of functional chimaeric mouse/human antibody. Nature, 312, 643-646.

Brüggemann, M. and Neuberger, M. (1988). Novel antibodies by DNA manipulation. Prog. Allergy, 45, 1-15.

Brüggemann, M., Free, J., Diamond, A., Howard, J., Cobbold, S. and Waldmann, H. (1986). Immunoglobulin heavy chain locus of the rat. Striking homology to mouse antibody genes. Proc. Natl. Acad. Sci. USA, 83, 6075-6079.

Brüggemann, M., Müller, H-J., Burger, C. and Rajewsky, K. (1986a). Idiotypic selection of an antibody mutant with changed hapten binding specificity, resulting from a point mutation in position 50 of the heavy chain. EMBO J., 5, 1561-1566.

Brüggemann, M., Williams, G.T., Bindon, C.I., Clark, M.R., Walker, M.R., Jefferis, R., Waldmann, H. and Neuberger, M.S. (1987). Comparison of the effector functions of human immunoglobulins using a matched set of chimaeric antibodies. J. Exptl. Med., 166, 1351-1361.

Bucchini, D., Reynaud, C-A., Ripoche, M-A., Grimal, H., Jami, J. and Weill, J-C. (1987). Rearrangement of a chicken immunoglobulin gene occurs in the lymphoid lineage of transgenic mice. Nature, 326, 409-411.

Cabilly, S., Riggs, A.D., Pande, H., Shiveley, J.E., Holmes, W.E., Rey, M., Perry, L.J., Wetzel, R. and Heynecker, H.L. (1983). Generation of antibody activity from immunoglobulin polypeptide chains produced in *Escherichia coli*. Proc. Natl. Acad., Sci. USA, 81, 3273-3277.

Cattaneo, A. and Neuberger, M. (1987). Polymeric immunoglobulin M is secreted by transfectants of non-lymphoid cells in the absence of immunoglobulin J chain. EMBO J., 6, 2753-2758.

Davies, J. and Jiminez, A. (1980). A new selective agent for eukaryotic cloning vectors. Am. J. Trop. Med. Hug. 29, 5, 1089-1092 (suppl.).

Deans, R.J., Denis, K.A., Taylor, A. and Wall, R. (1984). Expression of an immunoglobulin heavy chain gene transfected into lymphocytes. Proc. Natl. Acad. Sci. USA, 81, 1292-1296.

Duncan, A.R. and Winter, G. (1988). The binding site for C1q on antibodies. Nature, 332, 738-740.

Duncan, A.R., Woof, J.M., Partridge, L.J., Burton, D.R. and Winter, G. (1988) Localization of the binding site for the human high affinity Fc receptor on IgG. Nature, 332, 563-564.

Graham, F.L. and van der Eb, A.J. (1973). A new technique for the assay of infectivity of human adenovirus 5 DNA. Virology, 52, 456-467.

Goodhardt, M., Cavalier, P., Akimenko, M.A., Lutfalla, G., Babinet, C. and Rougeon, F. (1987). Rearrangement and expression of rabbit immunoglobulin κ light chain gene in transgenic mice. Proc. Natl. Acad. Sci. USA, 84, 4229-4233.

Grosschedl, R., Weaver, D., Baltimore, D. and Costantini, F. (1984). Introduction of an α immunoglobulin gene into the mouse germ line. Specific expression in lymphoid cells and synthesis of functional antibody. Cell, 38, 647-658.

Hale, G., Swirsky, D., Waldmann, H. and Chan, L.C. (1985). Reactivity of rat monoclonal antibody CAMPATH-1 with human leukaemia cells and its possible application for autologous bone marrow transplantation. British J. Haematol., 60, 41-48.

Helm, B., Marsh, P., Vercelli, D., Padlan, E., Gould, H. and Geha, R. (1988). The mast cell binding site on human immunoglobulin E. Nature, 331, 180-183.

Jones, P.T., Dear, P.H., Foote, J., Neuberger, M.S. and Winter, G. (1986). Replacing the complementarity-determining regions in a human antibody with those from a mouse. Nature, 321, 522-525.

Kenten, J., Helm, B., Ishizaka, T., Cattini, P. and Gould, H. (1984). Properties of a human immunoglobulin ε-chain fragment synthesized in *Escherichia coli*. Proc. Natl. Acad. Sci. USA, 81, 2955-2959.

Köhler, G. and Milstein, C. (1975). Continuous culture of fused cells secreting antibody of pre-defined specificity. Nature, 256, 495-497.

Kurokawa, T., Seno, M., Sasada, R., Ono, Y., Onda, H., Igarashi, K., Kikushi,

M., Sugino, Y. and Honjo, T. (1983). Expression of human immunoglobulin E ε chain cDNA in *E. coli*. Nucl. Acids Res., 10, 3077-3085.

Lui, A.Y., Robinson, R.R., Hellström, K.E., Murray, E.D.Jr., Chang, C.P. and Hellström, I. (1987). Chimeric mouse-human IgG1 antibody that can mediate lysis of cancer cells. Proc. Natl. Acad. Sci. USA, 84, 3439-3443.

McCutchan, J.H. and Pagano, J.S. (1968). Enhancement of the infectivity of simian virus 40 deoxyribonucleic acid with diethyl-aminoethyl-dextran. J. Natl. Cancer Inst., 41, 351-357.

Morrison, S.L. and Oi, V.T. (1984). Transfer and expression of immunoglobulin genes. Ann. Rev. Immunol., 2, 239-256.

Morrison, S.L., Johnson, M.J., Herzenberg, L.A. and Oi, V.T. (1984). Chimeric human antibody molecules: mouse antigen-binding domains with human constant region domains. Proc. Natl. Acad. Sci. USA, 81, 6853-6855.

Mulligan, R.C. and Berg, P. (1981). Selection for animal cells that express the *Escherichia coli* gene coding for xanthine-guanine phosphoribosyltransferase. Proc. Natl. Acad. Sci. USA, 75, 2072-2076.

Neuberger, M.S. (1983). Expression and regulation of an immunoglobulin heavy chain gene transfected into lymphoid cells. EMBO J. 2, 1373-1378.

Neuberger, M.S. (1985). Making novel antibodies by expressing transfected immunoglobulin genes. Trends in Biochem. Sci., 10, 347-349.

Neuberger, M.S., Williams, G.T. and Fox, R.O. (1984). Recombinant antibodies possessing novel effector functions. Nature, 312, 604-608.

Neuberger, M.S., Williams, G.T., Mitchell, E.B., Jouhal, S.S., Flanagan, J.G. and Rabbitts, T.H. (1985). A hapten-specific chimaeric immunoglobulin ε antibody with human physiological effector function. Nature, 314, 268-270.

Neuberger, M.S. and Williams, G.T. (1985). Protein engineering of antibody molecules. In:Protein engineering application in science, medicine and industry. (eds. M. Inouye and R. Saruna), Vol.19, Academic Press, New York, pp311-317.

Ochi, A., Hawley, R.G., Hawley, T., Shulman, M.J., Traunecker, A., Köhler, G. and Hozumi, N. (1983). Functional immunoglobulin M production after transfection of cloned immunoglobulin heavy and light chain genes into lymphoid cells. Proc. Natl. Acad. Sci. USA, 80, 6351-6355.

Ochi, A. Hawley, R.G., Schulman, M.J. and Hozumin, N. (1983a). Transfer of a cloned immunoglobulin light chain gene to mutant hybridoma cells restores specific antibody production. Nature, 302, 340-342.

Oi, V.T., Morrison, S.L., Herzenberg, L.A. and Berg, P. (1983). Immunoglobulin gene expression in transformed lymphoid cells. Proc. Natl. Acad. Sci. USA, 80, 825-829.

Pellicer, A., Wigler, M., Axel, R. and Silverstein, S. (1978). The transfer of stable integration of the HSV thymidine kinase gene into mouse cells. Cell, 14, 133-141.

Pepe, V.H., Sonenshein, G.E., Yoshimura, M.I. and Shulman, M.J. (1986). Gene transfer of immunoglobulin light chain restores heavy chain secretion. J. Immunol., 137, 2367-2372.

Potter, H., Weir, L. and Leder, P. (1984). Enhancer-dependent expression of human κ immunoglobulin genes introduced into mouse pre-B lymphocytes byelectroporation. Proc. Natl. Acad. Sci. USA, 81, 7161-7165.

Rice, D. and Baltimore, D. (1982). Regulated expression of an immunoglobulin κ gene introduced into a mouse lymphoid cell line. Proc. Natl. Acad. Sci. USA, 79, 7862-7865.

Riechmann, L., Clark, M., Waldmann, H. and Winter, G. (1988). Reshaping human antibodies for therapy. Nature, 332, 323-327.

Ritz, J. and Schlossman, S.F. (1982). Utilization of monoclonal antibodies in the treatment of leukemia and lymphoma. Blood, 59, 1-11.

Roberts, S., Cheetham, J.C. and Rees, A.R. (1987). Generation of an antibody with enhanced affinity and specificity for its antigen by protein engineering. Nature, 328, 731-734.

Sahagan, B.G., Dorai, H., Saltzgaber-Muller, J., Toneguzzo, F., Guindon, C.A., Lilly, S.P., McDonald, K.W., Morrissey, D.V., Stone, B.A., Davis, G.L., McIntosh, P.K. and Moore, G. (1986). A genetically-engineered murine-human chimaeric antibody retains specificity for human tumor-associated antigen. J. Immunol., 137, 1066-1074.

Sharon, J., Gefter, M.L., Manser, T., Morrison, S.L., Oi, V.T. and Ptashne, M. (1984). Expression of V_HC_κ chimaeric protein in mouse myeloma cells. Nature, 309, 364-367.

Sharon, J., Gefter, M.L., Manser, T. and Ptashne, M. (1986). Site-directed mutagenesis of an invariant amino acid residue at the variable-diversity segments junction of an antibody. Proc. Natl. Acad. Sci. USA, 83, 2628-2631.

Storb, U. (1987). Transgenic mice with immunoglobulin genes. Ann. Rev. Immunol., 5, 151-174.

Storb, U., O'Brien, R.L., McMullen, M.D., Gollahon, K.A. and Brinster, R.L. (1984). High expression of cloned immunoglobulin κ gene in transgenic mice is restricted to B lymphocytes. Nature, 310, 238-241.

Sun, L.K., Curtis, P., Rakowicz-Szulczynska, E., Ghrayeb, J., Morrison, S.L., Chang, N. and Koprowski, H. (1986). Chimaeric antibodies with 17-1A-derived variable and human constant regions. Hubridoma, 5, suppl. 1, 17-20.

Weidle, U.H., Borgya, A., Mattes, R., Lenz, R. and Buckel, P. (1987). Reconstitution of functionally active antibody directed against creatine kinase from separately expressed heavy and light chains in non-lymphoid cells. Gene, 51, 21-29.

Whittle, N., Adair, J., Lloyd, C., Jenkins, L., Devine, J., Schlom, J., Raubitschek, A., Colcher, D. and Bodmer, M. (1987). Expression in COS cells of a mouse-human chimaeric B72.3 antibody. Protein Engineering, 1, 499-505.

Williams, G.T. and Neuberger, M.S. (1986). Production of antibody-tagged enzymes by myeloma cells: application to DNA polymerase I Klenow fragment. Gene, 43, 319-324.

Wood, C.R., Boss, M.A., Patel, T.P. and Emtage, J.S. (1984). The influence of messenger RNA secondary structure on expression of an immunoglobulin heavy chain in *Escherichia coli*. Nucl. Acids Res., 12, 3937-3950.

Wood, C.R., Boss, M.A., Kenten, J.H., Calvert, J.E., Roberts, N.A. and Emtage, J.S. (1985). Synthesis and *in vivo* assembly of functional antibody in yeast. Nature, 314, 446-449.

Yamamura, K-I., Kudo, A., Ebihara, T., Kamino, K., Araki, K., Kumahara, Y. and Watanabe, T. (1986). Cell-type-specific and regulated expression of a human γ1 heavy-chain immunoglobulin gene in transgenic mice. Proc. Natl. Acad. Sci. USA, 83, 2152-2156.

Zimmermann, U. and Vienken, J. (1982). Electric field-induced cell-to-cell fusion. J. Membr. Biol., 67, 165-182.

The Location of Effector Function Sites on Immunoglobulin G

Alexander R. Duncan

Department of Biology, Q-063, University of California, San Diego, La Jolla, CA 92093, USA

ABSTRACT

A mouse IgG2b antibody with specificity for the hapten 3-iodo-4-hydroxy-5-nitrophenylacetate (NIP), and the ability to fix complement, has been subjected to surface scanning mutations in order to determine the attachment site for (1) C1q and (2) the high affinity receptor for IgG on human monocytes (FcRI). The C1q binding motif was defined as Glu(318)-x-Lys(320)-x-Lys(322) and removal of the polar side chains from these three critical amino acids rendered the antibody unable to fix complement. Since the motif is common to most human and rodent antibodies its removal has obvious benefits for *in vivo* serotherapy; we have also shown that a peptide based on the motif can inhibit lysis by complement and may have a clinical role in complement mediated disease. The mouse antibody was unable to inhibit the binding of human IgG1 to FcRI on human monocytes, but we have used site-directed mutagenesis to alter Glu(235) in the γ2b heavy chain to Leu(235) and shown that the mutant behaves essentially as human IgG1. This engineering of a rodent antibody also has implications for the potential of monoclonal antibodies in tumour therapy.

KEYWORDS

Antibodies; protein engineering; complement; FcR.

INTRODUCTION

Antibodies act as adaptor molecules in immune defence. The N-terminal variable domains of an antibody molecule bind antigen and trigger the binding and activation of effector molecules through the C-terminal constant regions. The exquisite specificity of the variable domains coupled with the ability to trigger a limited number of effector functions, even across different species, has prompted their use in therapy. Indeed, xenogeneic antiserum has been used for the prevention and treatment of human infections such as tetanus, diphtheria, gas gangrene and others, and in the treatment of toxic snake and insect bites, since the turn of the century. However, it is limited by the response of the recipient to the foreign protein and can result in allergic (Type I) or immune-complex (Type III) hypersensitivity. Interest in the field of immunotherapy was given new impetus with the introduction of the hybridoma technique by Kohler and Milstein (1975), and clinical trials using monoclonal antibodies, mostly of rodent origin, have been reported since the early 1980's (Nadler *et al.*, 1980; Epenetos *et al.*, 1985; Hale and Waldmann, 1985). Further advances seem likely with the development of molecular biological techniques, such as gene transfection and protein engineering. Thus, chimaeric antibodies have been produced by 'exon shuffling' of human and mouse antibody genes and transfection into lymphoid cells. Antibodies with mouse variable regions and human constant regions have been expressed and secreted which both bind antigen and trigger effector functions (Morrison *et al.*, 1984; Boulianne *et al.*, 1984; Neuberger *et al.*, 1985). These reagents may become even more useful for *in vivo* serotherapy if the location of the effector function binding sites were determined. Recently, the mast cell binding site on human IgE has been mapped to a 76 amino acid segment of polypeptide chain at the junction of the second and third constant domains (Helm *et al.*, 1988). The localization of effector function binding sites on IgG for the first subcomponent of complement, C1q, and the high affinity Fc receptor for IgG(FcRI) on human cells is now described.

1. Binding Site for C1q

The first step in the complement cascade is the binding of C1q to the Fc region of the antibody in the CH2 domain (Kehoe and Fougereau, 1969; Connell and Porter, 1971; Colomb and Porter, 1975). C1q has six globular heads connected by collagen like stems to a central stalk. C1q interacts with antibody via its heads, but the isolated heads bind to antibody rather weakly, with an affinity of 10^4 M^{-1}. Since isolated Fc or CH2 domains bind to C1q with comparable affinities (Yasmeen *et al.*, 1976), it appears that each head interacts separately with a single antibody molecule. However, binding of antibody to multiple epitopes arrayed on an antigen surface facilitates the binding of several C1q heads. This leads to enhanced binding of C1q, with an affinity of about 10^8 M^{-1} (see review by Burton, 1985), and triggering of complement lysis.

Several sites on the CH2 domain have been proposed for binding of C1q, based on chemical modification, inhibition of complement lysis by synthetic peptides and oligosaccharides, and sequence comparisons of antibody isotypes (Burton, 1985). Taken together, most of the solvent accessible surface of the CH2 domain has been invoked as part of a C1q binding site. Recently, the binding site for tRNA on the tyrosyl tRNA synthetase was localised by 'surface scanning': systematically removing side chains (at the level of the gene) and scoring the mutants for altered binding of tRNA (Bedouelle and Winter, 1985). We have used this approach to localise the C1q binding site in mouse IgG2b. The interaction of C1q and antibody is sensitive to ionic strength, and appears to be highly conserved throughout evolution since C1q reacts with IgG from different sources (Burton, 1985). We therefore resected polar side chains from conserved surface residues.

A chimaeric antibody (Neuberger and Williams, 1986) with the heavy chain of mouse IgG2b isotype which binds to the hapten NIP (3-iodo-4-hydroxy-5-nitrophenylacetate) and fixes complement, was used as a model for surface-scanning. 23 point mutations of the mouse γ2b gene were constructed with synthetic mismatch primers in M13K19 (Carter *et al.*, 1985), mostly cutting the side chains back to alanine. The gene was reassembled with the heavy chain variable domain in a pSV2-gpt vector and transfected into the myeloma

J558L which supplies the mouse λ-light chain (Neuberger and Williams, 1986). Transfectants secreting mutant antibody were selected and cloned, and mutant antibody harvested on an affinity column of NIP-Sepharose (Duncan and Winter, 1988).

The mutant antibodies were tested both for complement lysis and for C1q binding (listed in Table 1 and shown diagrammatically in Fig. 1; Duncan and Winter, 1988). Most of the mutant antibodies lysed red cells coated with the hapten NIP at the same titre (3 μg/ml) as the wild type antibody. However, four mutants, Asn297→Ala, Glu318→Ala, Lys320→Ala and Lys322→Ala, were non-lytic. Asn297 is the carbohydrate attachment site, and Glu318 - Lys320 are located close together on one side of a strand of β sheet (Fig. 1). From previous work, removing carbohydrate from antibody by glycosidases or by treating the cells with glycosylation inhibitor tunicamycin, destroys the ability of the mouse γ2b isotype to fix complement (Nose and Wigzell, 1983). It therefore seems the mutant Asn297→Ala is non-lytic due to lack of carbohydrate. All four mutants bound to hapten, to protein A, and apart from Asn297→Ala, were able to rosette mouse macrophages by binding to the low affinity FcRII receptor. This indicates that there were no major changes in other functions of the antibody. The mutant antibodies were also scored for binding to the first component of complement C1q. Antibody was aggregated by binding to a solid support (Affigel) derivatised with the hapten NIP, binding of ^{125}I labelled C1q measured, and the affinity determined (Table 1; Duncan and Winter, 1988). In this assay, the affinity of the mouse γ2b isotype is 10^8 M^{-1}, and similar to that of the human γ1 and γ3 isotypes, and mouse γ2a (Leatherbarrow and Dwek, 1984). For the four mutants, the assay showed that the affinity of the Asn297→Ala antibody for C1q was three fold weaker than the wild type γ2b, confirming previous work with aglycosylated protein (Leatherbarrow *et al.*, 1985). The affinities of the three other mutants, Glu318→Ala, Lys320→Ala and Lys322→Ala for C1q were estimated at least 30 fold lower than the wild type (results not shown). Thus, the residues 318, 320 and 322 appear to be critical for binding of C1q to aggregated antibody.

This site proves to be part of the site proposed by Dwek and colleagues (Burton *et al.*, 1980). By identifying the conserved, surface accessible residues characteristic of CH2 domains, these authors proposed the last two (C-terminal β strands) of the CH2 domain as a C1q binding site, hinting at residues Glu318, Lys320, Lys322 on one strand, Pro331 at the loop, and Glu333, Thr335 and Ser337 on the other strand. This work confirmed the involvement of the first strand and ruled out that of the second.

The motif Glu318-X-Lys320-Y-Lys322 is conserved in all the human γ isotypes, rat γ2b and γ2c isotypes, and mouse γ2a, γ2b and γ3 isotypes. However, in mouse γ1, rat γ1 and rat γ2a there are substitutions of Arg322, Thr318 and Arg320, and Thr320 respectively (Burton, 1985; for rat sequences M. Brüggeman, personal communication). To evaluate the effect of these substitutions, the mutants Glu318→Thr, Lys320→Arg and Lys322→Arg were constructed in the mouse γ2b gene, but the mutant antibodies proved identical to wild type in complement lysis or binding (Table 1). Thus, the site can employ Glu or Thr at 318, and Lys or Arg at both 320 and 322 in binding C1q. To dissect the requirements for the C1q site further, the mutants Glu318→Val, Lys320→Gln and Lys322→Gln were constructed. These proved non-lytic (Table 1). Since valine (non-lytic) differs from threonine (lytic) only by the substitution of a methyl for a hydroxyl group, and glutamine (non-lytic) lacks the basic group of lysine or arginine (lytic), requirements for lysis could involve a hydrogen bond from 318 and electrostatic interactions from 320 and 322.

Table 1. Complement assays of mouse IgG2b mutants

IgGs	haemolysis titre (υg/ml)	Ka nM^{-1}
MoIgG2b	3	
MoIgG2b Glu235-Leu	3	10
MoIgG2b Ser239-Ala	3	
MoIgG2b Lys248-Ala	3	
MoIgG2b Ile253-Ala	3	9
MoIgG2b Asp265-Ala	3	
MoIgG2b Ser267-Ala	3	
MoIgG2b Asp270-Ala	3	
MoIgG2b Gln274-Ala	3	
MoIgG2b Glu283-Ala	3	
MoIgG2b His285-Ala	3	12
MoIgG2b Gln290-Ala	3	11
MoIgG2b Glu294-Ala	3	
MoIgG2b Asn297-Ala	X	31
MoIgG2b Lys317-Ala	3	
MoIgG2b Glu318-Ala	X	>300
MoIgG2b Glu318-Thr	3	11
MoIgG2b Glu318-Val	X	
MoIgG2b Lys320-Ala	X	>300
MoIgG2b Lys320-Arg	3	11
MoIgG2b Lys320-Gln	X	
MoIgG2b Lys322-Ala	X	>300
MoIgG2b Lys322-Arg	3	11
MoIgG2b Lys322-Gln	X	
MoIgG2b Lys326-Ala	3	
MoIgG2b Pro331-Gly	3	12
MoIgG2b Glu333-Ala	3	12
MoIgG2b Thr335-Ala	3	10
MoIgG2b Ser337-Ala	3	11
MoIgG2b Lys340-Ala	3	

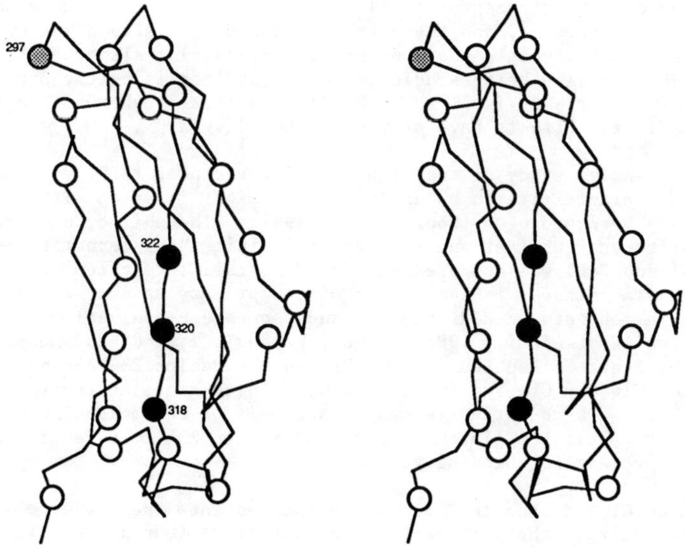

Fig. 1. Stereo pair of the CH2 domains from pooled human Fc (Deisenhofer, 1981). Those residues in which exposed side chains have been resected to alanine (or, in the case of proline 331, to glycine) are shown. The empty circles show mutants which are lytic; mutants unable to lyse cells are shown as filled circles: the carbohydrate attachment residue, Asn297 is shaded; Glu318, Lys320 and Lys322 are solid. The mutant EL235 is not shown since its position in the crystal structure is not known.

Synthetic peptides have been used to mimic parts of proteins and since the residues 318-322 are located in a short linear section of polypeptide chain, a peptide mimic should inhibit the complement cascade by binding to C1q. The peptide, or derivatives, could be valuable for therapeutic intervention in disease states in which complement fixation is symptomatic, such as in hereditary angiodema or autoimmune diseases. In previous work simple organic and inorganic ions (Bing, 1971) and peptides (Prystowsky *et al.*, 1981; Lukas *et al.*, 1981) have been found to inhibit the binding of C1q. For example, one peptide inhibited lysis at concentrations of 10^{-4} M (Lukas *et al.*, 1981). We designed a peptide based on the Glu-X-Lys-Y-Lys motif, with alanine residues as spacers and blocked termini: AcAla-Glu-Ala-Lys-Ala-Lys-Ala-NH2. This peptide also inhibited lysis at concentrations of approximately 1.7 x 10^{-4} M, similar to free antibody.

2. Binding Site for FcRI

Antibodies bind to cells through their Fc region and trigger a number of important and diverse biological responses including phagocytosis, release of inflammatory mediators, antibody-dependent cell-mediated cytotoxicity, placental transfer, and control of immunoglobulin production (Burton, 1985). Three types of IgG Fc receptor (FcR) in both human and mouse have been partially characterised, and these are expressed on distinct but overlapping haematopoetic cell types (Anderson and Looney, 1986). Although all these various receptors are being studied attempts at localising the structural features of an antibody involved in binding to these receptors and triggering their diverse effects have proved inconclusive.

In a comprehensive study of the high affinity receptor (FcRI) on human monocyte cells a single site on human IgGs which might interact with this receptor has been proposed (Woof *et al.*, 1984). The ability of a number of IgGs of different subclass and species to inhibit the interaction between human IgG1 and FcRI was compared and found to fall into three groups, tight, medium and weak binders. Examination of the primary sequences revealed a good correlation between FcRI binding and sequence conservation in the lower hinge region (Woof *et al.*, 1986). Thus, human IgG1, IgG3 and mouse IgG2a contain the sequence Leu Leu Gly Gly Pro in the region 234-238 but this becomes Leu Glu Gly Gly Pro in mouse IgG2b, which binds at least 100-fold more weakly. We used site-directed mutagenesis of the mouse IgG2b heavy chain gene to alter the glutamic acid at position 235 to a leucine and tested the mutant protein for binding to FcRI.

The mutation Glu235 → Leu (EL235) was introduced into the γ2b gene in an analogous manner to the surface-scanning mutations (Duncan *et al.*, 1988; Duncan and Winter, 1988). The purified wildtype mouse γ2b antibody did not inhibit the binding of ^{125}I labelled human IgG to the high affinity Fc receptor on U937 cells. The mutant γ2b EL235, however, shows inhibitory ability comparable to pooled human IgG. Direct binding of ^{125}I-labelled EL235 to U937 cells was quantified using Scatchard analysis and gave a mean Ka value of $3.13 \pm 0.58 \times 10^8$ M^{-1} with $3.6 \pm 0.35 \times 10^3$ receptors per cell, and shows that the mutant EL235 behaves essentially as human IgG1 (Duncan *et al.*, 1988).

Here we have shown that a single point mutation in a mouse IgG molecule can generate full Fc receptor binding activity from a wildtype antibody which does not bind human cells. This demonstrates the importance of the lower hinge region for binding to this Fc receptor. The mutant retains protein A and C1q binding and it seems unlikely that the mutation causes any structural change, so as to reveal the binding site. Furthermore, Fab arm obstruction of this site is not implicated since the isolated Fc of IgG2b also exhibits very weak affinity for FcRI and finally, sequence differences in the lower hinge region of other non-FcRI binders are diverse and unlikely to produce a common perturbation and suggests that this residue is a major determinant in FcRI binding.

By scanning the surface of the CH2 domain and looking for loss of activity (down mutants) and by testing a prediction and looking for gain of activity (up mutant) two binding sites on IgG constant regions have been localised. Thus, we have made an antibody non-complement lytic by removing the C1q binding motif. Since the antibody retains the ability to bind to cell receptor, this could be useful for serotherapy. For example, the administration of rodent antibodies to humans can result in immune complex hypersensitivity, in which the foreign antibody forms aggregates with host antibody, triggers complement lysis resulting in release of chemotactic C5a, recruitment of monocytes and inflammation. Mutations such as EL235 in the FcRI site demonstrate that protein engineering will not just aid the identification of binding sites but will also allow the turning on or off of effector functions. This will enable us to dissect the involvement of this and other FcRs in immune function. It, too, has potential implications for use in therapy. For example, monoclonal antibodies are being used for guided localisation of malignant lesions in patients with a number of solid tumours. However, their general use has been limited by several major problems (Epenetos *et al.*, 1985). The use of human monoclonal antibodies may also give high backgrounds due to non-specific binding to high affinity receptors in the lymphatics, liver and spleen. A human monoclonal antibody with the converse to the mutation described here (i.e. LE235) and which does not bind to this high affinity receptor may improve tumour localisation or therapy by enhancing specific uptake of the antibody while decreasing the background due to non-specific binding to FcR.

ACKNOWLEDGEMENTS

This work was carried out at the MRC Laboratory of Molecular Biology, Cambridge, U.K. I thank Michael Neuberger, Dennis Burton and especially Greg Winter for help and advice.

REFERENCES

Anderson, C.L. and Looney, R.J. (1986). Immunol. Today, 7, 264.
Bedouelle, H. and Winter, G. (1986). Nature, 320, 371.
Bing, J. (1971). Immunol., 107, 1243.
Boulianne, G.L., Hozumi, N. and Shulman, M.J. (1984). Nature, 312, 643.
Burton, D.R. (1985). Molec. Immunol., 22, 161.
Burton, D.R., Boyd, J., Brampton, A., Easterbrook-Smith, S., Emanuel, E.J., Novotny, J., Rademacher, T.W., van Schravendijk, M.R., Sternberg, M.J.E. and Dwek, R.A. (1980). Nature, 288, 338.
Carter, P., Bedouelle, H. and Winter, G. (1985a). Nucleic Acids Res., 13, 4431.
Colomb, M. and Porter, R.R. (1975). Biochem. J., 145, 177.
Connell, G.E. and Porter, R.R. (1971). Biochem. J., 124, 53.
Deisenhofer, J. (1981). Biochemistry N.Y., 20, 2361.
Duncan A.R. and Winter, G. (1988). Nature, in press.
Duncan, A.R., Woof, J.M., Partridge, L.J., Burton, D.R. and Winter, G. (1988) Nature, in press.
Epenetos, A.A., Hooker, G., Durbin, H., Bodmer, W.F., Snook, D., Begent, R., Oliver, R.T.D. and Lavender, J.P. (1985). Lancet, i, 350.
Hale, G. and Waldmann, H. (1985). In: Hybridoma Technology in the Biosciences and Medicine (ed. T. Springer), Plenum Press, New York.
Kehoe, J.M. and Fougereau, H. (1969). Nature, 224, 1212.
Kohler, G. and Milstein, C. (1975). Nature, 256, 465.
Leatherbarrow, R.J. and Dwek, R.A. (1984). Molec. Immunol., 21, 321.
Leatherbarrow, R.J., Rademacher, T.W., Dwek, R.A., Woof, J.M., Clark, A., Burton, D.R., Richardson, N. and Feinstein, A. (1985). Molec. Immunol., 22, 407.
Lukas, T.J., Muños, H. and Erickson, B.W. (1981). J. Immunol., 127, 2555.
Morrison, S.L., Johnson, M.J., Herzenberg, L.A., Oi, V.T. (1984). Proc. Natl. Acad. Sci. USA, 81, 6851.
Nadler, L.M., Stashenko, P., Hardy, R., Kaplan, W.D., Button, L.N., Kute, D.W., Antman, K.H. and Schlossman, S.F. (1980). Cancer Res., 40, 3147.
Neuberger, M.S., Williams, G.T., Mitchell, E.B., Jouhal, S.S., Flanagan, J.G. and Rabbitts, T.H. (1985). Nature, 314, 268.
Neuberger, M.S. and Williams, G.T. (1986). Phil. Trans. R. Soc. Lond.,A 317, 425.

Nose, M. and Wigzell, H. (1983). Proc. Natl. Acad. Sci. USA, 80, 6632.
Prytowsky, M.B., Kehoe, M. and Erickson, B.W. (1981). Biochemistry, 20, 6349.
Woof, J.M., Jafaar, M.I., Jefferis, R. and Burton, D.R. (1984). Molec. Immunol., 21, 523.
Woof, J.M., Partridge, L.J., Jefferis, R. and Burton, D.R. (1986). Molec. Immunol., 23, 319.
Yasmeen, D., Ellerson, J.R., Dorrington, K.J. and Painter, R.H. (1976). J. Immunol., 116, 518.

Structure–Function Relationships for IgG Subclasses

F. Shakib
Midlands Asthma and Allergy Research Association,
12 Vernon Street, Derby DE1 1FT, UK

ABSTRACT

The differentiation of IgG into four subclasses has allowed the efficiency of a given effector function to be maximised within an individual subclass and broadened the conditions under which effector mechanisms may be switched on. This is best demonstrated in IgG1 and IgG3 and less so in IgG2. IgG4, on the other hand, has probably evolved to regulate the effector functions of other subclasses.

KEYWORDS

IgG subclasses, IgG1, IgG2, IgG3, IgG4.

Introduction

The humoral defence system in man deploys an impressive array of immunoglobulin classes, subclasses, allotypes, isoallotypes and idiotypes. Within this highly heterogenous population of protein molecules, no antibody can perform all effector functions and no effector function is restricted to one antibody. The system, therefore, relies on recruiting immunoglobulins, which although products of distinct structural genes, can perform similar effector functions but with variable efficiency. Such clever working mechanisms will obviously safeguard against having total failure of a particular effector function consequent to selective immunoglobulin deficiency. It is, therefore, not surprising that total IgG deficiency has been reported in healthy individuals.

Human IgG subclasses have emerged relatively recently in evolution, as indicated by the high degree of sequence homology between the constant parts of their heavy chains. They may, therefore, not be critical to survival. However, it is possible that the divergence of IgG into four subclasses has allowed the efficiency of a given effector function to be maximised within an individual subclass and amplified the conditions under which effector mechanisms may be triggered. In this review, I will be discussing the structural differences between IgG subclasses in relation to their biological functions and disease activity.

Structure of IgG Subclasses

IgG subclasses are defined by the C_H genes encoding Cγ1, hinge, Cγ2 and Cγ3 domains. The constant domains of IgG subclasses have similar amino acid sequences - on average two subclasses show more than 95% identity. The greatest differences are found in the hinge in terms of the number of residues it contains and the number of disulphide bonds bridging the heavy chains (Table 1).

Table 1. The number of amino acid residues and the number of disulphide bridges in the hinge region of the four IgG subclasses

Subclass	No. of Residues	No. of Disulphide Bridges
IgG1	15	2
IgG2	12	4
IgG3	62	11
IgG4	12	2

Electron microscopic, hydrodynamic and small-angle X-ray scattering studies indicate that IgG3 has an extended hinge (Fig. 1) and that all of the hinge Cys residues being in disulphide bridge formation. The immunogenicity of IgG3 is, therefore, very dependent on the structure of the hinge region. From a functional point of view, a characteristic feature of IgG3 molecules is that the antigen binding sites are remote from the effector function sites and are not expected to interfere with one another (Burton *et al.*, 1986).

The hinge-deleted IgG1 Dob protein assumes a conformation in which the Fab arms are essentially colinear, ie. in the same plane as one another and spanning 180°. For intact IgG1 proteins the Fab arms are not in the same plane and span an angle of 40° - 60° between them. IgG4 is best fit by a model which resembles the IgG1 Dob but does not have the Fab arms colinear. IgG2 is effectively hingeless with folded-back Fab arms.

Structure - Function Relationships

In the immunoglobulin C_H gene locus both γ3 and γ1 chains and γ2 and γ4 chains are encoded sequentially (Fig. 2) and IgG effector functions and antibody activities generally tend to segregate into IgG3/IgG1 and IgG2/IgG4 (Table 2). Such gene arrangement may represent an evolutionary pathway in which a single γ gene duplicated giving rise to a common gene for γ3 and γ1 and a common gene for γ2 and γ4. I will now consider the interaction of IgG subclasses with complement, Fc-receptors and staphylococcal protein A.

Complement. In comparing the complement activation capacity of IgG sub-

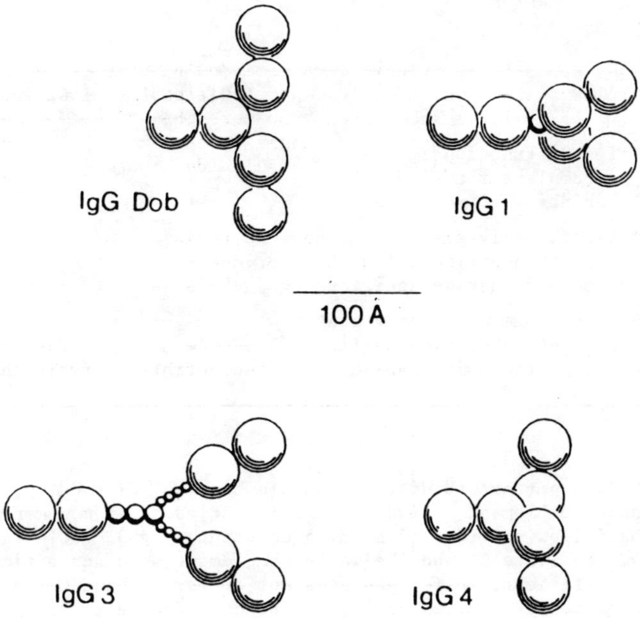

Fig. 1. Comparison of the structures of the subclasses of IgG. (Reproduced with permission from Burton *et al.*, 1986.)

$5' \ldots J - \mu - \delta \ldots \gamma 3 - \gamma 1 - \psi \varepsilon 1 - \alpha 1 \ldots \gamma 2 - \gamma 4 - \varepsilon - \alpha 2 \ldots 3'$

Fig. 2. C_H gene sequence in human.

Table 2. Subclass segregation in the effector functions, antibody activities and physico-chemical properties of IgG

	IgG1/IgG3	IgG2/IgG4
Effector functions		
complement activation	+ +	-
binding to monocyte Fc-R	+ +	-
Antibody activities		
protein antigens	+ +	-
polysaccharide antigens	-	+ + (IgG2)
Physico-chemical properties		
electrophoretic mobility	slow	fast
digestion with papain	susceptible	resistant

classes, it is important to distinguish C1q binding, C1 binding, C1 activation and whole complement activation. All subclasses in monomeric form bind C1q with the following order of binding constants: IgG3 > IgG1 > IgG2 > IgG4. IgG3 and IgG1 activate C1 and the whole complement sequence efficiently. IgG2 is less efficient, while IgG4 does not appear to bind C1 and does not activate complement. Interestingly, isolated Fc fragments of IgG4 do bind C1 and initiate the full complement sequence, implying that IgG4 has a C1 binding site but this is shielded in the intact IgG4 molecule by the Fab arms. This is in keeping with the proposed model for IgG4 (Fig. 1) where the Fab arms make a close approach to the Cγ2 domains. Thus, IgG4 and the hinge deleted IgG1 Dob protein both adopt a T shaped conformation and neither is able to activate complement. For IgG1 and particularly IgG3 the Fab arms are more distant from the Cγ2 domains and complement activation is efficient (Burton *et al.*, 1986).

It is worth pointing out though that studies of complement activation capacity of IgG subclasses employ aggregated myeloma proteins and the results obtained may, therefore, not reflect the physiological events *in vivo*. Thus, a study of the biological activity of IgG subclass antibodies to *Haemophilus influenzae* type b capsular polysaccharide using an *in vitro* complement-mediated bactericidal assay and an *in vivo* passive protection assay in infant rats, revealed that IgG2 antibodies are just as effective as those of IgG1 (Weinberg *et al.*, 1986). There is, therefore, a good case for using more such functional assays.

Fc-receptors. Fc-receptors on monocytes, neutrophils, platelets, lymphocytes and macrophages show a similar subclass specificity pattern in binding affinity, ie. IgG1 and IgG3 > IgG4 > IgG2. There is no good evidence for the commonly held view that the Cγ3 domain is involved in Fc-receptor binding to such cells. Certainly for the monocyte Fc-receptor ample evidence seem to favour Cγ2 as the domain involved. For example, in one study a panel of monoclonal antibodies against epitopes on the Cγ1, Cγ2, Cγ3 and Cγ2 - Cγ3 interface regions were tested for their capacity to inhibit the interaction between IgG and monocytes (Partridge *et al.*, 1986). Of all the monoclonal antibodies employed only those against the Cγ2 domain, were inhibitory. The

Cγ2 domain is, therefore, involved in two major effector functions, ie. complement activation and Fc-receptor binding. To date no effector function activity has been definitively attributed to the Cγ3 domain but it has been suggested that its function is to maintain the unusual disposition of the Cγ2 domains, as well as to facilitate Fc - Fc interaction (Burton , 1986). It has recently been proposed that the monocyte Fc-receptor interacts with the region between the hinge disulphide and the folded Cγ2 domain (Burton *et al.*, 1986).

A recent study of the ability of monoclonal IgG1 and IgG3 antibodies specific for blood group D (Rh) antigen to mediate red cell binding to macrophages, found that IgG3 opsonizes at a density of about 100 molecules/red cell, whereas IgG1 was only active at a level of 10,000 molecules/cell (Wiener *et al.*, 1987). It is suggested that the more potent opsonic activity of IgG3 is due to the long hinge, leading to greater accessibility to the Fc-receptor binding site on Cγ2. The lower affinities of IgG4 and IgG2 for monocyte Fc-receptors do not appear to be due to Fab arm obstruction, since Fc fragments have similar affinities to the parent IgG2 and IgG4 molecules. Thus, in monocyte Fc-receptor system, the subclass affinity order is a reflection of sequence changes in the region connecting the hinge disulphide bridge to the Cγ2 domain, rather than a direct Fab arm obstruction effect (Gregory *et al.*, 1987). For IgG4, therefore, the Fab arms prevent access to C1 but do not affect Fc-receptor binding. This may be due to a difference in molecular size between C1 (MW 920,000) and monocyte Fc-receptor (MW approximately 70,000) as suggested recently (Gregory *et al.*, 1987).

Data concerning the presence of Fc-receptors for IgG4 on mast cells and basophils are still not convincing. For instance, experiments in which polyclonal and monoclonal anti-IgG4 reagents were employed to show IgG4 bound to such Fc-receptors are no longer considered to be valid (Shakib, 1989). Thus the detection of IgG anti-IgE in human sera, particularly from those with allergic manifestations, would suggest that the IgG4 molecules detected on these cells using anti-IgG4 reagents are attached to IgE rather than to specific Fc-receptors (Fig. 3). In fact, this is highly likely to be the case, since IgG anti-IgE antibodies found in allergic sera are predominantly of the IgG4 subclass. Furthermore, unlike IgE, IgG4 lacks an extra pair of Fc domains, and the sequence within the Cγ3 domain in an analogous position to that in the Cε4 domain thought responsible for providing a triggering signal to the mast cell does not contain the requisite primary structure (ie. a basic N-terminal region and hydrophobic C-terminal region) (Stanworth, 1986).

Protein A. Interaction of IgG with staphylococcal protein A, although not an effector function, is the one most clearly understood in molecular terms. Fragment B of protein A binds at a site between the Cγ2 and Cγ3 domains. This binding is restricted to IgG1, IgG2, IgG4 and those IgG3 proteins bearing the G3m(s) and G3m(t) markers which are characteristic of the Mongoloid populations. In Caucasian IgG3 proteins, the protein A contact residue His 435 of Mongoloid IgG3 is substituted by Arg, which cannot be accommodated in the complex, and protein A does not bind (Burton *et al.*, 1986). There is apparently a close parallelism between the structural requirements for reactivity of human IgG with protein A and rheumatoid factor. Thus, it has been shown (Jefferis *et al.*, 1984) that a monoclonal rheumatoid factor, with specificity similar to that of general polyclonal rheumatoid factor, is reactive with IgG1, IgG2, IgG4 and IgG3 of variants G3m(s) and G3m(t) only - all of which have a His residue in position 435 (Stanworth, 1986).

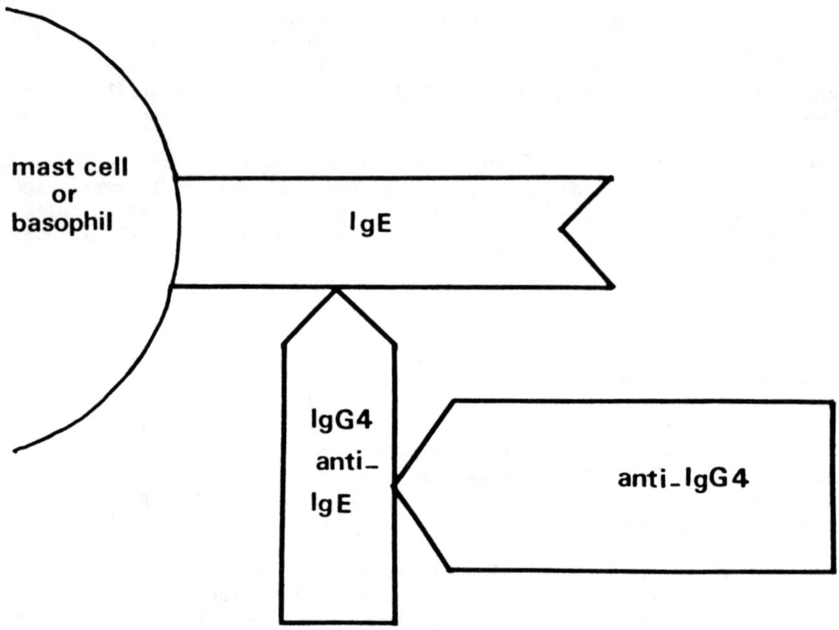

Fig. 3. Diagrammatic depiction of the mechanism by which anti-IgG4 reagents bind to mast cells and basophils.

Pathological Consequences of Subclass Variations in IgG Antibody Response

From previous discussions of structure-function relationships for IgG subclass, it is clear that the effector function efficiency of IgG is largely determined by its subclass composition. Since IgG antibody responses are often restricted to one or two subclasses, an appreciation of the subtle pathological consequences of IgG subclass antibody variations is relevant and revealing.

The question of whether the subclass profile of an IgG antibody response determines the type and the severity of the immunopathological lesion induced, is best dealt with by considering responses to a single antigen. Roberts *et al.* (1983) have elegantly demonstrated two contrasting morphologic forms of glomerular lesion in patients with systemic lupus erythematosus, in which the subclass compositions of IgG anti-native DNA antibody were characteristically different. Evidence indicating that differing effector function properties of these IgG subclasses may explain the inflammatory appearance of proliferative glomerulonephritis and the bland morphology of membranous glomerulonephritis are shown in Table 3. This situation offers the potential of using a serological test to distinguish between two different types of glomerular disease. Other pathological examples of IgG subclass variations are discussed elsewhere (Shakib, 1987).

Table 3. Some contrasting immunopathological features of proliferative glomerulonephritis (PGN) and membranous glomerulonephritis (MGN) in systemic lupus erythematosus patients with anti-native (n) DNA antibodies (Roberts *et al.*, 1983)

Contrasting Features	PGN	MGN
1. Renal biopsy deposits		
a - predominant IgG subclass	IgG3 and IgG1	IgG4
b - fixation of C1q, C3bp and C3c	high	low
2. Serum immune-bound anti-nDNA		
a - predominant IgG subclass	IgG3 and IgG1	IgG4
b - fixation of C3c	yes	no
3. Serum cryoglobulins		
a - concentration of C1q, C4 and C3	high	low
b - concentration of complement regulatory proteins	high	low
4. Serum complement components		
a - concentration of C1q	significantly low	average
b - concentration of C4 and C3	low	average

Physiological Role of IgG4

It will have become apparent from earlier considerations that IgG4 is a poorly constructed antibody which plays little or no part in immune elimination of foreign invasion. In this section I intend to develop the concept that the IgG4 subclass has evolved to become a mechanism for damping effector functions of other immunoglobulin classes and IgG subclasses.

IgG4 constitutes about 3% of total IgG in sera of healthy individuals, but it is often the predominant subclass during IgG antibody production. IgG4 restricted antibody response occurs with every possible route of immunisation and in response to a varied range of antigens (Table 4). Currently held views suggest that chronic antigenic exposure produces a switch from a predominant IgG1 to a predominant IgG4 antibody response. The phenomenon is likely to be the result of sequential "down-stream" switch of genes along the C_H gene locus (Fig. 2). Follow-up studies of IgG subclass antibody responses in patients undergoing desensitisation with dust mite, pollen and bee venom conform to this pattern (Nakagawa, 1986). It is, therefore, important to recognise that an IgG4-restricted antibody response is a time-related phenomenon (ie. occurs late during IgG response), and the term does not imply that other subclasses are not produced at an earlier stage. The tendency of prolonged antigenic stimulation to induce a shift towards maximum production of IgG4, a subclass which is not effective in activating

Table 4. Routes of immunisation and representative examples of antigens which induce IgG4 restricted antibody response (Reproduced with permission from Shakib, 1987).

Ingestion
  eg. ovalbumin

Inhalation
  eg. dust mite, pollen

Injection
  eg. desensitisation with dust mite, pollen and bee venom

Infusion
  eg. clotting factor VIII

Infection
  eg. schistosomiasis, filariasis

Autoimmunisation
  eg. thyroglobulin, DNA

effector functions, would suggest that IgG4 serves to control tissue damage induced by IgG1 and IgG3. Implicit in such hypothesis is that IgG antibody responses which do not culminate in predominant production of IgG4 would induce a more severe form of tissue destruction (Table 3). Thus, the presence of IgG4 antibody in association with those of other immunoglobulin classes and IgG subclasses at the site of an inflammatory reaction, will block effector functions of such other antibodies. IgG4 antibodies may, therefore, play a crucial role in preventing or minimising excessive tissue necrosis during the terminal stage of immune elimination.

Deleterious Role of IgG4

Apart from its lack of complement activation and binding to Fc-receptors on phagocytic cells, IgG4 exhibits functional defects in the Fab regions - namely monovalency (Aalberse *et al.*, 1983) and weak antigen binding affinity (Devey *et al.*, 1985). Studies in mice have shown that the production of low affinity antibody is associated with failure to eliminate antigen from the circulation and an increased susceptibility to immune complex membranous glomerulonephritis (Devey *et al.*, 1984) (see Table 3 for a parallel situation in humans). It has, therefore, been suggested (Devey *et al.*, 1985) that a predominantly low affinity antibody response (of which IgG4 may be an example) may be considered as a form of immunodeficiency and, as well as predisposing to chronic immune complex disease, may increase susceptibility to infection. For instance, parasitic infestations, like schistosomiasis and filariasis, are prolonged by high levels of IgG4 anti-parasite antibodies. The selective production of IgG4 antibodies in parasitic diseases is probably one further mechanism by which parasites evade or modify host immunity (Catty *et al.*, 1986). Thus, IgE mediated responsiveness is rarely seen in schistosomiasis despite high levels of parasite-specific IgE, and *in vitro* evidence suggest that IgG blocking antibodies are responsible for the modulation of this type of reaction (Hofstetter *et al.*, 1982).

Concluding Remarks

There is considerable interest in delineating the precise relationships between structure and function of IgG subclasses, and some real progress has been achieved in recent years. This has come in large part from the availability of monoclonal antibodies specific for IgG subclasses and their functional domains. Studies of this kind will have great therapeutic applications.

REFERENCES

Aalberse, R.C., van der Gaag, R. and van Leeuwen, J. (1983). Serologic aspects of IgG4 antibodies. I. Prolonged immunisation results in an IgG4 restricted response. J. Immunol., 130, 722-726.

Burton, D.R. (1986). Is IgM-like dislocation a common feature of antibody function? Immunol. Today, 7, 165-167.

Burton, D.R., Gregory, L. and Jefferis, R. (1986). Aspects of the molecular structure of IgG subclasses. In: Basic and Clinical Aspects of IgG Subclasses (F. Shakib, Ed.) Monogr. Allergy, Vol. 19, pp. 7-35, Basel, Karger.

Catty, D. Jassim, A., Hassan, K. and Raykundalia, C. (1986). IgG subclasses in parasitic infestations. In: Basic and Clinical Aspects of IgG Subclasses (F. Shakib, Ed.) Monogr. Allergy, Vol. 19, pp 144-155, Basel, Karger.

Devey, M.E., Bleasdale, K.M., French, M.A.H. and Harrison, G. (1985). The IgG4 subclass is associated with a low affinity antibody response to tetanus toxoid in man. Immunol., 55, 565-567.

Devey, M.E., Bleasdale, K., Stanley, C. and Steward, M.W. (1984). Failure of affinity maturation leads to increased susceptibility to immune complex glomerulonephritis. Immunol., 52, 377-383.

Gregory, L., Davis, K.G., Sheth, B., Boyd, J., Jefferis, R., Nave, C. and Burton, D.R. (1987). The solution conformations of the subclasses of human IgG deduced from sedimentation and small angle X-ray scattering studies. Mol. Immunol., 24, 821-829.

Hofstetter, M., Poindexter, R.W., Ruiz-Tiben, E. and Ottesen E.A. (1982). Modulation of the host response in human schistosomiasis: III. Blocking antibodies specifically inhibit immediate hypersensitivity responses to parasite antigens. Immunol., 46, 777-785.

Jefferis, R., Jaafar, M.I. and Steinitz, M. (1984). Immunogenic and antigenic epitopes of immunoglobulins. VIII. A human monoclonal epitope determined by Histidine/arginine interchange as residue 435 on immunoglobulin G. Immunol. Lett., 7, 191-194.

Nakagawa, T. (1986). IgG subclass changes in response to desensitisation. In: Basic and Clinical Aspects of IgG Subclasses (F. Shakib, Ed.) Monogr. Allergy, Vol. 19, pp. 253-261, Basel, Karger.

Partridge, L.J., Woof, J.M.,Jefferis, R. and Burton, D.R. (1986). The use of anti-IgG monoclonal antibodies in mapping the monocyte receptor site on IgG. Mol. Immunol., 23, 1365-1372.

Roberts, J.L., Wyatt, R.J., Schwartz, M.M. and Lewis, A.J. (1983) Differential characteristics of immune bound antibodies in diffuse proliferative and membranous forms of lupus glomerulonephritis. Clin. Immunol. Immunopathol., 29, 223-241.

Shakib, F. (1987). Aspects of the pathological significance of IgG4.

Exp. Clin. Immunogenet., 4, 193-200.

Shakib, F. (1989). Antiglobulin induced leucocyte mediator release. In: Autoantibodies to Immunoglobulins (F. Shakib, Ed.) Monogr. Allergy, Basel, Karger (in press).

Stanworth, D.R. (1986). The molecular pathology of IgG4. In: Basic and Clinical Aspects of IgG Subclasses (F. Shakib, Ed.) Monogr. Allergy, Vol. 19, pp. 227-240, Basel, Karger.

Stanworth, D.R. (1986). IgG subclasses in rheumatoid joint disorders. In: Basic and Clinical Aspects of IgG Subclasses (F. Shakib, Ed.) Monogr. Allergy, Vol. 19, pp. 188-197, Basel, Karger.

Weinberg, G.A., Granoff, D.M., Nahm, M.H. and Shackelford, P.G. (1986). Functional activity of different IgG subclass antibodies against type b capsular polysaccharide of *Haemophilus influenzae*. J. Immunol., 136, 4232-4236.

Wiener, E., Atwal, A., Thompson, K.M., Melamed, M.D., Gorick, B. and Hughes-Jones, N.C. (1987). Differences between the activities of human monoclonal IgG1 and IgG3 subclasses of anti-D (Rh) antibody in their ability to mediate red cell-binding to macrophages. Immunol, 62, 401-404.

Discussion: Session Two

Chairman: James Watson

1. Questions to David Marsh

Robert Hamburger. David, does it bother you that the ragweed antigens, *Amb a* V and *Amb t* V exhibit a high degree of amino acid sequence homology without any cross-reactivity?

David Marsh. No, not at all. Structural studies of other proteins indicate that the core region is highly conserved and that variability primarily involves the outer surface loops. From molecular modelling studies, which I did not have time to present, the homologous sequence to which you refer is most probably in the core region and therefore unavailable for recognition by antibodies.

Robert Hamburger. But that sequence still acts as a T-cell recognition site?

David Marsh. Yes, that appears to be correct. It is important to remember that (1) the T-cell epitope is highly conserved and since it is probably in the core region, will only be exposed during antigen presentation, e.g. by macrophages, and (2) the B-cell epitope is not highly conserved because it is on the surface which in turn means that is freely available for *in vitro* cross-reaction studies.

Lawrence Schwartz. From this type of analysis, can you tell whether either IgG or IgE is to be expressed?

David Marsh. We were emphasizing IgE responses, but we have found a very good relationship between specific IgG and specific IgE antibody levels in individuals who had not been receiving immunotherapy. Following this "artificial" exposure there is still a good correlation, although levels of IgG usually increase much more than for IgE, where the elevation is usually no more than two-fold.

If I may get on my hobby-horse for a moment: since 1973 I have been saying that there is nothing unique about allergens, nothing special about allergenic determinants; they are just the protein structures which in one case induce an IgE response and in another an IgG response. Most likely it is

Advances in the Biosciences Vol. 74

Printed in Great Britain

their presentation in the localised site of IgE production, e.g., the respiratory tract, which induces IgE responses.

Lawtence Schwartz. But doesn't that contradict what you said earlier that there is a familial tendency for some people to form IgE, even though everybody is exposed to the same allergens?

David Marsh. There certainly is a tendency for allergy to be inherited, but this is associated not only with the Ir genes. For example, in the mouse non-MHC genes have been clearly shown to regulate IgE production, although the case is less well proven in man. It may be that these same genes regulate other immunoglobulin classes as well.

2. Questions to Birgit Helm

Alex Duncan. Do you believe the receptor-site on your epsilon chain to be a linear determinant?

Birgit Helm. No, you certainly need a tertiary conformation, most probably involving determinants on the CH_2 and CH_3 regions. We did produce a polypeptide which comprised the putative sites in this area which were computed to be exposed to solvent, i.e., amino acids 304-314, 324-334 and 362-367. Unfortunately the molecule did nothing, so you really do need some conformational folding.

Alec Sehon. Regarding this question of conformation: I don't believe that it is essential for the fragment to have a conformation identical to the intact molecule, because it should be possible for it and the receptor to mould themselves into an acceptable fit.

Birgit Helm. I doubt it, but further experimentation is required to solve this point.

3. Questions to Marianne Brüggemann

Alec Sehon. The technology is most elegant and the points which you have covered speak for themselves, however, I would like you to comment on the immunogenicity of the humanized antibodies. I know of a number of people who have found the chimaeric antibodies to be highly immunogenic in rabbits, and this is not too surprising considering that (a) the idiotype is still foreign and (b) the hinge region is unlike that of a normal antibody.

Marianne Brüggemann. To my knowledge chimaeric antibodies have not yet been used in humans, but in theory the situation could be worse in humans than in rabbits. In addition to the reasons you cited, humans, because they are not inbred, could also respond to altotypic differences presented by the "humanized" chimaeric antibodies. However, on the brighter side, we have some preliminary data concerning the reverse situation, i.e., "mousified" chimaeric antibodies injected into mice, and it would seem that a foreign antibody is much more immunogenic than a less-foreign antibody. Nonetheless as long as chimaeric antibodies have not been tested in man we cannot say, or do, much about it.

Martin Chapman. Many of us are interested in looking at IgE and IgG binding epitopes on allergens, and one of the main problems is getting enough Fab antibody material with which to work. You have demonstrated that gene segments can be switched around, but you are still restricted by the need to isolate DNA from B-cells which in man circulate in very low numbers. I wonder if there are any more efficient strategies involving human hybridoma technology which have been developed recently and which might be applied in these situations.

Marianne Brüggemann. The problem as I see it is that immortalized human B-cell lines secrete very little antibody, this is apparently not due to gene deletion, but to cessation of antibody production. If a means could be found whereby the human B-cells could be revived so as to produce antibody, the problem could be solved. However, it might be more rewarding in the longer term to construct human antibodies in mouse myeloma or hybridoma cell lines, because antibody yields can be very high - we have one such line secreting more than 20 mg antibody per litre of culture supernate. Moreover, the antibodies could easily be purified in one step by filtering the supernate through a small affinity column. In the longer term, it will be possible to produce human antibodies when large gene segments can be inserted into transgenic mice.

4. Question to Farouk Shakib

Birgit Helm. Is anything known about the domain specificity of the gamma 4 anti-IgE antibodies, and the anaphylactic type response?

Farouk Shakib. The specificity is for the epsilon CH_2 domain. It has been shown *in vitro* that IgE autoantibodies can "fire-off" mast cells, unless the autoantibody was of the IgG4 subclass. This is because IgG4 behaves as a monovalent antibody and so cannot cross-link cell surface IgE molecules.

Session Three

CLONING OF ALLERGENS

Chairman: J. Watson

Cloning and Sequencing Vespid Venom Allergens: Sequence Similarity and Antigenic Cross Reactivity of Hornet Antigen 5 and Tobacco Leaf Protein

Te Piao King

The Rockefeller University, 1230 York Avenue,
New York, NY 10021, USA

ABSTRACT

Hornets, yellowjackets and wasps are known collectively as vespids. They have similar venom compositions. Their three main protein components are hyaluronidase, phospholipase A_1 and antigen 5 of unknown biologic function, and they are allergens in man. The homologous venom proteins of vespids have varying extents of antigenic cross reactivity.

The primary structures of two forms of antigen 5 from white face hornet (*Dolichovespula maculata*), also designated as *Dol m V*, have been determined by cDNA and protein sequencings. These two forms with 204 and 205 amino acid residues differ in 23% of their sequences and they are antigenically similar.

Dol m V is found to have sequence similarity with a pathogenesis-related protein from tobacco or tomato leaf. In a 130-residue overlap of these proteins, there is about 28% identity. Antigenic cross reactivity of *Dol m V* and tobacco pathogenesis-related protein was detected by enzyme immunoassay with rabbit antibodies specific for tobacco protein. This finding may be relevant to the clinical observation that about 10% or more of our population who are not allergic to insects have venom-specific IgE.

KEYWORDS

Insect; vespid; hornet; allergen; tobacco; pathogenesis-related protein.

INTRODUCTION

The most common type of insect sting allergy is that caused by honey bees followed by vespids which include hornets, yellowjackets and wasps [1,2]. It is generally accepted that the allergic symptoms which occur after a sting result from an IgE-mediated reaction. That is, reaction of cell-bound specific IgE antibodies with venom allergens (antigens) leads to formation and release of mediators of anaphylaxis. Insect venoms are widely used for

Advances in the Biosciences Vol. 74

Printed in Great Britain

the diagnosis and treatment of insect sting allergy. In the course of a year's treatment, mg amounts of venoms may be injected into a person to obtain immunity against subsequent stings. The immunity is believed to be associated with increased levels of venom-specific IgG and decreased levels of IgE.

Bee venom contains two major protein allergens, phospholipase A_2 (16 kd) and hyaluronidase (45 kd), and one minor allergen, melittin [3]. Melittin is a peptide of 26 amino acid residues yet surprisingly it is allergenic in about one-third of bee-sensitive patients. The primary structures of bee venom phospholipase A_2 and melittin are known [4]; the tertiary structure is also known for melittin [5].

Hornets, yellowjackets and wasps have similar venom compositions [6-12]. Their venoms each contain three major venom allergens, phospholipase A_1 (35 kd), hyaluronidase (43 kd) and antigen 5 (23 kd) of as yet unknown biologic function. Different vespid antigen 5s have significant structure similarity as suggested by their partial N-terminal sequence data [13]. Vespid phospholipase A_1 probably also have sequence homology as they yielded similar patterns of fragments on protease digestions [14].

As indicated in Table 1, there are many species of hornets (genus *Dolichovespula*), yellowjackets (genus *Vespula*) and wasps (genus *Polistes*) in N. America [15]. Tests with more antibodies have shown that the homologous proteins within each genus are usually antigenically indistinguishable but there are exceptions [6,8,10]. Tests with mouse antibodies have also shown that the homologous antigen 5s and hyaluronidases from vespids of different genera have varying levels of antigenic cross reactivity but the phospholipases do not [9,13]. On the other hand, tests with sera from untreated yellowjacket-sensitive patients revealed cross reactivity of antigen 5s of different vespids and to a lesser degree of phospholipases as well [14]. These comparisons on antigenic cross reactivity were made with homologous proteins from white face hornet (*D. maculata*), yellow hornet (*D. arenaria*), yellowjacket (*V. vulgaris*) and wasp (*P. annularis*). The two hornets were chosen as their homologous proteins are antigenically dissimilar [6]. The species of yellowjacket and wasp were chosen because of their wide distribution in this country [15].

In addition to insect venoms there are many other sources of allergens, such as pollens, foods, animal danders and mites. The primary structures of some of these allergens are known. It is not clear whether or not allergens are different from other protein antigens and have peptide sequence which can enhance IgE responses in some manner. For example, the lymphokine interleukin 4 (B cell stimulatory factor 1) can stimulate IgE response [16]. Observation of model antigens in mice indicate that specific IgE and IgG responses occur concomittantly and that only the magnitude and kinetics of these two responses differ. IgE response is more readily influenced by the antigen dose, the adjuvant and the genetic background of mice but thus far it is not known to be influenced by the nature of protein antigen [cf. 17].

Primary Structure of *Dol m V*. In order to have a better knowledge of insect allergens and allergens in general, we started sequencing studies on vespid venom allergens. At present we have only completed the sequencing of one of the venom allergens from white face hornet (*Dolichovespula maculata*), antigen 5 [18]. According to a recently proposed system of allergen nomenclature [19], this protein is designated as *Dol m V*. The primary structure of *Dol m V* was determined by protein and cDNA sequencings. Venom-specific mRNAs are

Table 1. Taxonomy of honeybees, hornets, wasps and yellowjackets.[1]

ORDER		**Hymenoptera**			
SUPERFAMILY		**Vespoidea**		**Apoidea**	
FAMILY		**Vespidae**			**Apidae**
SUBFAMILY	**Vespinae**			**Polistinae**	**Apinae**
GENUS	**Vespula Thomson**		**Vespa**	**Polistes**	**Apis**
SUBGENUS	**V. sensu strictu**	**Dolichovespula**			
SPECIES	**V. germanica** **V. maculifrons** **V. vulgaris** **V. squamosa**	**D. maculata** **D. arenaria**	**V. crabo**	**P. annularis** **P. exclamans** **P. fuscatus**	**A. mellifera**
COMMON NAME[2]	**Yellowjackets**	**Hornets**		**Wasps**	**Honeybees**

[1] A partial list of known species of different insects is given.

[2] The common names of *D. maculata*, *D. arenaria* and *V. crabo* are white (bald) face hornet, yellow hornet and European hornet respectively.

present in the acid glands of hornets, and these glands are also known as poison glands [15]. We used the crude RNAs from acid glands for constructing a cDNA library in the expression vector λgt11. Immunoscreening the library with venom-specific polyclonal antibodies led to the isolation of *Dol m V* encoding clones. But we were not able to detect the presence of clones encoding hyaluronidase or phospholipase A_1.

On the basis of their DNA sequence data we constructed a map of 9 of these clones given in Fig. 1A. Position 0 of the indicated map corresponds to the N-terminal residue of antigen 5 and position 615 corresponds to the C-terminal residue. Five of these clones encode the full length of antigen 5 containing 205 amino acid residues while four clones encode the C-terminal

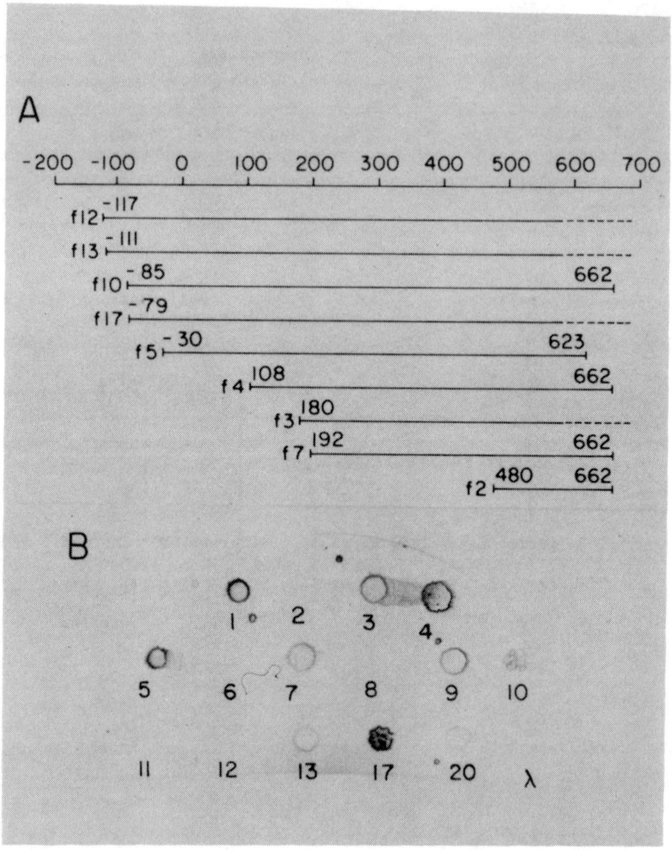

Fig. 1. A. A map of cDNAs encoding *Dol m V* (white face hornet antigen 5). The complete cDNA sequences of f5 and f10 have been determined [19], and only partial sequence data are known for others.

B. Immunoscreening of recombinant phages encoding *Dol m V*. About 10^5 plaque forming units of recombinant phage or control λgt11 were seeded on a lawn of Y1090 cells. The agar plates were incubated, overlaid with nitrocellulose paper then immunoscreened with mouse polyclonal antibodies.

45-169 residues of the molecule. The immunoscreening results in Fig. 1B suggest that the antigenic determinants of *Dol m V* are distributed throughout the linear sequence of the molecule as the clones which encode fragments of *Dol m V* were immunoreactive. The determinants present in the fragments of *Dol m V* are probably of the sequential type and they are not the immunodominant ones. The immunodominant determinants of intact *Dol m V* are believed to be of the topographic type, as mouse antibodies specific for the native protein bound weakly the chemically modified *Dol m Vs* which include the reduced and S-carboxymethylated, or the CNBr cleaved, derivatives [13].

Several of the cDNAs of the *Dol m V* clones in Fig. 1A show extensive sequence variation. On the basis of their partial or complete sequence data they can be divided into three groups. Group A contains f3, f12, f13 and f17, group B contains f2, f4, f7 and f10 and group C contains f5. *Dol m V* can be isolated in three distinct forms by ion exchange chromatography [6]. Comparison of the translated amino acid sequences of groups B and C with the partial sequence data of CNBr peptides of *Dol m V* establish that they represent chromatographic forms 2 and 3 of *Dol m V* respectively. These amino acid sequence data are given in Fig. 2. In order to get maximal alignment of the sequences, a gap at position 151 is added for form 2. The boxed residues indicate where the sequences of the two forms differ, and there are 49 boxed residues or about 1/4 of the molecule. The variations occur throughout the entire sequence with the highest density in the C-terminal half of the molecule.

```
Ag5-2  N N Y C K I K C R K  G I H T L C K F G T  S M K P N C G R N V  30
Ag5-3  N N Y C K I K C S R  G I H T L C K F G T  S M K P N C G S K I

Ag5-2  V K A Y G L T N D E  K N E I L K R H N D  F R Q N V A K G L E  60
Ag5-3  V K V H G V S N D E  K N E I V N R H N Q  F R Q K V A K G L E

Ag5-2  T R G K P G P Q P P  A K N M N V L V W N  D E L A K I A Q T W  90
Ag5-3  T R G N P G P Q P P  A K N M N V L V W N  D E L A K I A Q T W

Ag5-2  P N Q C D F N H D D  C R N T A K Y Q V G  Q N I A I S S T T A  120
Ag5-3  A N Q C S F G H D Q  C R N T E K Y Q V G  Q N V A I A S T T G

Ag5-2  T Q F D R P S K L I  K Q W E D E V T E F  N Y K V G L Q N S N  150
Ag5-3  N S Y A T M S K L I  E M W E N E V K D F  N P K K G T I G D N

Ag5-2  - F R K V G H Y T Q  M V W G K T K E I G  C G S I K Y I E D N  180
Ag5-3  N F S K V G H Y T Q  M V W G K T K E I G  C G S V K Y I E N N

Ag5-2  W Y T H Y L V C N Y  G P G G N D F N Q P  I Y E R K  (205)
Ag5-3  W H T H Y L V C N Y  G P A G N Y M D Q P  I Y E R K
```

Fig. 2. Amino acid sequences of *Dol m V* forms 2 and 3. The sequences were translated from the cDNA data and about half of the deduced sequences are supported by Adman degradation of their CNBr peptides [19]. The residues which differ in the two forms are enclosed by boxes.

In spite of the 25% sequence difference of *Dol m V* forms 2 and 3, they are indistinguishable on solid-phase enzyme immunoassay with mouse polyclonal antibodies. But these two forms are distinguishable with some preparations of mouse monoclonal antibodies. These results are shown in Fig. 3.

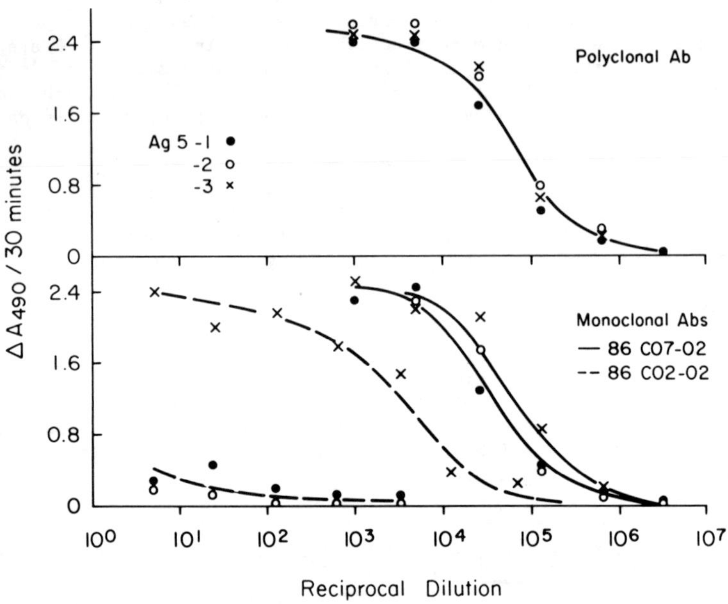

Fig. 3. Comparison of the binding of specific mouse antibodies to three forms of *Dol m V* by solid-phase enzyme immunoassay. One sample of polyclonal antibodies was tested in the upper pattern, and two samples of monoclonal antibodies were tested in the lower pattern.

Sequence Similarity of *Dol m V* and Other Proteins. We have previously noted that the N-terminal residues of vespid antigen 5s have limited sequence similarity with a segment of scorpion neurotoxin; 6 of the 20 residues compared were identical [13]. When we compared the complete sequence of *Dol m V* with other known proteins, we found sequence similarity with a class of proteins known as pathogenesis-related (PR) proteins which are present in leaves of different plants. In Fig. 4 is shown the sequence similarity of residue 74-202 of *Dol m V* with residue 20-138 of tobacco PR protein, which has a total of 138 amino acid residues [20]. The identical residues are enclosed by boxes, and hyphens indicate gaps added to obtain maximal similarity. For the 130-residue stretch compared there is about 29% sequence similarity of the two proteins. Very similar results were obtained when comparison was made with a homologous protein from tomato [21].

PR-proteins are normal leaf proteins present in a wide variety of plants like bean, celery, cucumber, tobacco, tomato and others [22]. Their levels increase markedly following infection with virus, fungi or bacteria, or irritation with chemicals. Some of these proteins are carbohydrate-specific

```
Ag5-2   M N V L V W N D E L A K I A Q T W A N Q C D F N H    98
PR-Pro  V E P L T W D N G V A A Y A Q N Y V S Q L A A D C    74
Ag5-3   M N V L V W N D E L A K I A Q T W A N Q C S F G H    98

Ag5-2   D D C R N T A K Y Q V G Q N - - I A I S S T T A T   121
PR-Pro  N L V H S H G Q Y - - G E N - - L A - - - Q G S G    92
Ag5-3   D Q C R N T E K Y Q V G Q N V A I A - - - S T T G   120

Ag5-2   Q - F D R P S K L I K Q W E D E V T E F N Y K V G   145
PR-Pro  D - F M T A A K A V E M W V D E K Q Y Y D H D S N   116
Ag5-3   N S Y A T M S K L I E M W E N E V K D F N P K K G   145

Ag5-2   L Q N S N - F R - - K V G H Y T Q M V W G K T K E   168
PR-Pro  - T C A Q G - Q V C - - G H Y T Q V V W R N S V R   137
Ag5-3   - T I G D N - N F S K V G H Y T Q M V W G K T K E   168

Ag5-2   I G C G S I K Y I E D N W - - - Y T H Y L V C N Y   190
PR-Pro  V G C A R V K C N N G G - - - - Y V - - V S C N Y   156
Ag5-3   I G C G S V K Y I E N N W H T H Y L - - V - C N Y   190

Ag5-2   G P G G N D F N Q P I Y                             202
PR-Pro  D P P G N V I G Q S P Y                             168
Ag5-3   G P A G N Y M D Q P I Y                             202
```

Fig. 4. Sequence similarity of *Dol m V* forms 2 and 3 and pathogenesis-related protein of tobacco leaf. Identical residues are enclosed by boxes. Amino acid residues are numbered at the right. The numbering for pathogenesis-related protein includes a 30 residue signal peptide.

hydrolases [23]. Some are of as yet unknown biological function, like the ones from tobacco and tomato which share sequence similarity with *Dol m V*.

<u>Antigenic Cross Reactivity of *Dol m V* and Pathogenesis-related Protein from Tobacco Leaf</u>. The sequence similarity of tobacco PR protein and *Dol m V* prompted us to test their antigenic cross reactivity by solid-phase enzyme immunoassay. Cross reactivity of these two protein was found with rabbit anti-sera specific for tobacco proteins but not with sera specific for hornet venom. As shown in Fig. 5 rabbit antibodies specific for tobacco PR protein bound solid-phase *Dol m V* with an affinity of about 1/20th of that for tobacco PR protein, but rabbit antibodies specific for hornet venom bound tobacco PR protein very weakly if at all.

This observation on the antigenic cross reactivity of *Dol m V* and tobacco PR protein may have clinical significance. It is known that 10% or more of the population who are not allergic to insects apparently have insect venom-specific IgEs [cf 24]. In Table 2 are given the skin test results of two separate groups of patients with *Dol m* venom, as reported by Reisman and

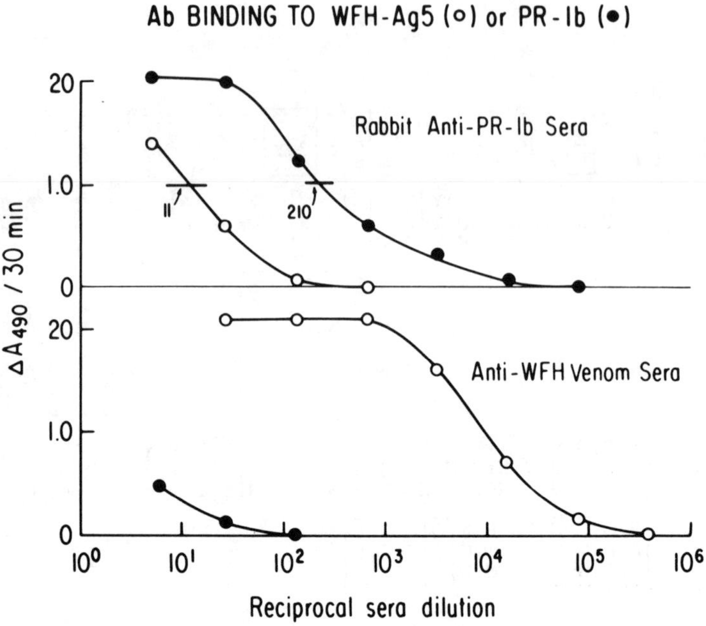

Fig. 5. Testing antigenic cross reactivity of *Dol m V* and tobacco PR protein by solid-phase enzyme immunoassay. Upper and lower patterns were obtained with rabbit antibodies specific for tobacco PR-protein and for whote face fornet (*D. maculata*) venom respectively.

Table 2. Distribution of positive skin tests endpoints ($\geq$+1) with white face hornet (*D. maculata*) venom in allergic and non-allergic individuals

Venom Conc.	Allergic Group 85 patients	Control Group 56 patients
0.01 μg/ml	24% +	2% +
0.1	26	2
1	13	23
10	(8)*	27
	71%	56%

Data are taken from Ref. 25.

*Only 9 of 85 patients were tested at 10 μg/ml concentration. The value given is an extrapolated one.

Georgitis [25]. The allergic group includes those who had large local or systemic reaction following insect sting and the control group includes those who had not been stung by insects previously. Of the allergic group, 71% showed positive skin reaction at venom concentrations of 10 μg/ml or less and this number is 54% for the control group. Some of the reactions at venom concentrations of >1 μg/ml may not be IgE antibody-mediated because venoms contain pharmacologically active substances like kinin and biogenic amines. Even after taking into account this possibility, one still concludes that a sizable number of the control group had venom-specific IgE. This raises the interesting possibility that some of the venom-specific antibodies in the control group may represent cross reactivity antibodies which are specific for plant proteins. Certainly most of us are very likely to have been exposed to many plant proteins through eating or smoking.

Future Studies. We are currently engaged in the cloning and sequencing of homologous antigen 5s of other vespids. These studies will be extended to another family of homologous vespid allergens, phospholipase A_1. Antigen 5 and phospholipase A_1 are different proteins with different biological functions and are antigenically unrelated, yet all insect allergic patients tested have antibodies to both proteins. Comparison of the sequence data of these two families of homologous venom allergens may facilitate the structure-function analysis of these molecules: namely, what are their B and T cell-specific epitopes? Therefore the sequence data can be useful for our understanding of vespid allergy.

ACKNOWLEDGEMENT

I would like to thank Dr L.C. van Loon of the Agricultural University, Netherlands, for gifts of pathogenesis-related protein from tobacco leaf and its specific rabbit antisera, and Mr Miles Guralnick of Vespa Laboratory, division of ALK, Inc. for gifts of venom samples. This research is supported in part by a US Public Health Service Grant AI-17021.

REFERENCES

1. Lichtenstein, L.M., Valentine, M.D. and Sobotka, A.K. (1979). Insect Allergy. State of the art. J. Allergy Clin. Immunol., 64, 5-12.
2. Reisman, R.D. (1985). Stinging insect allergy. J. Allergy Clin. Immunol., 75, 553-555.
3. King, T.P., Sobotka, A.K., Kochoumian, L. and Lichtenstein, L.M. (1976). Allergens of honey bee venom. Arch. Biochem. Biophys., 172, 661-671.
4. Habermann, E. (1972). Bee and wasp venoms. Science, 177, 314-322.
5. Terwillinger, T.C. and Eisenberg, D. (1982). The structure of melittin. J. Biol. Chem., 257, 6016-6022.
6. King, T.P., Sobotka, A.K., Alagon, A., Kochoumian, L. and Lichtenstein, L.M. (1978). Protein allergens of white face hornet, yellow hornet and yellowjacket venom proteins. Biochemistry, 17, 5165-5174.
7. King, T.P., Alagon, A.C., Kuan, J., Sobotka, A.K. and Lichtenstein, L.M. (1983). Immunochemical studies of yellowjacket venom proteins. Mol. Immunol., 20, 2997-3008.
8. King, T.P., Kochoumian, L. and Joslyn, A. (1984). Wasp venom proteins: Phospholipase A_1 and B. Arch. Biochem. Biophys., 230, 1-12.
9. King, T.P., Joslyn, A. and Kochoumian, L. (1985). Antigenic cross-reactivity of venom proteins from hornets, wasps and yellowjackets. J. Allergy Clin. Immunol., 79, 611-613.

10. King, T.P. (1987). Sensitivity of immunoassays for detecting cross reactivity of homologous venom proteins of yellowjackets. J. Allergy Clin. Immunol., 75, 621-628.
11. Gadde, J., Golden, D.B.K., Kagey-Sobotka, A., Valentine, M.D., Lichtenstein, L.M. and King, T.P. (1986). Sensitivity to purified vespid venom proteins. J. Allergy Clin. Immunol., 57, 621-628.
12. Hoffman, D.R. (1985). Allergens in Hymenoptera venom. XV. The immunologic basis of vespid venom cross-reactivity. J. Allergy Clin. Immunol., 75, 611-613.
13. King, T.P., Kochoumian, L. and Lam, T. (1987). Immunochemical observations of antigen 5, a major venom allergen of hornet, yellowjackets and wasps. Mol. Immunol., 24, 857-864.
14. Unpublished observations.
15. Akre, R.D. and Reed, H.C. (1984). Biology and distribution of social hymenoptera. In Handbook of Natural Toxins, Vol.2 (Tu, A.T., ed.) M. Dekker, Inc, New York, pp. 3-47.
16. Finkelman, F.D., Katona, I., Utban, J.F., Snapper, C.M., Ohara, J. and Paul, W.E. (1986). Suppression of *in vivo* polyclonal IgE responses by monoclonal antibody the lymphokine BSF1. Proc. Natl. Acad. Sci. USA, 83, 9675-9679.
17. Ishizaka, K. (1984). Regulation of IgE synthesis. Ann. Rev. Immunol. 2, 158-182.
18. Fang, K.S.F., Vitale, M., Fehlner, P. and King, T.P. (1988). cDNA cloning and primary structure of a white-face hornet venom allergen, antigen 5. Proc. Natl. Acad. Sci. USA, 85, 895-899.
19. Marsh, D.G., Goodfriend, L., King, T.P., Lowenstein, H. and Platts-Mills, T.A.E. (1986). Allergen nomenclature. WHO Bulletin, 64, 767-770.
20. Cornelissen, B.J.C., Hooft van Huijsduijnen, R.A.M., Van Loon, L.C. and Bol, J.F. (1986). Molecular characterization of messenger RNAs for pathogenesis related proteins 1a, 1b and 1c produced by TMV infection of tobacco. EMBO J., 5, 37-40.
21. Lucas, J., Camacho Henriquez, A., Lottspeich, F. and Henschen, A. (1985). Amino acid sequence of the pathogenesis-related leaf protein p14 from viroid-infected tomato reveals a new type of structurally unfamiliar proteins. EMBO J., 4, 2475-2749.
22. Van Loon, L.C. (1985). Pathogenesis-related proteins (1985). Plant Mol. Biol., 4, 111-116.
23. Kombrink, E., Schroder, M. and Hahlbrock, K. (1988). Several pathogenesis-related proteins in potato are 1,3-β-glucanases and chitinases. Proc. Natl. Acad. Sci. USA, 85, 782-786.
24. Zora, J.A., Swanson, M.C. and Yunginger, J.W. (1988). A study of the prevalence and clinical significance of venom-specific IgE. J. Allergy Clin. Immunol., 81, 77-82.
25. Reisman, R.E. and Georgitis, J.W. (1985). Frequency of positive venom skin tests in insect-allergic and non-allergic populations. J. Allergy Clin. Immunol., 76, 803-807.

Analysis and Expression of cDNA Clones Coding for House Dust Mite Allergens

W. R. Thomas, K. Y. Chua, G. A. Stewart, C. R. Doyle, K. J. Turner, R. J. Simpson and R. J. Dilworth

Clinical Immunology Research Unit, Princess Margaret Hospital, Perth, Western Australia

ABSTRACT

Reactivity of IgE in mite allergic serum is directed to a large extent to the major allergens *Der p I* and *Der p II* for the case of *Dermatophagoides pteronyssinus* and to the equivalent proteins from *D. farinae* and *D. microceras*. Another 3 components produce reactivity in about 20% of sera while reactitivity to other components is sporadic. To further study these allergens and provide reagents for determining their interactions with the immune system, a project to produce and analyse cDNA clones coding for mite allergens was commenced. To date cDNA clones coding for the allergens *Der p I* and *Der p II* have been isolated and sequenced. This has revealed that *Der p I* is a cysteine protease produced with a proenzyme sequence. *Der p II* codes for a protein of 129 amino acids with a molecular weight of 14,131 and has no known homology with other proteins. It is anticipated that these clones will permit many studies in mite allergy, particularly T cell function where relatively large amounts of pure allergen are required for meaningful analysis.

KEYWORDS

Allergen; cDNA cloning; house dust mite; cysteine proteases.

INTRODUCTION

The advantages that can be expected to develop from molecular cloning of allergens could be listed in the general areas of (a) the definition and chemical characterisation of allergens, (b) the production of large quantities of defined allergens for experimental and clinical use and (c) the mapping of structures of the allergens which interact with components of the immune system such as products of immunoglobulin and T cell-receptor V-genes and major histocompatibility complex molecules. The actual use of molecular cloning to produce diagnostic and therapeutic reagents will depend on the number of allergens to be cloned for each allergy and the cost of their production. While some difficulties in producing and purifying products from

Advances in the Biosciences Vol. 74

cloned cDNA can be anticipated it should be remembered that this technology is in its infancy and is rapidly developing. In addition, it is essential to consider the current effectiveness of mite immunotherapy, the varying and often low allergen concentrations in different extracts [1] and if many allergens are actually present in effective concentrations. The use of genetic engineering could also produce new products such as modified allergens for safe immunotherapy. The role of cDNA cloning in T cell studies should be emphasised. Since effective stimulation of T cells in culture requires about 10 μM of antigen [2] (250 μg/ml for the major mite allergens *Der p I*) and precise dose-response curves to isolated allergens, few meaningful experiments have been performed even for the major allergens. Even in experiments conducted with purified major allergen the dose of antigen used, presumably because of availability, was below plateau levels for responsiveness [3]. Thus products from cDNA clones should open the way to documenting and perhaps understanding T cell function in allergy and desensitisation [4]. While this report will restrict itself to describing cDNA cloning of house dust mite allergens it should be pointed out that similar advances have been made with vespid venom antigen 5 [5].

HOUSE DUST MITE ALLERGENS

Three species of mite, *Dermatophagoides pteronyssinus*, *Dermatophagoides farinae* and *Dermatophagoides microceras* are important in producing allergic disease. Their relative distribution varies in different geographical regions, but all produce similar allergens which, as far as is documented, have both shared and individual antigenic determinants. Most mite allergic serum contains IgE antibody which reacts with two major allergens *Der p I* and *Der p II* in the case of *D. pteronyssinus* [6-10]. Not only do most people react with these allergens, but most of the IgE in a given serum usually reacts with these allergens, particularly *Der p I* [9,10]. Another three allergens react with IgE in 20-30% of sera when examined by cross radio-immunoelectrophoresis or Western blotting. This is illustrated in miniature in Fig. 1 with 5 sera taken from mite-allergic people visiting a clinic in Melbourne, Australia. When examined for the reactivity of their IgE with electrophoresed house dust mite extract by Western blotting all showed reactivity to a band consistant with the position of *Der p I* (26 Kd) and 4/5 in this figure had reactivity in position corresponding to *Der p II* and *Der p III* (14 Kd, 31 Kd). It should be noted that reactivity to *Der p I* is diminished in these assays because of its sensitivity to reduction and denaturation. The sera from "Coc" and "Van" (Fig. 1) had two additional reactivities. Serum from "And" reacted strongly with many bands. In our experience this pattern is found in severe cases of allergy. The relationship between the number of mite allergens recognised by a particular serum and the degree of allergy needs further documentation. Perhaps skin tests to minor mite allergens (made available by molecular cloning) may be more useful in recognising severe mite allergy than responses to major allergens which dominate reactivity to mite extracts. It has also been apparent that while the crossed radioimmunoelectrophoresis and Western blotting techniques have their usefulness, there is often difficulty in interpreting patterns from gel to gel. From Fig. 1, for example, it is clear that there are at least two allergens with Mr of about 14,000 (see panel for "Coc"). To avoid confusing these and similar close bands the availability of a panel of cloned allergens which could be used *in vitro* or *in vivo* would help analysis.

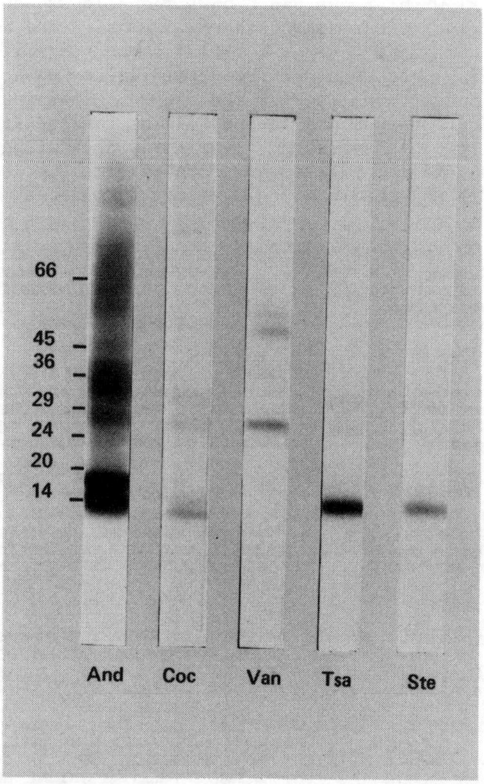

Fig. 1. Western blotting of mite extracts with allergic human IgE. Mite extract was electrophoresed on an 8-18% SDS-polyacrylamide gradient and transferred to nitrocellulose. Strips were reacted with neat allergic serum blocked by skim milk and reactivity developed by ^{125}I-monoclonal anti IgE and autoradiography. This procedure only produces reactivity in allergic sera. Allergic sera ("And" to "Ste") were from patients visiting an allergy clinic. The bands at 14 Kd are equivalent to *Der p II* and the *Der p I* and *III* migrate at 26 and 31 Kd under our conditions.

ISOLATION AND EXPRESSION OF cDNA CLONES CODING FOR *DER P I* AND *DER P II*

A cDNA library was established by standard techniques in lambda gt11 from mRNA isolated from adult *D. pteronyssinus* provided by Commonwealth Serum Laboratories, Melbourne. As a test for the integrity of the mRNA and its ability to code for allergens, *in vitro* translation products were produced in reticulocyte lysates and were shown to include many allergens including *Der p I* [11]. A clone, lambda gt11 pI (13T) was isolated by a combination of plaque immunoassay with a specific rabbit anti *Der p I* and oligonucleotide hybridisation [12] with probes based on amino acid sequences obtained from N-terminal and tryptic peptides of faecal *Der p I*. The nucleotide sequence of this clone reported by Chua *et al.* [13] showed that it was homologous with a group of cysteine proteases and shows particular conservation around the active site and contact residues of the enzyme. This analysis was possible because papain from the paw paw and actinidin from the Kiwi fruit were amongst the homologous enzymes and their X-ray crystallographic structure has been examined in detail. Fig. 2 shows the alignment of some of these cysteine proteases and active sites or contact residues marked by arrows.

```
                                            10        20            30        40        50
                      Der p 1:  TNACSINGNAPAEIDLRQMR..TVT..PIRMQGGCGSCWAFSGVAATESAYLAHRNQS
            Rat cathepsin H:            YPSSMDWRKKG.NVVS..PVKNQGACGSCWTFSTTGALESAVAIAS.GK
Chinese gooseberry actinidin:           LPSYVDWRSAG..AVV..DIKSQGECGGCWAFSAIATVEGINKITS.GS
              Papaya papain:            IPEYVDWRQKG..AVT..PVKNQGSCGSCWAFSAVVTIEGIIKIRT.GN
          Human cathepsin B:            LPASFDAREQWPQCPTIKEIRDQGSCGSCWAFGAVEAISDRICIHTN..
                                                             ▲   ▲▲▲

                                      60        70        80        90
                      Der p 1:  LD..LAEQELVDC...ASQHGCHGDTIPRGIEYIQHNGVVQ................
            Rat cathepsin H:  MM.TLAEQQLVDCAQNFNNHGCQGGLPSQAFEYILYNK....GI.............
Chinese gooseberry actinidin:  LI.SLSEQELIDCGRTQNTRGCDGGYITDGFQFIINDG....GI.............
              Papaya papain:  LN.QYSEQELLDC..DRRSYGCNGGYPWSALQL.VAQY....GI.............
          Human cathepsin B:  VSVEVSAEDLLTCCGSMCGDGCNGGYPAEAWNFWTRKGLVSGGLYESHVGCRPYSIPP

                                             100       110       120        130
                      Der p 1:  ......ESY..YRYVAREQSCRRPNAQRFGISNYCQIYPPNANK...IRE.ALAQPQR
            Rat cathepsin H:  ...MGEDSY.PYI..GKNGQCKFNPEKA...VAF.VKNVVNITL...NDEAAMVEAVA
Chinese gooseberry actinidin:  ...NTEENY.PYT..AQDGDCDVALQDQKY.VTI.DT.YENVPY...NNEWALQTAV.
              Papaya papain:  ...H.YRNT.PYY.EGVQRYCRSR.EKGPY.AAKTDGVRQVQPY...NQG.ALLYSIA
          Human cathepsin B:  CEHHVNGSRPPCTGEGDTPKCSKICEPGYSPTYKQDKHYGYDSYSVSNSEKDIMAEI.

                                 140        150       160            170       180
                      Der p 1:  YCRHYWT....IKDLDAFRHYDGRTIIQRDNGYQ.PN..Y.HAVNIVGYSNAQGVDYW
            Rat cathepsin H:  LY..NPVSFAFEV.TEDFMMYKS.GVYSSNSCHKTPD.KVNHAVLAVGYGEQNGLLYW
Chinese gooseberry actinidin:  TY..QPVSVALDAAGDAFKQYAS.GIFTGP.C...GT.AVDHAIVIVGYGTEGGVDYW
              Papaya papain:  N...QPVSVVLQAAGKDFQLYRG.GIFVGP.C...GN.KVDHAVAAVGY..NPG..YI
          Human cathepsin B:  .YKNGPVEGAFSV.YSDFLLYKS.GVYQ....HVTGEMMGGHAIRILGWGVENGTPYW
                                                                       ▲

                                 190       200       210       220
                      Der p 1:  IVRNSWDTNWGDNGYGYFAANIDLMMIEEYPYVVIL
            Rat cathepsin H:  IVKNSWGSNWGNNGYFLIERGK....NMCGLAA..CASYPIPQV
Chinese gooseberry actinidin:  IVKNSWDTTWGEEGYMRILRNVG.GAGTCGIAT..MPSYPVKYNN
              Papaya papain:  LIKNSWGTGWGENGYIRIKRGTGNSYGVCGLYT..SSFYPVKN
          Human cathepsin B:  LVANSWNTDWGDNGFFKILRGQ....DHCGIESEVVAG..IPRTD
                                   ▲▲▲
```

Fig. 2. Composite alignment of the amino acid sequences of *Der p I* and other cysteine proteases [13]. Gaps were added to maximise homology and arrows indicate residues in the active site of papain and actinidin as deduced by X-ray crystallography.

Furthermore, the conservation of tertiary structure between actinidin and papain [14] has given some insights into the possible structure of *Der p I*. Previously we speculated that the cDNA clone codes for the mature protein and a short proenzyme sequence beginning with a phenylalanine at -13 which was preceded by a truncated leader sequence. This was based on the anology with other cysteine proteases which have pre and proenzyme sequences and the lack of a typical leader peptide and Kozak sequences in our clone. Position -13 was chosen as the possible residue for the beginning of the proenzyme because it follows a typical (ala-x-ala) signal peptidase cleavage site. These features are illustrated in the full nucleotide sequence of the *Der p I* clone in Fig. 3. It now appears that the proenzyme sequence extends beyond this site and beyond the 23 preamino terminal residues of the current cDNA clone. As recently discussed by Ishidoh and colleagues [15] most cysteine proteases examined have about 120 preproenzyme residues with cathepsin B being the smallest with 80. In addition a highly conserved area in the pro-enzyme sequence of cysteine proteases has been recognised and this shows

```
          -23     -20          *                           -10
GAATTC    AAA AAC CGA TTT TTG ATG AGT GCA GAA GCT TTT GAA CAC CTC AAA ACT
Eco RI    Lys Asn Arg Phe Leu Met Ser Ala Glu Ala Phe Glu His Leu Lys Thr
                        -1  +1                                  +10
CAA TTC GAT TTG AAT GCT GAA ACT AAC GCC TGC AGT ATC AAT GGA AAT GCT CCA GCT
Gln Phe Asp Leu Asn Ala Glu Thr Asn Ala Cys Ser Ile Asn Gly Asn Ala Pro Ala
                            +20                                      +30
GAA ATC GAT TTG CGA CAA ATG CGA ACT GTC ACT CCC ATT CGT ATG CAA GGA GGC TGT
Glu Ile Asp Leu Arg Gln Met Arg Thr Val Thr Pro Ile Arg Met Gln Gly Gly Cys
                                +40                                     +50
GGT TCA TGT TGG GCT TTC TCT GGT GTT GCC GCA ACT GAA TCA GCT TAT TTG GCT CAC
Gly Ser Cys Trp Ala Phe Ser Gly Val Ala Ala Thr Glu Ser Ala Tyr Leu Ala His
                                    +60
CGT AAT CAA TCA TTG GAT CTT GCT GAA CAA GAA TTA GTC GAT TGT GCT TCC CAA CAC
Arg Asn Gln Ser Leu Asp Leu Ala Glu Gln Glu Leu Val Asp Cys Ala Ser Gln His
+70                                     +80
GGT TGT CAT GGT GAT ACC ATT CCA CGT GGT ATT GAA TAC ATC CAA CAT AAT GGT GTC
Gly Cys His Gly Asp Thr Ile Pro Arg Gly Ile Glu Tyr Ile Gln His Asn Gly Val
  +90                                     +100
GTC CAA GAA AGC TAC TAT CGA TAC GTT GCA CGA GAA CAA TCA TGC CGA CGA CCA AAT
Val Gln Glu Ser Tyr Tyr Arg Tyr Val Ala Arg Glu Gln Ser Cys Arg Arg Pro Asn
      +110                                    +120
GCA CAA CGT TTC GGT ATC TCA AAC TAT TGC CAA ATT TAC CCA CCA AAT GCA AAC AAA
Ala Gln Arg Phe Gly Ile Ser Asn Tyr Cys Gln Ile Tyr Pro Pro Asn Ala Asn Lys
                            +130                  +140
ATT CGT GAA GCT TTG GCT CAA CCA CAG CGC TAT TGC CGT CAT TAT TGG ACC ATC AAA
Ile Arg Glu Ala Leu Ala Gln Pro Gln Arg Tyr Cys Arg His Tyr Trp Thr Ile Lys
              +150                                    +160
GAT TTA GAC GCA TTC CGT CAT TAT GAT GGC CGA ACA ATC ATT CAA CGC GAT AAT GGT
Asp Leu Asp Ala Phe Arg His Tyr Asp Gly Arg Thr Ile Ile Gln Arg Asp Asn Gly
                  +170                                    +180
TAC CAA CCA AAC TAT CAC GCT GTC AAC ATT GTT GGT TAC AGT AAC GCA CAA GGT GTC
Tyr Gln Pro Asn Tyr His Ala Val Asn Ile Val Gly Tyr Ser Asn Ala Gln Gly Val
                      +190                                    +200
GAT TAT TGG ATC GTA CGA AAC AGT TGG GAT ACC AAT TGG GGT GAT AAT GGT TAC GGT
Asp Tyr Trp Ile Val Arg Asn Ser Trp Asp Thr Asn Trp Gly Asp Asn Gly Tyr Gly
                          +210                                    +220
TAT TTT GCT GCC AAC ATC GAT TTG ATG ATG ATT GAA GAA TAT CCA TAT GTT GTC ATT
Tyr Phe Ala Ala Asn Ile Asp Leu Met Met Ile Glu Glu Tyr Pro Tyr Val Val Ile
+222 *
 CTC TAAACAAAAAGACAATTTCTTATATGATTGTCACTAATTTATTTAAAATCAAAATTTTTAGAAAATGAATAAA
 Leu     3' non-coding region
 TTCATTCACAAAAATT - Poly A tail
```

Fig. 3. Nucleotide and predicted amino acid sequence of cDNA lambda gt11 p1 [13]. Numbers to the right are nucleotide positions and those above the sequence are amino acids. Positive numbers are mature faecal *Der p I* starting at threonine. Negative numbers are proenzyme sequences. Residues 52-54 are an N-glycosylation site. The polyadenylation site is underlined.

homology with residues -13 to -23 of *Der p I* (Fig. 4). This area is located between about -26 and -36 for the other cysteine proteases but as previously shown in the homology alignment, mature *Der p I* (Fig. 2) has an additional 9 N-terminal amino acids which permit it to adopt similar folding of its proenzyme structure. Attempts to isolate clones with a full proenzyme sequence will be made because this is important to achieve the correct folding of the mature protein.

The difference in structure between *Der p I*, *Der f I* and *Der m I* will be of interest because of the desirability of creating common recombinant class I mite allergen and because differences and similarity in their antigenicity

			* * *** ***		
<u>Der p</u> I			KNRFL-MSAEAF	-13	...
Cathepsin H	-114	......	LNQFSDMSFAEI	-25	......
Cathepsin L	-114	......	MNAFGDMTNEEF	-26	......
Papain	-133	......	LNVFADMSNDEF	-29	......
Aleurain	-144	......	INRFSDMSWEEF	-27	......
Cathepsin B	- 80	......	RN-FYNV-DISF	-26	......

The number on the left shows the size of the preprosequence and the number on the right shows the location of the conserved proenzyme area.

***conserved residues.**

Fig. 4. Conserved propeptide region of cysteine proteases.

may pinpoint epitopes. Previous published comparison of *Der p I* and *Der f I* homology showed only 11/20 homologous residues in the N-terminal [12,16]. Further data for *Der f I* [17] shows that most of the non-homologous residues were in extreme N-terminal and the overall homology will be about 80% (Chapman *et al.*, unpublished). Fig. 5 shows the homology of the first 30 residues of *Der p I* and *Der f I*. The allergen *Der m I* has a primary structure very similar to *Der f I* [18].

Southern blot analysis of genomic DNA from *D. pteronyssinus*, probed with the cDNA coding for *Der p I*, has shown that the DNA containing the cDNA sequence is located on fragments of less than 2.5 Kd. This is sufficient for only one gene hence there is only one *Der p I*-like molecule produced by the mite and secondary modifications must account for any heterogeneity in structure.

The allergen *Der p II* was cloned by screening by plaque immunoassay with allergic human serum and monoclonal anti IgE. The clone producing the allergen lambda gt11-pII (c1) reacted with about 70% of allergic serum tested [Doyle, C.R., Chua, K.Y., Turner, K.J., Stewart, G.A. and Thomas, W.R. in preparation]. It has been sequenced and shown to contain a 597 bp mite sequence coding for a 129 amino acid protein with a derived molecular weight of 14,131. A scheme of this cDNA structure is presented in Fig. 6 [Chua, K.Y. Dilworth, R.J., Doyle, C.R., Simpson, R.J., Stewart, G.A. and Thomas, W.R. in preparation]. The protein has a typical 17 residue leader peptide and has no N-glycosylation sites. No significant homology with other proteins has been found except to *Der f II* [17]. Our cDNA and peptide data [16] has shown homology in 31/35 of residues available for comparison (Fig. 7). It also reacts with a monoclonal anti *Der p II* supplied by Dr R.C. Aalberse. This high degree of antigenicity contrasts with our *Der p I* clone which reacts well with rabbit anti *Der p I*, but poorly with human IgE. This is probably related to the finding that the *Der p II* is expressed in our cDNA clone by itself while the *Der p I* is expressed as a fusion protein. Several modifications are now being made to both increase and decrease antigenicity.

FUTURE PERSPECTIVES

The study presented here is the first stage in developing products from cDNA cloning which will assist our investigation of mite allergy. Studies to determine the B and T cell epitopes are in progress and will be partially

```
                       10       20      30
Der p I    TNACSIN-GNAPAEIDLRQ-MRTVTPIRMQGG
            *  *  ** * * *    *

Der f I    TSACRINSVNVPSELDLRSLMRTVTPIRM
```

Fig. 5. Comparison of N-terminal sequences of *Der p I* and *Der f I*. Asterisks show non-homologous residues.

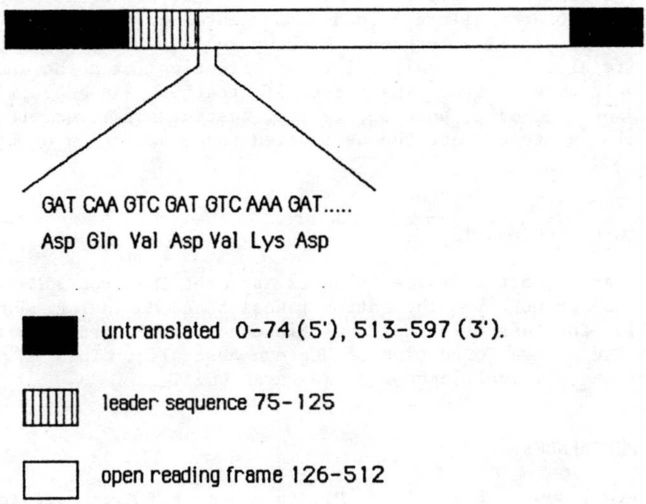

Fig. 6. Schematic representation of *Der p II* cDNA sequence from clone lambda gt11-pII (cl). The identity of the clone was recognised by its ability to react frequently with mite allergic sera and its N-terminal homology with the known *Der p II* sequence [16]. It codes for 129 amino acids with an estimated MW of 14,131 Kd.

```
                   10        20        30
Der p II   DQVDVKDCANHEIKKVLVPGCHGSEPCIIHRGKPF
                     *     * *     *

Der f II   DQVDVKD?ANNEIKKVMVDG?HGSDP?IIHRGKPF
```

Fig. 7. Comparison of N-terminal sequences of *Der p II* and *Der f II*. Asterisks show non-homologous residues.

reported elsewhere in this volume. It is anticipated that the availability of these reagents will stimulate the study of T cell function in the allergic process and provide material for experimentation in delivery systems and modified allergens. The potential use of recombinant molecules in clinical practice is a real possibility for house dust mite allergens since for many people the bulk of the allergic load is presented by class I and II and to a lesser degree class III allergens. With up to 5% of many populations suffering from mite allergy the availability of effective immunotherapeutic reagents would have a large impact even if treatment was appropriate in only half the cases. To date, however, we must content ourselves with having used cDNA cloning to advance the definition and description of mite allergens.

ACKNOWLEDGEMENTS

This study was supported by the Princess Margaret Children's Medical Research Foundation and grants from the National Health and Medical Research Council of Australia, the Australian National Biotechnology Program Research Grants Scheme and the Asthma Foundation of Western Australia. This is publication No. 2899 of the Clinical Immunology Research Unit.

REFERENCES

1. Ford, A.W., Rawle, F.C., Lind, P., Spieksma, F.T.M., Lowenstein, H. and Platts-Mills, T.A.E. (1985). Standardisation of *Dermatophagoides pteronyssinus*: Assessment of potency and allergen content in the coded extracts. Int. Arch. Allergy Appl. Immunol., 76, 58-67.
2. Ashwell, J.D., Fox, B.S. and Schwartz, R.H. (1986). Functional analysis of the interaction of the antigen-specific T cell receptor with its ligands. J. Immunol., 136, 757-768.
3. Rawle, F., Michel, E.B. and Platts-Mills, T.A.E. (1984). T-cell responses to the major allergen from the house dust mite *Dermatophagoides pteronyssinus* antigen PI: Comparison of patients with asthma atopic dermatitis and perennial rhinitis. J. Immunol., 133, 195-201.
4. Rocklin, R.E. (1983). Clinical and immunologic aspects of allergen-specific immunotherapy in patients with seasonal allergic rhinitis and/or allergic asthma. J. Allergy Clin. Immunol., 72, 323-334.
5. Fang, K.S.F., Vitale, M., Fehlner, P. and King, T.P. (1988). cDNA cloning and primary structure of a white-face hornet venom allergen antigen 5. Proc. Natl. Acad. Sci., 85, 895-899.
6. Krilis, S., Baldo, B.A. and Basten, A. (1984). Antigens and allergens from the common house dust mite *Dermatophagoides pteronyssinus*. Part II.

Identification of the major IgE binding antigens by cross radioimmuno-electrophoresis. J. Allergy Clin. Immunol., 74, 142-146.

7. Tovey, E.R. and Baldo, B.A. (1987). Comparison by electroblotting of IgE-binding components in extracts of house dust mite bodies and spent mite culture. J. Allergy Clin. Immunol., 79, 93-102.
8. Stewart, G.A., Butcher, A., Lees, K. and Ackland, J. (1986). Immuno-chemical and enzymatic analyses of extracts of the house dust mite *Dermatophagoides pteronyssinus*. J. Allergy Clin. Immunol., 77, 14-24.
9. Tovey, E.R., Chapman, M.D. and Platts-Mills, T.A.E. (1981). Mite faeces are a major source of house dust allergens. Nature, 289, 592-593.
10. Chapman, M.D. and Platts-Mills, T.A.E. (1980). Purification and characterisation of the major allergen from *Dermatophagoides pteronyssinus* antigen PI. J. Immunol., 125, 587-592.
11. Stewart, G.A. and Thomas, W.R. (1987). *In vitro* translation of Messenger RNA from the house dust mite *Dermatophagoides pteronyssinus*. Int. Arch. Allergy Appl. Immunol., 83, 384-389.
12. Thomas, W.R., Stewart, G.A., Simpson, R.J., Chua, K.Y., Plozza, T.M., Dilworth, R.J., Nisbet, A. and Turner, K.J. (1988). Cloning and expression of DNA coding for the major house dust mite allergen *Der p I* in *Escherichia coli*. Int. Arch. Allergy Appl. Immunol., 85, 127-129.
13. Chua, K.Y., Stewart, G.A., Thomas, W.R., Simpson, R.J., Dilworth, R.J., Plozza, T.M. and Turner, K.J. (1988). Sequence analysis of cDNA coding for a major house dust mite allergen *Der p I* in homology with cysteine proteases. J. Exp. Med., 167, 175-182.
14. Baker, E.N. and Drenth, J. (1987). The Thiol proteases structure and mechanism. In: Biological Macromolecules and Assemblies, Vol.3, (Ed. F. Jurnack) A. McPherson, pp.314-368.
15. Ishidoh, K., Imajah, S., Emore, Y., Ohno, S., Kawasaki, H., Minami, Y., Kominami, E., Katunuma, N. and Suzuki, K. (1987). Molecular cloning and sequencing of cDNA for rat cathepsin H. Homology in pro-peptide regions of cysteine proteases. FEBS Letters, 226, 33-37.
16. Stewart, G.A., Simpson, R.J., Moritz, R.L., Thomas, W.R. and Turner, K.J. (1988). Physico-chemical characterisation of allergens *Der p I* (Dpt 12) Dpt 22 and Dpt 36 from the house dust mite. In: Mite Allergy, (Eds. A.L. de Weck, A. Todt), The UCB Institute of Allergy, Brussels, pp.35-38.
17. Chapman, M.D., Heymann, P.W. and Platts-Mills, T.A.E. (1988). Epitope mapping of major dust mite (*Dermatophagoides*) allergens using monoclonal antibodies. In: Mite Allergy (Eds. A.L. de Weck, A. Todt), The UCB Institute of Allergy, Brussels, pp.27-29.
18. Lind, P. and Horn, N. (1988). Extensive analogy between the epitope specificity of mouse hybridoma and human IgE antibodies to *D. pteronyssinus* major allergen *Der p I*. In. Mite Allergy (Eds. A.L. de Weck, A. Todt), The UCB Institute of Allergy, Brussels, pp.30-34.

Molecular and Immunological Approaches to Epitopes of Selected Airborne Pollen Allergens

David G. Klapper, Susan Woods,
Joseph Olson, Robert Esch, James J. Smith
and Thorunn Rafnar

Department of Microbiology and Immunology,
University of North Carolina School of Medicine,
Chapel Hill, NC 27599, USA

ABSTRACT

Current methodology for defining epitopes of natural airborne allergens is described in this report. The grass Group I system has proved to be amenable to traditional approaches such as isolation and characterization of peptides which retain antigenic activity. Antigen E of short ragweed has not yielded useful information from similar approaches and this report describes the tools which are being used to clone the structural gene for antigen E to enable the use of molecular biology techniques to address the question of epitopes on this allergen.

KEYWORDS

Grass Group I; amino acid sequence; synthetic oligonucleotide.

INTRODUCTION

The ability to fractionate crude extracts of allergens and isolate highly purified proteinaceous materials which react with human IgE has led to many recent advances in the field of allergy and allergens. With highly purified allergens, it has become possible to prepare monospecific antibodies and begin to standardize materials and define cross-reactive molecules. Through the use of monoclonal antibodies, it is now becoming possible to perform epitope mapping of discrete allergens [1-9] and design synthetic peptides which will mimic these epitopes [10]. These approaches to understanding the interaction between the immune system and selected allergens depend upon the ability to isolate sufficient highly purified specific allergen to be able to perform at least limited structural studies and various immunoassays. Another level of investigation aimed at providing virtually unlimited quantities of allergen and the ability to manipulate the structures of these allergens will require the latest in technology, namely the cloning, characterization, and expression of the genes coding for these molecules.

This laboratory has long been interested in the structure of allergens and

Advances in the Biosciences Vol. 74

AB 74-F

has attempted to apply recent advances in technology to this area of investigation. In this report, two systems on which we are currently working will be described as models for different approaches to determining the B cell epitope structures of selected, naturally occurring airborne allergens.

MATERIALS AND METHODS

Isolation of Allergens. Protocols for isolation of grass Group I allergens and antigen E from short ragweed have been published in detail [5,8]. Briefly, pollen is soaked in neutral, aqueous buffer overnight at 4°C, the pollen stroma centrifuged out, and the allergen of interest isolated by a series of molecular sieve and ion exchange chromatographic steps. Fractions from these columns are assayed using specific antisera and purification is monitored by SDS-PAGE.

Modification of Allergens. Reduction and alkylation and citraconylation with subsequent tryptic digestion of the allergen were performed as described [10]. Analytical Procedures: An Applied Biosystems 470A gas phase sequencer with a Waters HPLC connected on-line was used for the bulk of the sequence analysis performed here. Peptide synthesis was performed with a Biosearch 9500 automated solid phase instrument using standard t-BOC chemistry protocols provided by the manufacturer. HPLC based amino acid analysis and purification of peptides was performed with Waters HPLC equipment using C18 reverse phase resins.

ELISA. This assay in the many forms in which we use it has also been well described [5]. In general, polyvinyl microtiter plates are used, Tween 20 is used to prevent non-specific binding, our usual label is alkaline phosphatase, and the plates are washed with a Biotek ELISA plate washer and read with a Biotek model 309 plate reader.

Monoclonal Antibodies. The monoclonal antibodies utilized for the studies reported in this manuscript have all been described in the literature [5,8, 10]. The standard fusion partner in this laboratory is P3X63-Ag8.653, a non secreting derivative of MOPC 21. Antigens, administered either in Freund's adjuvant or as alum precipitates, are either native or have been reduced and alkylated. Typically, 3 rounds of immunization (50 micrograms of antigen) spaced 10 days apart are given, the animal is rested 10-14 days and a dose of antigen not in any adjuvant is given. Three days later, the mice are sacrificed and spleen cells recovered for the fusion [11]. Plating, feeding, assaying, cloning in soft agar, etc. are now fairly routine procedures. Large amounts of antibody are recovered either by processing large volumes of tissue culture supernates or from ascites fluids in pristane treated mice [12] using high speed, high recovery techniques previously described [13].

Oligonucleotide Probes and Hybridization. Two sets of oligonucleotides, corresponding to two AgE peptide sequences, were synthesized on an automated DNA synthesizer. The first set contained a mixture of 32 sequences, each 18 nucleotides long, corresponding to all possible codon combinations coding for the peptide sequence Trp Glu Asn Phe Lys Lys (oligonucleotide sequence 5' T/CTT-T/CTT-G/AAA-G/ATT-T/CTC-CCA 3'). The second set of oligonucleotides contained all possible codon combinations coding for the peptide sequence Lys Asp Gly Asp Ile Tyr. Since this sequence is highly degenerate, the probes were divided into three pools to simplify screening. The first pool contained the sequences 5' G/ATA - TAT - G/ATC - T/G/ACC - G/ATC - T/CTT 3'; the second and third pools had identical sequences except for position 4 where pool 2 contains a G and pool 3 contains and A. The oligonucleotides were end labeled and used as hybridization probes to screen a cDNA library made from ragweed flower heads. The hybridization conditions of Wood *et al.* [14] were used and the washes performed at temperatures 4°C below the calculated Tm.

RNA Isolation and cDNA Library Construction. Membrane bound polysomal RNA was isolated as described by Larkins and Hurkman [15] and polyA RNA selected on an oligo dT cellulose column. cDNA was synthesized using AMV reverse transcriptase and RNase H, attached to EcoR1 linkers, and ligated into lambda gt10. The library was amplified on *E. coli* strain C600Hf and plated out for screening on the same strain.

Oligonucleotide Primer Extensions. cDNA was synthesized using the oligonucleotides as primers under conditions described by Noyes *et al.* [16] except instead of end labeling the oligonucleotides, the cDNA strand was labeled by adding ^{32}P-dCTP (10 microcurie) to the reaction mixture. After the first strand synthesis was complete, the reaction was stopped by adding EDTA to a final concentration of 12 mM. The cDNA was separated from unincorporated label by chromatography on a G-50 spun column and precipitated with sodium acetate and ethanol. The pellet was dissolved in 0.1 M NaOH/1 mM EDTA and the RNA hydrolyzed for 30 minutes at 70°C. Sample preparation and gel electrophoresis were as described [17] through 5% acrylamide in 8 M urea gels at 25 watts for 5-6 hours. The gel was wrapped in several layers of plastic film and exposed to X-ray film at -20°C.

RESULTS

Grass Group I allergens. Well over 100 monoclonal antibodies have been produced and evaluated in this laboratory over the last several years. These antibodies can be conveniently placed into three categories of activity in terms of their ability to recognize the proband Group I and to cross react with other Group I molecules from other grasses [8]. Some of the monoclonals (at least one for each Group I allergen) react very specifically with the immunizing Group I allergen and do not show any appreciable cross reactivity with any other Group I allergen. By far, the majority of the monoclonal antibodies show some degree of cross reactivity with other Group I allergens with tremendous variation in the pattern of reactivity. Some might recognize both the proband and one other Group I, some might recognize the proband and several other (but not all) Group I allergens, and some might recognize the proband and one or two others with high affinity, but other Group I allergens with low affinity. The third category of monoclonal antibodies are those which cannot distinguish between the proband and any other Group I allergen, that is, they are extremely cross reactive. The former monoclonal antibodies (highly specific) have potential for use in standardization, identification, and confirmation assays for particular grass pollen extracts. The latter antibodies (highly cross reactive) are useful for characterizing epitopes among Group I allergens responsible for their known clinical cross reactivity.

Among the monoclonal antibodies which fall into this highly cross reactive category, it has previously been determined that each is capable of inhibiting the binding of human IgE from grass Group I allergic individuals to Group I allergen. In addition, from additivity assays, it has been demonstrated that at least 2 epitopes defined by these antibodies must exist [18]. These epitopes have been designated A and D and site A is defined by at least 5 monoclonal antibodies while site D is defined by a single monoclonal antibody. Any of the anti site A antibodies in the presence of the anti site D antibody will cause additive inhibition of the binding of human IgE to any Group I molecule. Any single anti site A antibody inhibits human IgE binding as well as any mixture of anti site A antibodies.

One of the rather fortuitous situations discovered early on in our studies is that many of the monoclonal antibodies in our collection, although raised against native Group I allergens, react quite well with reduced and alkylated Group I allergens. Among the monoclonal antibodies against site A, several have this remarkable ability and have been used to isolate and characterize the peptide which carries this epitope [19].

When assayed by either direct ELISA or inhibition of ELISA, the mixture of peptides from a complete tryptic digest of meadow fescue Group I allergen contained no immunoreactive material. However, if the Group I molecule was first treated with citraconic anhydride before the tryptic digest and subsequently decitraconylated after the trypsin had been inactivated, the total digest could be used to coat a polyvinyl chloride microtiter plate and an anti site A antibody bound to it. In addition, the digestion mixture could inhibit the ability of an anti site A antibody to bind to purified, native or reduced and alkylated Group I allergen. Using an anti site A monoclonal antibody to screen fractions for the presence of an immunoreactive peptide, the total digest was fractionated first by size, then by charge, and finally by reverse phase HPLC. A single peak which was shown to be a single peptide was isolated as shown in Figure 1 and its complete amino acid sequence determined (Fig. 2). The amino acid sequence agreed quite well with the amino acid composition (not shown). Also not shown here are the results of solid phase peptide synthesis of this entire 28 residue peptide. The synthetic material reacted with the anti site A monoclonal antibody and when used to immunize several mice, the antisera produced bound the synthetic peptide and either native or reduced and alkylated Group I allergen. When other Group I allergens were used as starting material for these structural studies, substantially similar results were obtained [19].

It is clear, then, that this very straightforward approach to defining B cell epitopes has the potential to work well in this particular system. The unique features of this system are that there are a set of related molecules which can be isolated from a variety of temperate grasses, there is considerable clinically relevant antigenic cross reactivity among these molecules, both human and murine antibodies recognize this cross reactivity, and the epitope(s) do not seem to be conformation dependent.

Short Ragweed Antigen E. This allergen, described by King in 1964 [reviewed, 20] is the major allergen in short ragweed and is perhaps the most common and clinically relevant allergen in the eastern United States. The molecule has been well described and consists of a single non glycosylated polypeptide chain with a molecular weight of 38000 daltons. During chromatographic isolation, antigen E (AgE) is cleaved into an alpha and beta chain by a trypsin-like enzyme but retains full allergenic and antigenic activity. In this 2 chain form, AgE is quite enzyme resistant. Both polyclonal and monoclonal antibodies have been made against AgE in this and other laboratories and the finding is that these antibodies recognize either the single or 2

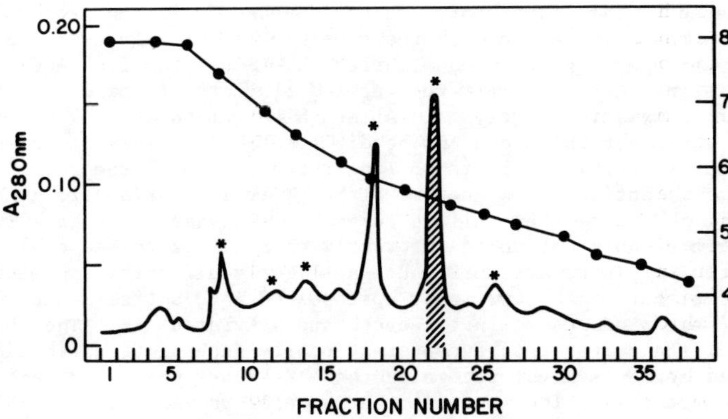

Fig. 1. DEAE ion exchange chromatography of a fraction from the tryptic digest of an extensively reduced and alkylated, citraconylated Group I molecule. See [19] for details. Elution was by an increasing salt, decreasing pH gradient and individual fractions were monitored in an ELISA assay for reactivity with a monoclonal antibody to a highly cross reactive (among other Group I allergens) epitope. The 'hatched' area represents the antibody positive material which was pooled, further purified, and sequenced. Asterisks are areas of the chromatogram which reacted with polyclonal antisera against grass Group I allergen.

Y T T E G G T K S E V E D V I P E G W K A D T S Y S A K

Fig. 2. Amino acid sequence of the 28 residue peptide which represents the highly homologous (among Group I allergens), highly cross reactive peptide isolated from the tryptic digest of citraconylated denatured Group I allergen.

chain form of AgE, but not modified AgE. The epitopes are apparently extremely conformationally dependent. Approximately 80% of the IgE antibodies in serum from allergic individuals can be inhibited from binding to AgE by a combination of 2 murine monoclonal antibodies which recognize 2 non repeated, non overlapping allergenic sites [5]. In fact, all sera (human, rabbit, and mouse) tested to date give the same somewhat surprising result –

there are only a few major epitopes on this relatively large molecule.

Since several monoclonal antibodies directed against the two major epitopes were generated in this laboratory, it was decided to attempt to isolate and characterize the epitopes in much the same way as was used to characterize the site A epitope of grass Group I allergen. Since "native" AgE is highly enzyme resistant, AgE was first reduced and alkylated to make it more sensitive to enzymatic hydrolysis. Although reduction and alkylation of AgE results in a product which can now be digested by a variety of enzymes, the modified (denatured) molecule is no longer reactive with the monoclonal (or any polyclonal) antibodies. And so, although it is possible to generate many fragments of AgE using chemical and enzymatic cleavage, none have retained the appropriate conformation to interact with antisera raised against the native molecule. In an attempt to use a slightly different approach to identifying either of the two major epitopes on AgE, antisera and monoclonal antibodies were produced against reduced and alkylated AgE. The thought was that even if the epitopes were conformationally dependent, antibodies to short linear segments might recognize those segments when they were part of the native structure. The hope was that some of the antibodies against the denatured molecule might inhibit the ability of the antibodies against the native molecule to bind to the native molecule. Toward this end, 17 monoclonal antibodies against reduced and alkylated AgE were raised and characterized [10]. Some of these were shown by Western blotting (Fig. 3) to bind to both the intact structure (38000 daltons) and either alpha or beta chains. None of the 17 monoclonals which were examined cross reacted or recognized overlapping antigenic sites. Unfortunately, none of the 17 recognized native AgE. Several of the monoclonal antibodies were used to isolate peptides from tryptic digests of citraconylated, reduced and alkylated AgE (as was done from grass Group I allergens) and antisera raised against synthetic peptides modeled from the amino acid sequence information gleaned from these isolated peptides reacted with reduced and alkylated AgE or with alpha or beta chains as appropriate. In no case did an antiserum against a peptide fragment react with native AgE. Unlike the grass Group I model system, this is a system where the conformation dependence of the epitopes renders a simple, straightforward approach unrewarding.

For a number of reasons, it was decided that cloning the gene for AgE might give rise to material which would be more amenable to our goal of defining the two major epitopes of the molecule. Since the complete amino acid sequence of AgE is not yet known, cloning the gene would give us that information. In an expression vector, it might be possible to express large fragments of the molecule and modify the sequence in ways which would help us to preserve and therefore elucidate the epitope structure. One approach to identifying and characterizing the structural gene coding for AgE might be the use of the panel of monoclonal antibodies against denatured AgE to screen a cDNA library in an expression vector. An alternative approach, and the one to be described in this report, is to synthesize oligonucleotides designed from our knowledge of some of the amino acid sequence of AgE. Two such oligonucleotide probes were synthesized (see Materials and Methods) and originally used to directly screen a cDNA library made from RNA isolated from ragweed flower heads which are actively synthesizing pollen and therefore AgE. Although hundreds of thousands of plaques were blotted and hybridized with these mixtures of radiolabeled, synthetic probes, we were concerned with the lack of reproducibility when duplicate blots were screened and when a putative positive plaque was expanded and reblotted. We became convinced that the probes were too short, too degenerate, and that no one probe had a high enough specific activity to unambiguously hybridize under sufficiently stringent conditions to identify a positive clone.

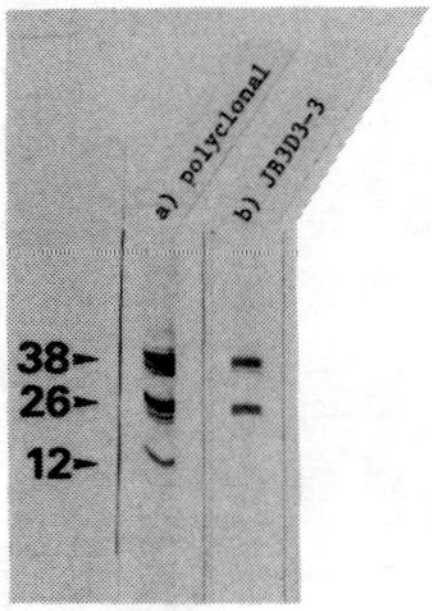

Fig. 3. Western blots of antigen E probed with polyclonal and monoclonal antisera to denatured antigen E. The polyclonal antiserum recognizes the intact allergen (38,000 daltons), alpha chain (26,000 daltons), and beta chain (12,000 daltons). The monoclonal antibody depicted here (JB3D3-3) is specific for an epitope on the alpha chain and recognizes both the alpha chain and the intact allergen (which contains both alpha and beta chains in their uncleaved form).

We turned, therefore, to an approach which we thought would give rise to a limited number of cDNA probes long enough to allow screening of our cDNA library under very stringent conditions. That approach is a modification of the primer extension method [16] and depends upon the ability of oligonucleotide mixtures to anneal (as best they can) to polyA RNA. Reverse transcriptase will begin to make a cDNA from 3' to 5' and that cDNA will have a unique oligonucleotide sequence since it is being copied from a unique template. If the successful oligonucleotide anneals with a complementary sequence near the 5' end of the message, the cDNA will be relatively small, but if it anneals near the 3' end of the message, there is the possibility that the cDNA will be quite large. The cDNA can be labeled either with a single radioactive base or multiple radiolabeled bases, and when recovered, should act as a highly specific probe for hybridization analysis. Although the relatively short synthetic oligonucleotide probes we are using may not be highly specific for the gene of interest, even in the worst case, they will anneal with only a limited number of RNA's and generate a limited number of cDNA's. Figure 4 shows a radioautograph of the results of a typical primer extension experiment. The left lane contains a series of size markers of known base lengths (HaeIII digest of ØX174), lane P1 contains the reaction products resulting from one of the mixtures of synthetic oligonucleotides and polysomal RNA, and lane A2 contains the other mixture of synthetic oligonucleotides and a total polyA RNA preparation. Although there are a number of faint background bands, there is at least one more prominent band in each lane. Since radioisotope continues to be added as the cDNA gets longer and longer, it might be expected that the larger cDNA's should give a more pronounced signal than the smaller cDNA's. In both cases, the band with the most significant signal is one of the lower molecular weight bands, suggesting that it was more efficiently synthesized and therefore might reflect a higher affinity for the synthetic oligonucleotide (and

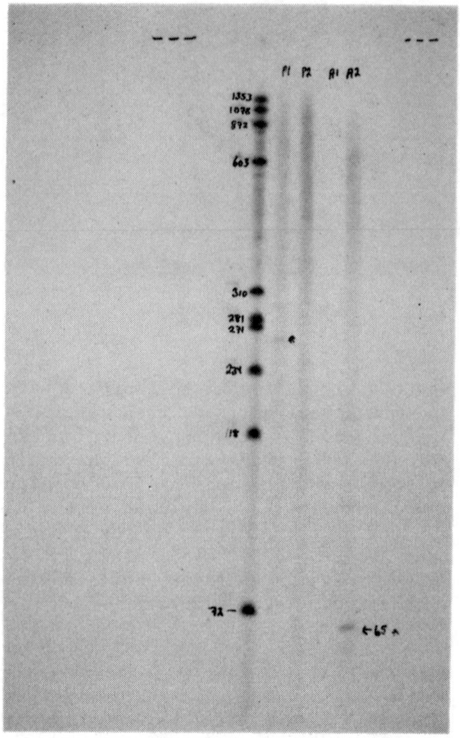

Fig. 4. Acrylamide/urea gel of the reaction products from a primer extension reaction. See text for details. The bands marked with an asterisk in lane P1 and marked as '65*' in lane A2 are thought to be specific for antigen E.

hopefully represent a cDNA from the target AgE mRNA). The band of approximately 260 nucleotides could represent as much as 25% of the amino terminal region of AgE while the 65 nucleotide cDNA would be expected to anneal closer to the 5' end of the mRNA. Studies are in progress to use these highly specific, radiolabeled cNDA's to probe the cDNA library.

DISCUSSION

Of the many model systems for studying epitopes and the immune response to them, natural airborne allergens seem particularly relevant. Current studies in this laboratory have focused on those epitopes recognized by B cells and their products, antibodies. In light of the fact that there is a strong association between class II histocompatibility antigens and allergic response to certain grass and ragweed allergens [21-25], it makes sense to study T cell epitopes of these antigens as well. In order to study epitopes, a number of things must be known about the antigen and a number of tools and assays developed. Structural information must be available in order to define peptides and subsequently synthesize them or derivatives of them. Antibodies (preferably monoclonal) are useful for assaying proteolytic fragments and peptides, as well as for inhibition studies to ultimately define those epitopes recognized by human reaginic antibodies. A full complement of monoclonal antibodies, both against native and denatured determinants has proved to be useful in our hands. With these tools, it is possible not only to define and characterize epitopes of allergens, but to isolate and characterize the genes which code for specific allergens. Once a gene is cloned, it is possible to use the powerful tools of molecular biology to mutate that gene, to express parts of the gene, to probe for related molecules within and between plant species, etc.

In the case of grass Group I allergens, it has been possible to define B cell epitopes using only monoclonal antibodies and protein sequence/peptide synthesis technology. The 28 residue peptide described in this report is highly homologous among the Group I allergens of all five temperate grasses under investigation in this laboratory [19]. In addition, a monoclonal antibody to another shared epitope among these Group I allergens is currently being used to isolate and characterize it. Such highly conserved peptides defined by these monoclonal antibodies have the potential to be the basis for much improved immunotherapy for grass Group I allergens. Although the complete amino acid sequence for any Group I allergen is not yet available, it is anticipated that the use of monoclonal antibodies and oligonucleotide probes modeled from the sequence of these highly cross reactive peptides will allow the structural genes coding for these allergens to be isolated and thoroughly characterized.

Antigen E, the major allergen of short ragweed, has been a very difficult model system to develop. For a relatively large molecule, there seem to be a sparcity of epitopes on the native molecule and these are extremely conformation dependent. In its native form, AgE is resistant to proteolytic degradation which has made the isolation of peptides representing these epitopes impossible, even with clever techniques such as peptide 'footprinting' [26,27]. It has not yet been possible to obtain crystals of this antigen (it may have too many isoelectric forms to permit crystal formation) and it is too large to be approached by current 2D nuclear magnetic resonance technology, so it may be a while before we have a three dimensional picture of this interesting and clinically important molecule. In the meantime, there are a number of approaches which may lead to an understanding of the epitope structure of the molecule. It may be that something in the primary amino acid sequence of AgE will be of use in predicting an epitope, and it may be that expression of structural genes or defined parts of structural genes designed using *in vitro* mutagenesis [28] will result in peptides containing one or more of the known native epitopes. To this end, the tools described in this report have been developed and current approaches to identifying the structural gene of AgE are well underway.

REFERENCES

1. Chapman, M.D., Sutherland, W.M., Platts-Mills, T.A.E. (1984). Recognition of two *Dermatophagoides pteronyssinus* specific epitopes on antigen P. by using monoclonal antibodies: binding to each epitope can be inhibited by serum from dust mite-allergic patients. J. Immunol., 133, 2488.
2. Ekramoddoullah, A.K.M., Kisil, F.T., Bundasen, P.G., Fischer, J.M.M., Rector, E.S., Sehon, A.H. (1984). Determinants of Rye grass pollen cytochrome C recognized by human IgE and murine monoclonal antibodies.

Mol. Immunol., 21, 375.

3. Ekramoddoullah, A.K.M., Kisil, F.T., Sehon, A.H. (1986). Partial characterization of an antigenic site of high molecular weight basic antigen, a Rye-grass pollen allergen, using a monoclonal antibody. Mol. Immunol., 23, 111.
4. Kahn, C.R., Marsh, D.G. (1986). Monoclonal antibodies to the major *Lolium perenne* (Rye-grass) pollen allergen *Lol p I* (Rye I). Mol. Immunol. 23, 1281.
5. Olson, J.R., Klapper, D.G. (1986). Two major human allergenic sites on ragweed pollen allergen antigen E identified by using monoclonal antibodies. J. Immunol., 136, 2109.
6. Chapman, M.D., Hermann, P.W., Platts-Mills, T.A.E. (1987). Epitope mapping of two major inhalant allergens *Der p I* and *Der f I* from mites of the genus *Dermatophagoides*. J. Immunol., 139, 1479.
7. Ekramoddoullah, A.K.M., Kisil, F.T., Cook, R.T., Sehon, A.H. (1987). Recognition of a site of a Kentucky Bluegrass pollen allergen by antibodies in the sera of allergic and non-atopic human and a murine monoclonal antibody. J. Immunol., 138, 1739.
8. Esch, R.E., Klapper, D.G. (1987). Cross-reactive and unique Group I antigenic determinants defined by monoclonal antibodies. J. Aller. Clin. Immunol., 79, 489.
9. Mazur, G., Becker, W.M., Baur, X. (1987). Epitope mapping of major insect allergens (chironomid hemoglobins) with monoclonal antibodies. J. Aller. Clin. Immunol., 80, 876.
10. Smith, J.J., Olson, J.R., Klapper, D.G. (1988). Monoclonal antibodies to denatured ragweed pollen allergen *Amb a I*: Characterization, specificity for the denatured allergen, and utilization for the isolation of immunogenic peptides of *Amb a I*. Mol. Immunol., 25, 355.
11. Kennett, R.H., Denis, K.A., Tung, A.S., Klinman, N.R. (1978). Hybrid plasmacytoma production: fusions with adult spleen cells, monoclonal spleen fragments, neonatal spleen cells, and human spleen cells. Curr. Top. Microbiol. Immunol., 81, 77.
12. Potter, M. (1972). Immunoglobulin-producing tumors and myeloma proteins of mice. Physiol. Rev , 52, 631.
13. Klapper, D.G., Osgood, S., Esch, R., Olson, J. (1986). Use of new HPLC resins to solve old problems. J. Liq. Chromat., 9, 1613.
14. Wood, W.I., Getschier, J., Lasky, L.A., Lawn, R.M. (1984). Base composition-independent hybridization in tetramethylammoniumchloride: A method for oligonucleotide screening of highly complex gene libraries. Proc. Natl. Acad. Sci., 82, 1585.
15. Larkins, B.A., Hurkman, W.J. (1978). Synthesis and deposition of zein protein bodies of maize endosperm. Plant Physiol., 62, 256.
16. Noyes, B.E., Mevarech, M., Stein, R., Agarwal, K.L. (1979). Detection and partial sequence analysis of gastrin mRNA by using an oligodeoxynucleotide probe. Proc. Natl. Acad. Sci., 76(4), 1770.
17. Maxham, A.M., Gilbert, W. (1977). A new method for sequencing DNA. Proc. Natl. Acad. Sci., 74(2), 560.
18. Esch, R.E., Klapper, D.G. Identification and localization of allergenic determinants on grass Group I antigens using monoclonal antibodies. J. Immunol. (in press).
19. Esch, R.E., Klapper, D.G. Isolation and characterization of a major cross-reactive grass Group I allergenic determinant. Mol. Immunol. (in press).
20. King, T.P. (1976). Chemical and biological properties of some atopic allergens. Adv. Immunol., 23, 77.
21. Marsh, D.G., Hsu, S.H., Roebber, M., Kautzky, E.E., Freidhoff, L.R., Meyers, D.A., Pollard, M.K., Bias, W.B. (1982). HLA-Dw2: a genetic marker for human immune response to short ragweed pollen allergen Ra5.

I. Response resulting primarily from natural antigenic exposure. J. Exp. Med., 155, 1439.

22. Marsh, D.G., Meyers, D.A., Freidhoff, L.R., Kautzky, E.E., Roebber, M., Norman, P.S., Hsu, S.H., Bias, W.B. (1982). HLA-Dw2: a genetic marker of human immune response to short ragweed pollen allergen Ra5. II. Response after ragweed immunotherapy. J. Exp. Med., 155, 1452.
23. Roebber, M., Klapper, D.G., Goodfriend, L., Bias, W.B., Hsu, S.H., Marsh, D.G. (1985). Immunochemical and genetic studies of *Amb t V* (Ra5G), an Ra5 homologue from giant ragweed pollen. J. Immunol., 134, 3062.
24. Goodfriend, L., Choudhury, A.M., Klapper, D.G., Coulter, K.M., Dorval, G., DelCarpio, J., Osterland, C.K. (1985). Ra5G, a homologue of Ra5 in giant ragweed pollen: isolation, HLA-DR associated activity and amino acid sequence. Mol. Immunol., 22, 899.
25. Marsh, D.G., Friedhoff, L.R., Ehrlich-Kautzky, E., Bias, W.B., Roebber, M. (1987). Immune responsiveness to *Ambrosia artemisiifolia* (short ragweed) pollen allergen *Amb a VI* (Ra6) is associated with HLA-DR5 in allergic humans. Immunogenetics, 26, 230.
26. Jemmerson, R., Paterson, Y. (1986). Mapping epitopes on a protein antigen by the proteolysis of antigen-antibody complexes. Science, 232, 1001.
27. Sheshberadaran, H., Payne, L.G. (1988). Protein antigen-monoclonal antibody contact sites investigated by limited proteolysis of monoclonal antibody-bound antigen: Protein "footprinting". Proc. Natl. Acad. Sci., 85, 1.
28. Kunkel, T.A. (1985). Rapid and efficient site-specific mutagenesis without phenotypic selection. Proc. Natl. Acad. Sci., 82, 488.

The Rye-Grass Pollen Allergen, *Lol p I*

R. Bruce Knox, Mohan B. Singh, T. Hough
and P. Theerakulpisut

School of Botany, University of Melbourne, Parkville,
Victoria 3052, Australia

ABSTRACT

Grass pollen is the principal outdoor airborne allergen in spring and early summer. Rye-grass pollen is the major single source of allergenic pollen in cool temperate climates. The most abundant allergenic protein in rye-grass is *Lol p I*, a glycoprotein of molecular weight 34,000. The gene *Lol p I*, which encodes the major allergenic protein from the pollen of rye-grass, has been isolated and cloned. The clones are selected from a lambda-gt11 cDNA library using a combination of specific monoclonal antibodies and specific IgE antibodies from allergenic patients. The tissue and organ specificity of *Lol p I* is exclusive to pollen as shown by slot blotting and Northern analyses. A partial nucleotide sequence and deduced amino acid sequence has been obtained, and no homology is apparent with other known proteins. The cDNA clones have several potential applications in biotechnology.

KEYWORDS

Rye-grass; *Lol p I*; cloning; nucleotide sequence; amino acid sequence; homology.

INTRODUCTION

What is an Allergen?

Allergens are those environmental antigens that can induce an immediate hypersensitivity reaction in man following interaction with specific IgE molecules in the airways, gastro-intestinal tract, skin or mucous membrane.

Where do Allergens Occur?

Allergens occur in a diversity of sources, including higher animals, plants, insects, mites, fungi and bacteria where they are part of the natural micro-fauna and flora of the atmosphere, or are primary or secondary chemicals in

Advances in the Biosciences Vol. 74

Printed in Great Britain

plants used as food or in animal defence mechanisms. This article deals with environmental allergens. The major indoor allergen of the human environment is house dust mite, and its allergens are reviewed by Stewart *et al.*, Baldo *et al.* (this volume). Here, the focus is on the major outdoor allergen - grass pollen. Its association with hayfever is first noted by Elliotson in 1831. He found that the worst of catarrh symptoms accompanied the maturing of the grasses during the haymaking season. Blackley (1873) not only demonstrated the role of grass pollen in initiating allergic disease, but also the presence of the allergenic pollens in the atmosphere. During the past century, there has been continued clearing of land around large cities for agriculture, usually for grazing of cattle for milk production. This has meant that the area of pastures and grasslands close to population centres has increased substantially.

What is the Prevalence of Allergic Disease?

Because of the increased presence of these environmental agents, hayfever and seasonal asthma continue to have significant morbidity and socio-economic impact on Western communities, despite advances made in pharmacology and immunology. Approximately 50% of the population are capable of producing IgE (Platts-Mills, 1987), but the proportion suffering severe symptoms of allergic disease is smaller. In USA, Broder *et al.* (1962) showed that 19% of children suffer from wheezing and asthma, while among adults 4.1% suffer from asthma and 6.3% from hayfever, a total of 10.4%. In Australia, Hall *et al.* (1972) showed that asthma may affect up to 16% of children, while among adults, 5% had asthma, and 16% hayfever, a total of 21%.

Why is Grass Pollen the Major Outdoor Allergen?

Grass pollen is the major outdoor airborne allergen in temperate climates (McNichol *et al.*, 1973; Marsh, 1975; Hill *et al.*, 1979). The grass pollen is present in the atmosphere as part of the pollination process. The pollen carries the male gametes, and is passed by air currents from the male part of the flower to the female. Most other types of plants use animals, especially bees as pollen vectors. Pollen calendars show that grass pollen is the major component of the airspora in spring and early summer when most of the grasses flower (Knox, 1979; Knox and Tuohy, 1982).

Is Rye-grass the Major Grass Pollen Allergen in Cool/Temperate Climates?

In cool temperate climates such as southern Australia, rye-grass pollen shows the greatest abundance of all types of grass pollen (Smart *et al.*, 1979). Its productivity is phenomenal - up to 460 Kg per hectare in rye-grass pasture that is not mown or grazed. Rye-grass is the major allergen for 67 and 75% of allergic patients, respectively, while house dust mite is the major indoor allergen for 71 and 79% of allergic patients (McNichol *et al.*, 1973; J. Hobday, unpublished data). In a warm temperate climate, eg. Sydney, the percentage attributable to rye-grass falls to 49% of adult allergic patients (A. Woolcock, unpublished data). Rye-grass is the major outdoor allergen in spring and early summer in other cool temperate climates, eg. northern California, USA and Northern Europe. In other parts of North America, Kentucky bluegrass, *Poa pratensis*, and in Europe, timothy, *Phleum pratense* are important grass pollen aeroallergens.

The above data, taken together, present a strong case for studying rye-grass pollen allergens, particularly their structure-function relationships as allergens.

What is a Grass Pollen Allergen?

The experiments of Noon (1911), showed that aqueous extracts of grass pollen contain the allergenically active components, producing skin reactions in sensitive patients. These allergens could be sterilized by autoclaving, showing their resistance to high temperatures. The pioneering studies of Johnson and Marsh (1965) resulted in isolation of the allergenic components from rye-grass pollen.

The first descriptions of the rye-grass allergenic proteins showed that they are immunochemically distinct, and are known as groups I, II, III etc. (Johnson and Marsh, 1965, 1966). Using the recently proposed International Union of Immunological Societies' (IUIS) nomenclature, these allergens are designated *Lol p I*, *Lol p II* etc. The major allergen, *Lol p I*, is an acidic glycoprotein of molecular weight ca. 34 kD and comprises four isoallergenic variants. Each one has been isolated in multiple electrophoretic forms, which in the case of *Lol p I* appear to differ in amino acid composition only in the ratio of glutamate to glutamine. The other minor allergens isolated from rye-grass pollen range in molecular weight from 10 to 76 kD (Ford and Baldo, 1986).

What are the Characteristics of *Lol p I*?

The allergen *Lol p I* constitutes ca. 5% of the total extracted pollen proteins and is a glycoprotein (Johnson and Marsh, 1966; Howlett and Clarke, 1981) containing a 5% carbohydrate moiety. Studies with carbohydrate splitting enzymes have demonstrated that the carbohydrate does not contribute to the allergic response. Allergenic activity is lost following proteolytic digestion (see review by Baldo, Sutton and Wrigley, 1982), but is resistant to heat treatment, eg. 100°C for 30 minutes at neutral pH (Marsh *et al*., 1966).

Lol p I is defined as an allergen because of its ability to bind to specific IgE antibody in sera of rye-grass-sensitive patients, to act as an antigen in IgG responses, and to trigger T-cell responses. The allergenic properties have been assessed by direct skin testing of grass pollen-sensitive patients. Results of tests in a US population study showed that 84% had a skin sensitivity to *Lol p I* (Friedhoff *et al*., 1986), demonstrating the primary importance of this protein as the major pollen allergen. Furthermore, 95% of patients demonstrated to be grass pollen-sensitive possessed specific IgE antibody that bound to *Lol p I*, as demonstrated by immunoblotting (Ford and Baldo, 1986).

Substantial allergenic cross-reactivity between grass pollens has been demonstrated using an IgE-binding assay, the radioallergo-sorbent test (RAST) eg. Marsh *et al*. (1970) and Lowenstein (1978).

The immunochemical relationships of *Lol p I* with other grass pollen antigens have been demonstrated using both polyclonal and monoclonal antibodies (eg. Smart and Knox, 1979; Singh and Knox, 1985). Table 1 provides a list of the various published cases where antibodies have been prepared to characterized grass pollen allergens, and *Lol p I* is the most frequent allergen studied. Monoclonal antibodies have been prepared to both purified protein and IgE-

Table 1. Selected list of monoclonal antibodies raised to pollen allergens

Species	Allergen	Comment	Reference
Lolium perenne	*Lol p I* and *Lol p II*	IgE binding by using ELISA	Smart *et al.* (1984)
Lolium perenne	*Lol p I*	Obtained 20 mAbs which can be divided into five families according to epitope specificity on *Lol p I* molecule	Bose *et al.* (1986)
Lolium perenne	*Lol p I*	Developed method for immunoaffinity purification of *Lol p I*	Kahn and Marsh (1986)
Lolium perenne	*Lol p I*	Checked cross reactivity of rye allergens with that of timothy grass by Western blotting	Mourad *et al.* (1986)
Festuca elatior	*Gp I*	Defined mAbs with specificities unique for the immunizing allergen and other mAbs that cross reacted with allergens from other grass species	Esch and Klapper (1987)
Parietaria judaica	*Pj10*	Used mAbs to purify allergen by immunoadsorption	Corbi *et al.* (1985)

binding components. These data demonstrate that the major allergen present in pollen of closely related grasses is immunochemically similar to *Lol p I* (Singh and Knox, 1985).

Antibodies to *Lol p I* allergenic protein are immunologically cross-reactive with other grass pollens, namely: Pooid (Festucoid) grasses. Group 1: Triticanae: *Bromus inermis*, smooth brome; *Agropyron repens*, English couch; *A. cristatum*; *Secale cereale*, rye; *Triticum aestivum*, wheat. Group 2: Poanae: *Dactylis glomerata*, orchard grass or cocksfoot; *Festuca elatior*, meadow fescue; *Lolium multiflorum*, Italian rye-grass; *Poa pratensis*, Kentucky blue-grass; *P. compressa*, flattened meadow grass; *Avena sativa*, oat; *Holcus lanatus*, velvet grass or Yorkshire fog; *Anthoxanthum odoratum*, sweet vernal grass; *Arrhenatherum elatius*, oat grass; *Agrostis alba*, red top; *Phleum pratense*, timothy; *Phalaris arundinacea*, reed canary grass. Panicoid Grass, *Paspalum notatum*, Bahia grass, Andopogonoid grasses: *Sorghum halepensis*, Johnson grass; *Zea mays*, maize. The question arises whether *Lol p I* is a generic pollen protein, and what are its relationships with pollen proteins from ragweed and other plants? The availability of both monoclonal antibodies and cloned gene sequences as probes should enable an answer in the future.

Lol p I is located in the outer wall of the rye-grass pollen grain, and also in the cytoplasm. This has been observed by immunocytochemistry with polyclonal antibodies of freeze-dried pollen using a post-embedding indirect staining method with both fluorescent- and ferritin-labelled antibodies (Knox *et al.*, 1980; Vithanage *et al.*, 1982). Freeze-drying of pollen samples proved to be essential to retain activity, as the allergen diffuses out of the samples in aqueous preparations. A novel membrane print technique to locate surface allergens in individual pollen grains has been developed in our laboratory (O'Neill *et al.*, 1986).

Molecular Cloning of *Lol p I*

The complete coding sequence of the cDNA clone 12R complementary to the full length rye-grass pollen allergen *Lol p I* mRNA is being determined (Singh *et al.*, 1988), and a partial nucleotide sequence has been obtained (R.B. Knox, M.B. Singh, T. Hough and P. Theerakulpisut, Rye-grass pollen allergen, Australian Provisional Patent, applied for March 1988). The clone 12R contains an 1100 base pair (bp) insert excluding the oligo (G) and poly (A) tails. The identification of the clone is based on the recognition of the protein expressed by *Escherichia coli* transformed with lambda-gt11 phage, using both specific monoclonal antibodies and specific serum IgE from grass pollen-sensitive patients. A cDNA clone, 6R, has also been isolated on the basis of its antibody specificity. However, 6R failed to show specific IgE binding, indicating that this recombinant clone lacks the DNA sequence that specificies allergenicity.

The pathway used for isolation of cDNA clones of *Lol p I* is shown in Table 2.

Preparation of rye-grass pollen mRNA. The original source of the genetic material is fresh rye-grass pollen, *Lolium perenne* L, collected from field sources near Melbourne. Pollen RNA is isolated according to the method of Herrin and Michaels (1984).

One gram of rye-grass pollen yielded 1 mg total RNA, and poly (A^+) RNA selected by affinity chromatography using standard methods. The integrity of poly (A^+) RNA is examined in terms of its ability to act as a template for synthesis of single-stranded cDNA as well as its translational activity in the rabbit reticulocyte system.

Isolation of cDNA clones of *Lol p I*. The *Lol p I* glycoprotein is seen electrophoretically as the major allergen present in the rye-grass pollen We have demonstrated the specific binding of monoclonal antibody FMC-A1 to the allergen. For cloning the *Lol p I* gene, the isolated mRNA is used as a template to synthesize double stranded complementary DNAs using the Amersham cDNA synthesis system. The cDNA library is prepared by cloning the double stranded cDNA into the phage lambda-gt11 at the EcoR1 site of the lacZ gene.

Screening the cDNA library for *Lol p I* clones. From this cDNA library of rye-grass pollen, recombinant lambda-gt11 phage containing the *Lol p I* insert are detected by screening the library with (1) specific monoclonal antibody FMC-A1; (2) specific IgE from sera of rye-grass-sensitive patients. The library of rye-grass pollen allergen-specific monoclonal antibodies used are those produced by Smart *et al.* (1983). The lambda-gt11 expression

Table 2. Flow chart of procedure used for isolation and cloning of *Lol p I*, the gene encoding the rye-grass pollen allergen

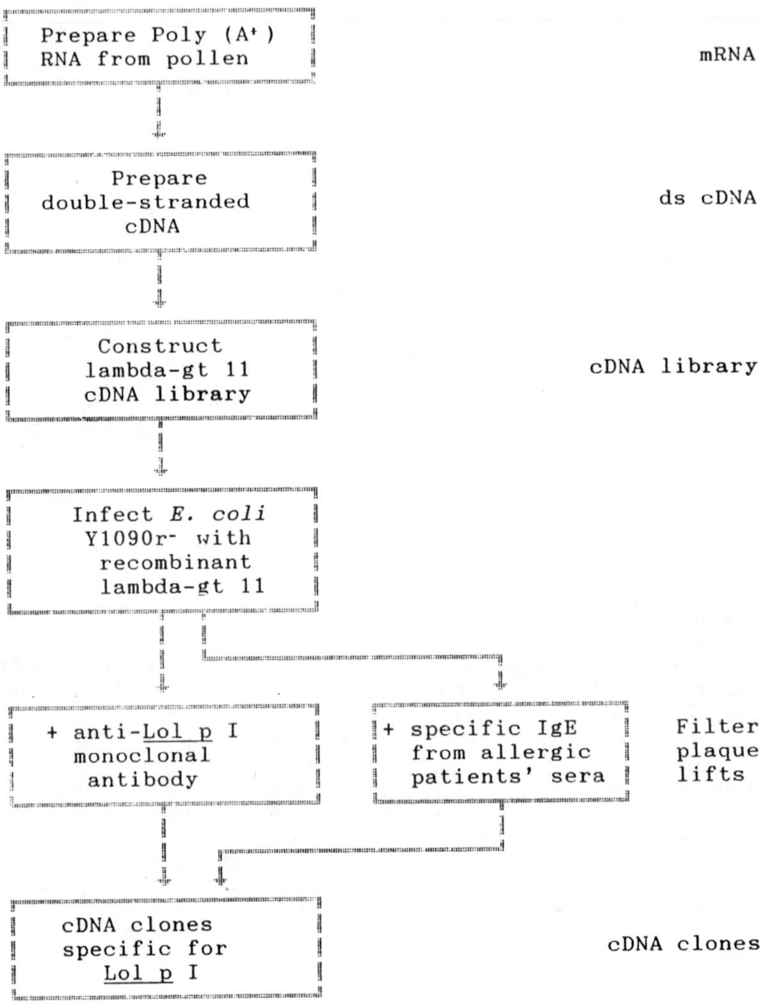

library is screened using a modification of the method of Huynh, Young and Davis (1985). Filter plaque lifts are dried and incubated in the monoclonal antibodies (ascites fluid) to the rye-grass pollen allergen. Specific antibody-binding is detected using peroxidase-conjugated affinity-purified anti-mouse IgG. The developed filter is used to locate specific plaque areas on the plate, corresponding to a positive signal. Positive phage plaques are collected and the eluted phage purified to individual antigen-positive lambda-gt11 clones by rescreening. Once plaque purification has been achieved, the ability of the recombinants to bind IgE from allergic patients, is detected using the same procedure, with the exception of incubation in allergic patients serum as the 1st antibody.

Analysis of *Lol p I*-beta-galactosidase fusion protein. The fusion protein has been isolated in preparative amounts from bacterial lysogens in which the operon repressor is inactivated, at 42°C, allowing the lac Z gene to be expressed and the allergenic proteins to be synthesized as a fusion protein with beta-galactosidase. The fusion protein is then fractionated by SDS-polyacrylamide gel electrophoresis, and the proteins transferred to nitrocellulose membranes for probing with allergic patients' sera and monoclonal antibodies. These antibodies recognise a protein with a molecular weight greater than the *E. coli* beta-galactosidase as would be expected of an allergen beta-galactosidase fusion protein.

Pollen-specificity of *Lol p I*. Selected cDNA clones were used to probe total RNA isolated from other rye-grass plant organs to test whether *Lol p I* allergen is pollen-specific or not. Slot-blotting and Northern analyses were employed. Total RNA is extracted from pollen, leaf, hydrated seed and root samples as previously described for pollen, and electrophoresed according to the method of Maniatis *et al.* (1982). The RNA is transferred to nitrocellulose filters, incubated in hybridization buffer containing the random primed *Lol p I* DNA probe, and exposed for autoradiography. No hybridization is detectable for total RNA from leaf, seed or root samples. These data indicate that *Lol p I* is expressed only in the pollen of the rye-grass plant.

Nucleotide sequence of clone 12R. Selected cDNA clones are ligated into M13 vector for sequencing. DNA restriction fragments to be sequenced are inserted into M13 mp14 (Messing and Vieira, 1982). M13 cloning and dideoxy chain termination DNA sequencing followed the method of Messing (1983). The amino acid sequence of a partial clone has been deduced. This sequence contains a wide range of amino acids, but is especially rich in valine, proline, serine, phenylalanine, arginine and leucine.

Applications of cDNA Clones to Pollen Allergens

Synthetic peptides. There are several interesting applications of the powerful techniques of gene cloning in biotechnology. Reliable amino acid sequences of biologically important proteins can be deduced from the determined nucleotide sequences. Ragweed pollen allergen Ra5, a 45 amino acid protein, has been synthesized (Choudhury and Goodfriend, 1983). The synthetic peptide retains complete biological activity, and analogues related to the allergen provide a potential test system for the molecular genetic analysis of allergic reactions (Hodges, Parker and Singh, 1986).

Synthetic peptides of this kind have made a major contribution to advances in molecular and cellular immunology. The methods for determining the regions in a protein responsible for allergenic reactions, ie. antigenic and allergenic epitopes, are few and time consuming. Accordingly, methods have been developed for the concurrent synthesis of all possible overlapping hexapeptides of a complete protein, eg. the coat protein of a virus (Geysen *et al.*, 1984). Alternatively, antigenic regions can be predicted by computer-based methods using: (1) hydrophilicity parameters of individual amino acids derived from HPLC retention times to predict surface residues (Guo *et al.*, 1985); (2) hydrophilicity recognition profiles of the proteins (Fraga and Singh, 1985). The latter group has used synthetic peptide antigens to elucidate T-cell recognition sites on poly 18, cytochrome c and insulin. A minimum of nine amino acids in a sequence are required for immunogenicity in such systems.

Synthetic peptides have been widely used to raise antibodies of known and selected specificities in attempts to design synthetic vaccines. Such methods have been applied to ragweed allergens, eg. the full profile of continuous antigenic sites of allergen Ra3 using synthetic overlapping peptides (Atassi and Atassi, 1986). Ten consecutive peptides each comprising 15 residues and overlapping each of its neighbours by 5 peptides have been synthesized. From these studies of ragweed pollen, we can conclude that:

- distinct epitopes are present for IgE, IgG and T-cell binding;
- immunodominant and silent regions exist in allergen molecules;
- the IgE and IgG epitopes are structurally similar.

The similarity of the IgE and IgG binding sites indicated by Atassi and Atassi (1986) may be dependent on the length of the peptides (15 amino acid residues). Possibly shorter peptides may show greater specificity. The potential of whether peptides corresponding to the immunosilent regions can elicit IgG formation needs to be investigated.

Peptide technology can also reveal some antigenic sites that could not otherwise have been found by the empirical hydrophilicity approach (Hopp and Woods, 1981). The use of synthetic peptides in the study of the structure-function relationships of grass pollen allergens will lead to a new understanding of the molecular basis of antigenicity, immunogenicity and allergenic activity.

Such methods offer the opportunity to answer the longstanding question of the structure of allergenic determinants. In particular, are structurally similar determinants shared among allergenic proteins originating from different grass species, other non-grass pollens, other allergenic plant proteins, eg. wheat endosperm, and other proteins of animal origin, eg. bovine serum albumin. In this connection, it is relevant that the major allergenic protein *Der p I* of house dust mite is structurally homologous to cysteine proteases, especially actinidin from kiwifruit and papain from papaya (Thomas *et al.*, 1988). These approaches have the desirable spin off that potentially potent synthetic vaccines could be developed in the future.

Pollen-specific promoters. The *Lol p I* gene has pollen-specific promoters (control regions) that can be isolated from genomic DNA libraries. There is the potential to develop hypoallergenic lines of rye-grass in which the allergen is not expressed in the pollen. This would provide strains of the grass free of the major allergen for planting in low allergen gardens and public parks. Male sterile lines in which pollen or anther development is prevented are a further potential application in plant biotechnology.

<u>cDNA clones to other grass pollen allergens</u>. The monoclonal antibodies used in the present work to screen the cDNA library for *Lol p I* clones show cross-reactivity with allergenic proteins from pollen of various related grass species. This shows there is a homology between allergenic proteins produced by these pollens with *Lol p I* allergen. This homology can be exploited to isolate DNA encoding other allergenic proteins without the need for protein microsequencing and oligo-nucleotide primers.

The cDNA clone 12R insert described here can be used as a hybridization probe for example in a colony hybridization screen. The cDNA insert can be labelled with ^{32}P by nick translation, end labelling or random priming.

<u>Immunodiagnostics</u>. The availability of these patented cDNA clones means that a ready source of purified rye-grass pollen *Lol p I* allergen protein can be guaranteed using the techniques of biotechnology. This pure monovalent material is available for the first time for use in immunodiagnostic test kits, rather than the presently used complex polyvalent pollen extracts.

<u>Immunotherapy</u>. Renewed interest has been shown in the immunotherapy of allergic disease, largely because the available spectrum of drugs have side-effects associated with long-term usage. Pollen preparations used as allergens are polyvalent and of poor quality. Consequently, concentrations used are frequently high in order to induce IgG responses, but may be lethal through triggering of anaphylaxis. Recent studies of ragweed pollen sensitivity showed that there is a correlation between allergen-specific IgG levels and relief from allergic symptoms (Lichtenstein *et al.*, 1983). Thus a goal is to stimulate allergen-specific IgG production in clinical treatment without inducing anaphylaxis. This can best be achieved through the use of pure allergen preparations and active peptides.

<u>Understanding sensitivity to allergens</u>. The pollen grains of related species contain unique as well as common allergens. Since allergic individuals may produce specific IgE antibodies to either one or other of these types of allergens, patients can show either clear cut sensitivity to one species, or to several related species because of the structural relationships among allergen molecules. The way is now open to address the question of why is a pollen allergen an allergen? The primary requirement is for molecular analysis of these pollen allergens, using gene cloning techniques. When the genes are isolated and cloned, nucleotide sequence data enable their amino acid sequence to be deduced. These data are used to look for sequence homologies with proteins of known function, and also to synthesize biologically active peptides useful in gaining a new understanding of the allergenic activity, and to define the role of these proteins in pollen biology.

<u>Plant recognition factors</u>. The pollen grain carries at its surface substances which are rapidly released upon contact with a moist surface. A large number of proteins are released, which include enzymes whose function is known, and species-specific recognition factors which are implicated in pollen-pistil interactions. Allergens are among the proteins released whose

function in pollen biology is unknown, but whose interactions with human cells has been very precisely characterised.

REFERENCES

Atassi, H. and Atassi, M.Z. (1986). Eur. J. Immunol., 16, 229.

Choudhury, A.M. and Goodfriend, L. (1983). J. Protein Chem., 2, 321.

Geysen, H.M., Melden, R.H., Barteling, S.J. (1984). Proc. Natl. Acad. Sci. USA, 81, 3998-4002.

Hodges, R.S., Parker, J.M. and Singh, B. (1981). TIBS, Elsevier Science Publishers, B.V.

Hopp, T.P. and Woods, K.R. (1981). Proc. Natl. Acad. Sci. USA, 78, 3824-3828.

Knox, R.B. and Tuohy, M. (1982). Proc. 6th Australian Weeds Conference, 1981, Vol.2, pp.125-142.

Lichtenstein, L.M., Norman, P.S., Kagey-Sobotka, A., Adkinson, N.F., Golden, D.B.K. (1983). In: XIth Int. Congress of Allergology and Clin. Immunol. (eds. J.W. Kerr and M.A. Ganderton) pp.285-289.

Smart, I.J., Tuddenham, W. and Knox, R.B. (1979). Aust. J. Bot., 27, 333-342

Thomas, W.R., Stewart, G.A., Simpson, R.J., Chua, K.Y., Plozza, T.M., Dilworth, R.J., Nisbet, A. and Turner, K.J. (1988). Int. Archs. Allergy Appl. Immunol., 85, 127-129.

Baldo, B.A., Sutton, R. and Wrigley, C.W. (1982). Grass allergens with particular reference to cereals. Prog. Allergy, 30, 1-66 (Karger, Basel).

Blackley, C.H. (1873). Experimental researches on the cause and nature of *Catarrhus Aestivus*. Bailliere, Tindall and Cox, London.

Broder, F. *et al.* (1962). J. Allergy, 33, 524-531.

Esch, R.E. and Klapper, D.G. (1987). J. Allergy Clin. Immunol., 79, 489-95.

Ford, S.A. and Baldo, B.A. (1986). A re-examination of ryegrass (*Lolium perenne*) pollen allergens. Int. Archs. Allergy Appl. Immunol., 81, 193-203.

Freidhoff, L.R., Ehrlich-Kantzky, E., Grant, J.H., Meyers, D.A. and Marsh, D.G. (1986). A study of the human immune response to *Lolium perenne* (Rye) pollen and its components, *Lol p I* and *Lol p II* (Rye I and Rye II). J. Allergy Clin. Immunol.

Gubler, V. and Hoffman, B.J. (1983). Gene, 25, 263.

Hall, G.J. *et al.* (1972). Int. J. Epidem., 1, 289-405.

Herrin, D. and Michaels, A. (1984). Isolation, fractionation and analysis of intact, translatable RNA from walled algal cells. Plant Mol. Biol. Reporter, 2, 24-29.

Hill, D.J., Smart, I.J. and Knox, R.B. (1979). Childhood asthma and grass pollen aerobiology in Melbourne. Med. J. Aust., 1, 426-429.

Howlett, B.J. and Clarke, A.E. (1981). Isolation and partial characterization of two antigenic glycoproteins from rye-grass (*Lolium perenne*) pollen. Biochem. J., 197, 695-706.

Howlett, B.J. and Knox, R.B. (1984). Allergic interactions. In: Encyclopaedia of Plant Physiology (eds. J. Heslop-Harrison and H.F. Linskens) Springer, Heidelberg, Vol. 17, pp.655-674.

Huynh, T.V., Young, R.A. and Davis, R.W. (1985). Constructing and screening cDNA libraries in lambda-gt10 and lambda-gt11. In: DNA cloning; a practical approach, Vol. I (ed. D.M. Glover), IRL Press, Oxford, pp.49-78.

Johnson, P. and Marsh, D.G. (1965). 'Isoallergens' from ryegrass pollen. Nature, 206, 935.

Johnson, P. and Marsh, D.G. (1966). Allergens from common ryegrass pollen (*Lolium perenne*). I. Chemical composition and structure. Immunochemistry 3, 91-100.

Kahn, C.R. and Marsh, D.G. (1986). Mol. Immunol., 23, 1281-1288.

Knox, R.B. (1979). Pollen and allergy. Studies in Biology, No.107, Edward Arnold, London, p.64.

Lowenstein, H. (1978). Quantitative immunoelectrophoretic methods as a tool for the analysis and isolation of allergens. Prog. Allergy, 25, 1-62, (Karger, Basel).

Maniatis, T., Fritsch, E.G. and Sambrook, J. (1982). Molecular Cloning; A Laboratory Manual. Cold Spring Harbor Laboratory.

Marsh, D.G., Haddad, Z.H. and Campbell, D.H. (1970). A new method for determining the distribution of allergenic fractions in biological materials: Its application to grass pollen extracts. J. Allergy, 46, 107-121.

Marsh, D.G. (1975). Allergens and the genetics of allergy. In: The Antigens Vol.3, (M. Sela), Academic Press Inc., London, New York, pp.271-359.

Marsh, D.G., Milner, F.M. and Johnson, P. (1966). The allergenic activity and stability of purified allergens from the pollen of common ryegrass. Int. Archs. Allergy, 29, 52.

McNichol, K.N. *et al.* (1973). Brit. Med. J., 4, 16-20.

Messing, J. (1983). New M13 vectors for cloning. In: Methods in Enzymology - Recombinant DNA techniques 101, Part C, pp.20-78.

Messing, J. and Vieira, J. (1982). A new pair of M13 vectors for selecting either DNA strand of double-digest restriction fragments. Gene, 19, 269-276.

Mourad, W., Pelltier, G., Boulet, A., Islam, N., Valet, J-P. and Herbert, J. (1986). J. Immunol. Methods, 89, 53-59.

Noon, L. (1911). Prophylactic inoculation against heyfever. Lancet, 1, 1972.

O'Neill, P.M., Singh, M.B. and Knox, R.B. (1986). Applications of a new membrane print technique in biotechnology. In: Biotechnology and ecology of pollen (eds. D.L. Mulcahy, G.B. Mulcahy and E. Ottaviano) Springer, New York, pp.203-208.

Singh, M.B. and Knox, R.B. (1985). Grass pollen allergens: Antigenic relationships detected using monoclonal antibodies and dot blotting immunoassay. Int. Archs. Allergy Appl. Immunol., 78, 300-304.

Smart, I.J., Heddle, R.J., Zola, M., Bradley, J. (1983). Development of monoclonal antibodies specific for allergenic components in rye-grass (*Lolium perenne*) pollen. Int. Archs. Allergy Appl. Immunol., 72, 243-248.

Smart, I.J. and Knox, R.B. (1979). Rapid batch fractionation of ryegrass pollen allergens. Int. Archs. Allergy Appl. Immunol., 62, 173-187.

Identification of cDNA Clones Representing Group I Pollen Allergens from Kentucky Bluegrass

T. W. Hatton, A. K. M. Ekramoddoullah, F. T. Kisil, A. H. Sehon and R. D. Hill

Departments of Plant Science and Immunology, University of Manitoba, Winnipeg, Manitoba, Canada

ABSTRACT

The group I pollen allergens are a group of proteins of about 33 KDa and 38 KDa that are potent allergens in extracts of Kentucky Bluegrass pollen. A monoclonal antibody (Mab60) has been produced which specifically detects these proteins on Western blot analysis. It has been used to screen a cDNA library of Kentucky Bluegrass mRNA in lambda gt11 to obtain complementary cDNA clones for the analysis of the structure of the allergens. Several antibody positive clones were detected. Other clones have been selected from the library by screening with IgE from grass-sensitive patients. The cDNA clones contain inserts ranging from about 200 bp to 800 bp and will be used to recover more full-length copies of the allergens from the library.

KEYWORDS

Kentucky Bluegrass pollen; monoclonal antibodies; cDNA.

INTRODUCTION

It is apparent that a greater understanding of the structure of pollen allergens will allow for more specific and efficacious treatment of allergic responses. For instance, if patients are primarily responsive to only one or a few pollen components, treatment with these specific compounds would be safer and more effective than using a complex mixture. Current treatment protocols risk sensitizing patients to allergens which they previously had not reacted. Moreover, a thorough structural analysis of allergens could reveal common structural motifs which might be exploited in diagnostic and treatment strategies.

For molecular analysis of allergen proteins, cDNA clones provide valuable experimental tools, especially when the encoded proteins are expressed at low levels or are difficult to purify. These clones can be used both for analysis, and as a template to investigate the effects of altered amino acid sequence on allergenicity.

Advances in the Biosciences Vol. 74

Printed in Great Britain

Several approaches may be used to obtain DNA clones representing proteins of interest: if some sequence data is available, then synthetic oligonucleotides can be created to provide a hybridization probe. This approach has the advantage of insensitivity to orientation and reading frame, thus allowing all cognate sequences to be detected: however, as a result of the degeneracy of the genetic code, multiple oligonucleotide sequences are required to account for the possible coding information. This can make identification of positive clones more difficult.

Alternatively, if no amino acid sequence data is availabe, then antibodies raised against the protein may be used in conjunction with lambda gt11 [1] or other expression vectors to detect cloned sequences expressed as fusion proteins. However, this approach requires that the insert at least be in the correct orientation and reading frame, thus one can detect only about one sixth of the available clones.

We have previously identified a number of allergens from Kentucky Bluegrass, and have prepared antibodies against several of these allergens. Thus, the objective of this work was to use these antibodies to obtain cDNA probes representing major pollen allergens from Kentucky Bluegrass.

MATERIALS AND METHODS

Pollen was obtained from Hollister-Stier Laboratory (Mississauga, Ontario, Canada). The pT7T319 multifunctional vector was obtained from Pharmacia LKB Inc. (Dorval, Quebec, Canada). The lambda gt11 cloning and screening kits were obtained from Promega Biotech (Madison, Wisconsin, USA) and Western blots were performed using reagents from Bio-Rad Laboratories (Mississauga, Ontario, Canada).

Extraction of RNA

Pollen was ground by hand in a mortar and pestle chilled on dry ice, then extracted in guanidine thiocyanate as described [2]. Reconstituted RNA was translated *in vitro* in the presence of 35S-methionine and newly synthesized proteins were separated on polyacrylamide gels [3], and autoradiographed.

Total RNA was passed over oligo-dT or poly U-Sepharose columns [4], and the recovered mRNA was re-passed over the columns to minimize ribosomal RNA contamination.

Western Blots

Pollen proteins were electrophoresed and transfered to nitrocellulose [5]. The blot was blocked with 1% gelatin, and incubated with a pool of monoclonal antibodies directed against known pollen allergens. Bound antibodies were revealed by colour development.

Construction and Screening of gt11 Library

Messenger RNA prepared as described above was used to construct double-stranded cDNA [6]. After methylation of Eco RI sites and blunting the ends of the cDNA, lambda gt11 arms were ligated to the cDNA at 15°C overnight. The library was amplified and then screened using Mab60 and the Promega kit, and positive phage appearing on duplicate filters were purified to homogeneity. Eco RI excised inserts were subcloned into pT7T319.

RESULTS AND DISCUSSION

Although we anticipated that the yield might be low, or that the RNA might not be intact, commercial pollen appeared to be a convenient source for RNA extraction. The extraction of mature blugrass pollen yielded about 0.5 mg of total RNA per gram of pollen. When this RNA was translated as total RNA in an *in vitro* reticulocyte system, the pattern of proteins synthesized closely followed the pattern of the extractable proteins, thus indicating that intact messenger RNA for pollen proteins was present.

We tested other plant tissues for the presence of allergens by extracting roots and leaves, including meristematic tissue, and performing Western blot analysis using the monoclonal antibodies raised against several of the allergens. These experiments indicated that the identified allergens appear to be pollen-specific. Thus, other plant tissues were not suitable for extraction of RNA for allergen proteins.

After separation of pollen mRNA on oligo-dT and poly U-Sepharose columns, the mRNA was translated *in vitro*, and using monoclonal antibodies, we attempted to immunoprecipitate newly synthesized allergen proteins. These attempts were not generally successful, however we were able to precipitate the smaller of the group I allergens (about 33 KDa).

A cDNA library in lambda gt11 was constructed from the mRNA prepared from pollen. The library contained about 300,000 phage, about 85% of which were independent recombinants. This library was amplified to facilitate screening with different antibodies. When the amplified library was screened with Mab60, directed against the group I allergens, we detected several positive phage on duplicate blots. These phage were purified to homogeneity and subcloned into the T3T7 vector for further analysis. The clones isolated thus far will be used to rescreen the library to recover more full-length clones representing the group I allergens.

In addition to using monoclonal antibody, we have screened the library with IgE from grass-allergic patients. We have purified a number of positive phage isolated in this fashion. It will be of interest to compare these clones to those isolated by Mab60.

The characterization of pollen allergens may enable us to understand more completely the nature of the allergic response, and may provide clues to more effective treatment regimes. We have begun to analyse the major pollen allergens from Kentucky Bluegrass (*Poa pratensis L.*) and have isolated cDNA clones representing the proteins. These clones will provide tools to determine important structural features of the allergens, as well as providing templates for directed mutagenesis to examine the effects of altered structure on allergenicity.

REFERENCES

1. Huynh, T.V., Young, R.A. and Davis, R.W. (1985). In: DNA Cloning (Ed. Glover, D.M.) IRL Oxford, Vol.1, pp.49-78.
2. Chirgwin, J.M., Przybyla, A.E., MacDonald, R.J. and Rutter, W.J. (1979). Biochemistry, 18, 5294-5299.

3. Laemmli, U.K. (1970). Nature, 227, 680-685.
4. Aviv, H. and Leder, P. (1972). Proc. Natl. Acad. Sci. USA, 69, 1408-1412.
5. Towbin, H., Staehelin, T. and Gordon, J. (1979). Proc. Natl. Acad. Sci. 76, 4350-4354.
6. Gubler, E. and Hoffman, B.J. (1983). Gene, 25, 263-269.

Discussion: Session Three

Chairman: James Watson

1. Questions to Wayne Thomas

Jim Braun. I am sorry but you lost me in the last couple of slides. The protein you produced from an isolated recombinant clone was not immunogenic and was able to block a regular response, yet before that you had identified it because it was immunogenic. Were they different clones?

Wayne Thomas. I am sorry, I switched clones on you; I went from *Der p I* to *Der p II* and back to *Der p I* for the final experiments. The *Der p I* clone is very sensitive to any sort of denaturation and is not antigenic, whereas the *Der p II* clone is exceptionally antigenic. The experiment to which you refer has not been performed with *Der p II*.

Jim Braun. I see, so the *Der p I* when it is denatured can simply derail a response even though it is physically present. Was that with the recombinant material?

Wayne Thomas. Yes, it was a fusion recombinant.

2. Questions to David Klapper

Herbert Mansmann. I was impressed by your slide depicting the sequencing of five grasses. Clinically those grasses would be expected to show considerable cross-reactivity in some grass pollen allergic individuals, and so I wonder if you have tested the synthetic peptides for biological activity?

David Klapper. Yes, we have done that, and yes they do cause pronounced wheal and flare reactions in susceptible individuals. However, because the peptides represent allergenic epitopes they behave functionally as monomers, and must be aggregated before this biological activity is evident. We used glutaraldehyde as the aggregating agent.

Martin Chapman. A question along similar lines David. Have you used the peptides to inhibit the binding of specific antibodies to rye grass?

Advances in the Biosciences Vol. 74

Printed in Great Britain

David Klapper. Yes, the peptides will inhibit the binding of rye grass-specific monoclonal antibodies.

Martin Chapman. What about rye grass-specific human IgE antibodies?

David Klapper. The peptides are in the region we call "human site one" and it turns out that there is four to five times more human IgE activity against "human site two". Consequently the peptides achieved a maximum inhibition of around 20% with human IgE sera.

Session Four

IMMUNOASSAY OF ANTIBODIES USING LABELLED ALLERGENS

Chairman: T. Merrett

Use of Antigen Binding Radioimmunoassays to Measure Class and Subclass Specific Antibody to Food Antigens in Atopic Dermatitis Patients

S. Rowntree*, T. A. E. Platts-Mills† and E. B. Mitchell*

*Children's Research Centre, Our Lady's Hospital for Sick Children, Crumlin, Dublin 12, Ireland

†University of West Virginia, Charlottesville, USA

ABSTRACT

In patients with atopic dermatitis IgE antibody responses to a major allergen of *D. pteronyssinus* (*Der p I*) and egg and milk proteins were monitored by antigen binding radioimmunoassay (ABA). These results were compared with those obtained by IgE-RAST, skin testing and Prausnitz-Kustner (P-K) reaction, and a good correlation was obtained between the test procedures for *Der p I* but not for the food antigens. High levels of IgG antibody to foods, but not inhalants such as *Der p I*, are normally seen, and when the ABA was repeated using IgG depleted specimens the results were found to agree with the other tests. It is concluded that the ABA for IgE class antibody to food is susceptible to interference by the proportionately greater amounts of IgG antibody. An ABA was developed for subclass IgG4 antibody and it confirmed that (a) IgG4 antibody is a normal (i.e. in atopics and non-atopics) component of the immune response to foods, and (b) is much increased compared to the proportion of total IgG which is total IgG4.

KEYWORDS

Antigen binding RIA; subclass specific antibody; food antigen; dermatitis.

An immune response to food proteins can occur in both normal and atopic subjects. In the latter subjects this response may include the production of specific IgE antibody. Assessment of this antibody using RAST may be difficult in view of the very high total IgE levels which may be present. The fact that food induced symptoms are often non-immediate may suggest that components of the immune response other than IgE antibody may be involved in food hypersensitivity. This is of particular relevance for atopic dermatitis. We have developed antigen binding radioimmunoassays to evaluate the antibody response to egg and milk proteins in subjects with atopic dermatitis and compared these results with immediate hypersensitivity skin testing and RAST [1].

Advances in the Biosciences Vol. 74

AB 74-G

The assay of IgE antibody using antigen binding RIA's was performed on 33-50% SAS fractions of serum, with IgE PS myeloma added as carrier to form a precipitate. In some cases gel chromatography (Sephadex G200) was subsequently undertaken. Enriched specific antibody fractions were prepared using affinity chromatography, the IgG antibody present in these fractions then being removed by immunoprecipitation with mono-specific goat anti-IgG.

The results indicated that approximately one third of all atopic dermatitis patients tested showed a positive skin test to egg protein. When house dust mite was used as eliciting antigen approximately 83% were positive. Quantitative intradermal skin testing showed that a higher mean concentration of ovalbumin (10^{-2} Hg/ml) and ovomucoid (10^{-2} Hg/ml) from egg protein was necessary to induce positive responses than was the case with house dust mite (*D. pteronyssinus*) *Der p I* (10^{-5} Hg/ml).

Using RAST, positive binding to egg white was found in 21/60 atopic subjects whereas only 6/60 showed a similar binding for milk whey. Most patients with a positive skin test to egg also had a positive RAST, however, three adults had positive RAST binding in the absence of a positive skin test. There was a poor correlation between the level of skin sensitivity to egg as determined by QID and the level of RAST binding ($r = 0.03$; $p = >0.1$). Of 10 patients with positive skin tests to milk only 3 showed positive RAST binding. RAST absorption experiments using sera from 8 dermatitic subjects known to have positive skin tests to house dust mite and egg suggested that 3.5% of the total IgG was specific for egg white whereas 8.9% was specific for house dust mite.

Using antigen binding RIA's to measure isotype specific responses to purified major allergens from egg white and milk whey demonstrated that IgG antibody was demonstrable in both atopic and non-atopic subjects. However, the binding activity was significantly greater in the atopic dermatitis group than the non-atopic control group (ovalbumin $p = <0.001$, ovomucoid $p = 0.01$, β-lactoglobulin $p = <0.001$). The level of IgG BA did not relate to the presence of clinical sensitivity, immediate skin tests or RAST binding. In contrast to responses seen with ingested allergens IgG antibody to *Der p I* was present only in atopic subjects and correlated well with the level of specific IgE antibody ($r = 0.69$, $p = <0.01$). The results of the binding assays to ovalbumin and ovomucoid suggested that a proportion of the patients with atopic dermatitis had IgE antibody to these allergens. This apparent activity did not correlate with the presence of immediate skin tests or RAST binding ($r = 0.1$, $p = >0.1$). The apparent ratio of IgG:IgE in these sera for both egg allergens was very high (>50:1) in comparison with *Der p I* (2:1). The preparation of IgG depleted fractions from the sera of 10 patients sensitive to house dust mite and egg protein demonstrated that removal of IgG resulted in a fall in the apparent level of IgE to ovalbumin but not *Der p I*. Therefore, in the absence of specific IgG, the level of egg specific IgE antibody in prick test positive subjects appeared to be low and at the threshold of sensitivity for the antigen binding RIA.

Prausnitz-Kustner (P-K) testing was performed using the sera from five egg and house dust mite prick test and RAST positive subjects. A strong P-K activity could be elicited in 3 of the 5 sera to *Der p I*. Only one sera showed a weak activity to egg proteins the remainder being negative. To confirm the presence of IgE to egg affinity purification was performed in two cases. The fractions possessed good P-K activity to egg protein. Depletion of IgG antibody in these fractions (removing IgG by greater than 98%) demonstrated accurately the levels of IgE antibody in these sera.

It has proven difficult to assay IgE antibody to egg using antigen binding RIA's in the presence of high IgG antibody. When a minor class of antibody represents only a small proportion of the specific antibody in a serum it becomes difficult to maintain antigen excess without diluting the serum to a level where the minor class is no longer detectable. The results for RAST appear to be more reliable. The good correlation between symptoms prick tests and RAST makes it unlikely that the results are in some way related to total IgE levels or high levels of IgG specific antibody. Measuring antibodies to inhalant allergens has indicated that the antigen binding assay is approximately as sensitive as RAST. However, IgE antibody was not detectable in patients who required $>10^{-3}$ Hg/ml of allergen to induce a positive intradermal test. The mean of skin sensitivity to egg protein in our patients was $>10^{-3}$ Hg/ml. RAST may be more sensitive if the IgE antibody is of low affinity or because of the large proportion of relevant antigen on the disc. The quantity of IgG antibody present may play a role in determining skin reactivity as the difficulty in obtaining P-K activity was unexpected. Overall, however, it would appear patients who develop immediate hypersensitivity responses to both ingested and inhaled allergens are less sensitive to foods than they are to inhalants [2].

IgG4 Subclass Antibody

An antigen binding radioimmunoassay to measure IgG4 antibody to egg protein and house dust mite antigen in these patients was developed [3], to assess the possible role of this IgG subclass in the pathogenesis of atopic diseases. The assays were used to directly compare IgG antibody (i.e. the activity of all four subclasses) and IgG4 antibody levels. In the latter assay mouse monoclonal anti-human IgG4 subclass reagents specific for the Fc domain of IgG4 [kind gift of Dr R. Jefferies (University of Birmingham)] were used. As in the class-specific antibody radiolabelled antigen 3-7 Ng (constant for a given antigen) was added to 100 µl of a serum dilution, and followed by 100 µl of monoclonal anti-IgG4 in 1/100 NMS. Subsequently rabbit anti-mouse IgG antiserum was added followed by overnight inhalation at 4°C. The precipitate formed was washed in BBS three times and the remaining radioactivity was counted in a gamma counter.

To ensure precipitation of IgG4 ab antigen complexes alone, antigen must remain in excess. This was achieved by using dilutions of sera that bound <20% of the total precipitable counts added. The counts bound were converted to units of BA by reference to a total IgG ab standard curve run in parallel.

In all adult atopic dermatitis subjects tested IgG4 subclass antibody to purified egg protein ovalbumin was present and comprised approximately 20% of the total ovalbumin specific IgG antibody. While only 6 out of 10 non-atopic subjects were shown to have IgG antibody responses to this antigen in all of these IgG4 antibody made a significant contribution to the ovalbumin specific antibody. The actual levels of both class and subclass specific antibody were significantly greater in the atopic dermatitis patients. Most patients had developed IgG antibody to β-lactoglobulin from cow's milk and approximately half also had demonstrable IgG4 which comprised 14% of this response. While the prevalence of IgG4 and IgG was greater in atopic dermatitis subjects no significant difference in the actual levels was apparent between atopic and non-atopic subjects. The prevalence of Rye 1 specific IgG4 antibody was less than that seen with food antigens (7/21) and in those positive subjects it comprised only a smaller percentage of the total Rye 1 specific IgG antibody. In the case of the perennial allergen house dust mite *Der p I* the situation was somewhat different with a majority of the atopic dermatitis patients having IgG subclass antibody. Again this contributes only a small percentage (6%) to the *Der p I* specific IgG antibody. In the subjects with antibody a good correlation between IgG and IgG4 was evident for 3 of the 4 antigens (ovalbumin $r = 0.935$, $p = <0.001$; Rye 1 $r = 0.57$, $p = <0.02$; *Der p I* $r = 0.647$, $p = <0.01$; β-lactoglobulin $r = 0.261$, $p = >0.1$). In the case of IgG4 and IgE antibody the relationship was not significant for *Der p I* ($r = 0.364$, $p = >0.05$) or ovalbumin ($r = 0.202$, $p = >0.1$), in contrast to the relationship between Rye 1 specific IgG4 and IgE antibody ($r = 0.891$, $p = 0.01$). There was no significant association between total serum IgE and specific IgG4 levels to any antigen studied ($p = >0.05$).

It is necessary to interpret increases in IgG4 antibody relative to the increase in IgG antibody before conclusions can be drawn about the significance of these IgG4 ab responses. The results presented confirm that IgG antibody to major food allergens is present in a majority of adult patients with atopic dermatitis, but that IgG4 antibody is also demonstrable in non-atopic individuals and represents a normal component of the immune response. The IgG4 ab response to these antigens is selectively increased relative to the IgG4 subclass contribution to total IgG immunoglobulin. However, although mean levels are higher in the atopic group the subclass antibody appears to represent a similar proportion of the IgG food specific antibody in both groups. It would appear from these results and other data (not shown) that the frequency and dosage of antigen may be a factor in determining the IgG4 antibody responses. However, the significance of IgG4 ab in disease expression remains unclear.

REFERENCES

1. Rowntree, S., Mitchell, E.B. and Platts-Mills, T.A.E. (1986). Assessment of skin reactivity and specific antibody to egg and milk proteins in patients with atopic dermatitis. Int. Arch. Allergy Appl. Immunol., 79, 132.
2. Rowntree, S., Cogswell, J.J., Platts-Mills, T.A.E. and Mitchell, E.B. (1985). Development of IgG and IgE antibodies to food and inhalant allergens in children at risk of allergic disease. Arch. Dis. Child., 60, 727.
3. Rowntree, S., Platts-Mills, T.A.E., Cogswell, J.J. and Mitchell, E.B. (1987). A subclass IgG4 specific antigen binding radioimmunoassay (RIA) - Comparison between IgG and IgG4 antibodies to food and inhaled antigens. J. All. Clin. Immunol., 80, 622.

Assessment of *in vitro* Testing to Diagnose IgE Mediated Allergic Disease

L. Perelmutter

Department of Pediatrics, Division of Allergy and Clinical Immunology, Thomas Jefferson University, Philadelphia, Pennsylvania, USA

ABSTRACT

The purpose of this article is to give an up-to-date assessment of *in vitro* allergy IgE testing to diagnose allergic disease. The emphasis is on clinical correlates, the use in clinical practice as well as discussions of the various assays being utilized today. A summary of various comparison studies including recommendations for users of *in vitro* IgE antibody assays and the use of gold standards to evaluate these assays, will be discussed. There will be further emphasis on the utilization of purer allergen fractions and recombinant DNA production in these assays.

KEYWORDS

In vitro; specific IgE; allergy; diagnosis; assays evaluation; pure allergen.

INTRODUCTION

Since the late 1960s sandwich-type radioimmunoassays (RIA), such as PRIST and RAST have been used extensively to detect total and specific IgE, respectively. The original form of these assays was a modified anti-globulin technique. Studies done in the early 1970s indicated that RAST correlated well with other diagnostic tests and clinical responses such as skin tests, bronchial challenge, histamine release from leukocytes and patient history. About the same time, the RAST inhibition assay was adapted to standardize allergen preparations, a method still used today to detect allergenic components in crude allergenic extracts. In addition, the PRIST and RAST tests were utilized to monitor seasonal changes of total and specific IgE antibody levels in patients with pollen allergies. Based on these early findings, Pharmacia marketed the first commercial PRIST and RAST kits. The ingenious use of the paper disc rather than the particle technique greatly simplified the assay and made the procedure practical for routine applications.

Advances in the Biosciences Vol. 74

Printed in Great Britain

Pharmacia developed the arbitrary RAST unit called the Phadebas RAST unit or PRU, and the RAST semiquantitative classification system was introduced. Thus, for the first time the RAST test had clinical significance when used together with the case history and skin tests in the diagnosis of allergy [1,2].

Initially the PRIST (total IgE) test was employed as a screening test for atopic or type 1 hypersensitivity allergy reactions, but it was soon noted that a number of symptomatic patients with so-called "normal" total IgE levels, had elevated levels of specific IgE [3].

As a result, other schemes using total IgE together with specific IgE levels were proposed for a more efficient diagnosis. It was implied that an individual with a total IgE level of less than 25 IU with only RAST class 1 scores would be unlikely to have atopic allergies, whereas one with a high total IgE level together with a RAST score of class 2 and above was likely to have atopic allergy [3].

From the mid-1970s, RAST expanded markedly and with this extended use certain problems emerged, the major of these being the sensitivity of the Phadebas RAST test and the cost effectiveness of the test in clinical diagnosis.

Many comparison studies arose dealing with the relationship between RAST and skin tests and/or patient history. Though the test was found to be very specific, the major complaints were the large number of false negatives that were obtained. The lower assay sensitivity of the Phadebas RAST led to the development of the modified RAST procedure by Fadal and Nalebuff [4]. They described procedural and interpretive modifications of Phadebas RAST. As part of their modifications, these investigators lowered the assay cut-off point. Much debate resulted in the modification to the RAST, the primary problem being that the lower cut-off would result in lower assay specificity, ie., increased number of false positives.

This did occur to some extent, but not to the detriment of the test [5,6]. Thus, when one employed the modified RAST, there appeared to be an increase in the sensitivity of the procedure without a significant loss in specificity. At the present time the modified assay is the one most widely used and has essentially "overcome" the initial criticism of the test's poor sensitivity.

In order to deal with the problem of cost effectiveness, the commercial laboratories and kit manufacturers suggested panels of allergen testing to meet the economic need of the user and patient. These panels took many forms, ie., pediatric panels, indoor panels, general panels, etc. Along with these selected panels, information for selection of panels based on patient history, geographically relevant allergens, allergen cross-reactivity, and *in vivo* testing results were released. The difficulty here was that there were a multitude of these schemes being proposed by the clinical laboratories, manufacturers, and physicians using the test in their office.

More recently, an efficient and cost-effective approach to allergy diagnosis using *in vitro* testing has been proposed. The use of a mini-screen consisting at least one highly prevalent representative of each major inhalent allergen group has been shown to detect up to 98% of those patients with demonstrable levels of allergen specific IgE [3,7]. This mini-screen could vary from between six to nine allergens depending on the geographic region.

The major inhalent allergen groups include grasses (timothy/Bermuda), weeds (short ragweed/English plantain), trees (oak/"other"), molds (*Alternaria*), animal danders (cat), and dust (dust mite, *D. farinae*). By combining the mini-screen with total IgE even more information can be obtained, for if the inhalent screen is negative while total IgE is elevated the possibility of atopic food allergy exists.

Another approach being used to reduce the cost of *in vitro* allergy testing is combining several allergens on a single, solid phase system. This multi-screen procedure, either using allergen mixes of the same group on a single disk (ie., trees) or by linking the mini-screen allergens on a single disk, the micro-screen [3] has been adopted by several manufacturers. This approach, while appearing promising, is still under active investigation.

Today the variety of assay kits for detecting total and specific IgE and the variation in their procedures are quite extensive. All these kits utilize one of the three available labelling systems: radioisotope, enzyme conjugate or fluorochrome. Some use monoclonal antibodies while others use polyclonal antibodies. Because of the multitude of tests available, it can be quite confusing to the perspective user of the kits as to which one to choose.

Pathophysiology

This section deals with the importance of circulating total and specific IgE in relation to various allergic disease states. IgE antibodies are triggered by small quantities of allergen. The interaction of cell-bound IgE antibody on target cells with allergen is known to cause release of mediators and hence type I hypersensitivity symptoms. How much IgE antibody is required to set off this chain of reactions and how meaningful is circulating (serum) IgE antibody levels with regard to this *in vivo* response? Although *in vitro* allergy testing agrees with the parameters of *in vivo* response such as skin tests and bronchial challenge, the use of these tests as they relate to symptomatology remains the subject of some controversy.

In interpreting the results of these tests, consideration should be given to the possible differences of an *in vitro* measurement compared with *in vivo* response. Furthermore, factors influencing total and specific IgE response should be considered. Both environmental and genetic factors are known to control IgE response; thus, infectious disease may alter total and specific IgE levels and should be considered as an influencing factor in the measurement of total and specific IgE. Another factor that influences these antibody levels is exposure to allergen as during pollen seasons with regard to pollen allergies. Extensive studies have shown that IgE antibody levels rise during the season but fall marginally off season. Of interest is the fact that circulating IgE-antibody-secreting B cells rise markedly during season with only slight changes in serum IgE antibody levels compared with preseasonal levels. Furthermore, patients in immunotherapy for several years experience a rather slow but consistent decrease in IgE antibody levels whereas their B cells from PBL secrete much less IgE antibody after short-term immunotherapy. As mentioned previously, the effects of IgE antibody response *in vivo* may not be reflected by their *in vitro* measurement. The presence of IgE antibody response *in vivo* may not be reflected by their *in vitro* measurement. The presence of IgE antibodies to food allergens does not necessarily prove that the patient is allergic to that food and food challenge may be warranted to prove specificity. Also, children placed on food substitutes like Vivonex continue to generate high IgE antibody levels to the food allergens for long periods of time after allergen exposure.

We need to know more about the mechanisms that control the synthesis of IgE and the interaction of this antibody on a cellular level with the target cells. As part of these studies, there is a growing interest in the role of IgG subgroups in allergic disease. Of particular importance is the anaphylactic and blocking antibody role of IgG4 levels in the clinical diagnosis and treatment of allergy [8]. It is the hope that the results of investigations of these subgroups will contribute to improved diagnosis to allergic disease.

Diagnostic and Therapeutic Considerations

As indicated earlier, *in vitro* IgE antibody methods have been used to diagnose allergic response to a variety of allergens including pollens, weeds, inhalants, animal danders, molds, foods, and some drugs. As questioned under "Pathophysiology", the value of the results should be considered with regard to the multiple mechanisms involved and the variety of allergens being employed. Of particular importance from regulation studies of IgE is the fact that serum IgE antibody levels do not always reflect *in vivo* response and exposure to allergen, and also the genetic predisposition of the patient is important. In addition to serum, total and specific IgE levels have been measured in saliva, nasal secretions, and tears. These latter studies have been useful in correlating these levels more closely to *in vivo* response compared with serum levels alone. Total and specific IgE levels have been extensively studied during and post-season in order to evaluate the appropriate time in which to sample for *in vitro* testing. The practice of single and not sequential testing may sometimes give erroneous false negative results. Another aspect in the utilizing of *in vitro* IgE testing, particular for routine testing, has been the possibility of IgG blocking antibody interference. This has been studied quite extensively by a number of investigators [9]. This problem has been found particularly with patients on immunotherapy but also in a small number of untreated patients. It is also known to occur in patients with food allergies where high levels of IgG antibodies to these allergens can be found. It has been concluded that the concentration of the allergen on the solid phase is critical to this effect and several research modifications to these *in vitro* tests have been introduced to eliminate or at least minimize the problem of IgG antibody interference [10]. Can these *in vitro* techniques be useful for following patients on immunotherapy? As part of the above studies dealing with IgG antibody interference on IgE antibody test, it was noted that it was imperative to know the IgG/IgE antibody levels. The proper determination of these two antibodies could be of value in determining effectiveness of immunotherapy; however, even today, the mechanism to explain immunotherapy is unknown. IgG antibodies or subgroups of IgG antibodies probably play some role. There has been extensive studies on IgG4 antibodies in relation to immunotherapy. This antibody appears to be associated with long-term or chronic exposure to antigen, whereas IgG1 is produced after short exposure to antigen. It may be that IgG1 and IgG4 antibody level response may be important to assess immunotherapy of a patient and the IgG1/IgG4 ratio could be the indicator in future. Although Pharmacia introduced IgG antibody kits to detect venoms, recently several manufacturers extended the antigen testing base to include a wide variety of food allergens. The value of such tests, however, whether using paper discs or microtiter plates, are still under investigation and such tests need further modification to eliminate non-specific binding of IgG. If such modifications are not done, the assays under the present conditions will produce an extensive number of false positives. The clinical application, therefore, of these IgG tests now being offered by some clinical laboratories for the diagnosis of food allergies should be regarded at this time as premature and relatively inaccurate. The value of relating IgG antibody levels to clinical improvement awaits studies indicating the role of antigen-specific suppressor T cells or anti-idiotypic antibodies generated during immunotherapy. It may be that in the future a combination of cellular and humoral response may be of importance to evaluate the patient and decide on the proper course of treatment.

Recently, more extensive comparison between the various kits commercially available for measuring specific IgE have been reported. Differences in sensitivity and specificity between the RAST, IP, MAST, and FAST tests were presented. This study illustrated not only wide but also individual intra-test differences allergen to allergen. Other studies have shown not only qualitative but significant quantitative class differences between the various test kits. This is of practical importance for those physicians who plan to determine the dosage for immunotherapy from the *in vitro* test results.

Differences found between these kits have led various committees of the American Academy of Allergy, the American College of Allergists, and the American Association of Clinical Immunologists and Allergists to call for better standardization and quality control for the reagents used by the kit manufacturer. Furthermore, the research total and specific IgE tests are presently undergoing close scrutiny by various groups of investigators involved in IgE regulation studies. Variations found in cell supernatants of IgE levels has prompted these groups of investigators to evolve primary IgE standards and more uniform assays and reagents for these studies. The principles involved in these studies will have an impact on total and specific IgE antibody detection for routine use. In the meantime, the users of the *in vitro* IgE antibody tests are interested in knowing which commercial kit to employ. I would like to propose a few questions that may help guide these users to a suitable testing system.

1. What is the source of the allergens used on the solid phase? Are these allergens standardized, and if so, by which procedure?

2. What is the source of anti-IgE? What myeloma was it raised against? Is it polyclonal or monoclonal?

3. Does the manufacturer provide evidence of quality control?

4. Has the manufacturer provided evidence of published data on relationship between specificity and sensitivity?

5. Does the assay have intra-test allergen to allergen consistency and reproducibility? Assay to assay reproducibility? Reliable semi-quantitative results on which to base immunotherapy?

6. Is the manufacturer providing evidence of sufficient R and D to back-up claims?

The problem that has been bothersome for both the kit manufacturer and the user of these assays, namely cost and sensitivity, is now close to being resolved. With the thrust of immunologic developments today newer innovative techniques will evolve to measure IgE antibodies. Immunologic reagents will be produced at a much lower cost by utilizing hybridoma and recombinant DNA technology. One such benefit will be reflected on techniques to measure IgE response. Also, the generation of monoclonal antibodies to IgE had had impact on development of *in vitro* methods giving us some insight into allotypic and anti-idiotypic determinants on the IgE molecule. In the future recombinant DNA prepared allergens will replace the monoclonal preparations used for *in vitro* diagnostics of IgE antibody response and provide proper matched allergens for treatment and diagnostic modes.

Finally, there is a need to normalize and standardize IgE antibody procedures. The College of American Pathologists have had a proficiency program available to each laboratory performing total and specific IgE antibody tests since 1986. This program is designed to help the laboratory in evaluating its ability to determine target values relative to other laboratories. The users of *in vitro* allergy tests should utilize this program.

In an attempt to further assess the various commercial kits it is imperative that comparison studies be performed from time to time to determine not only specificity, sensitivity and overall efficiency, but also class to class comparisons. Such studies should include multi-allergen procedures to dipsticks and multi-screens.

A re-evaluation of skin-test procedures need to be addressed in order to obtain more homogenous results. Incorporation of provocation (bronchial and nasal) challenge should also be included in order to determine a proper gold standard to compare with the *in vitro* tests.

REFERENCES

1. Evan, R., Reisman, R.E. and Wypyeh, J.I. (1972). An immunologic evaluation of ragweed sensitive patients by newer techniques. J. Allergy Clin. Immunol., 49, 285.
2. Fadal, R.G. and Nalebuff, D.J. (1978). Tools of the allergist: old and new. Contin. Educ. Family Phys., 10, 37.
3. Nalebuff, D.J. (1985). Use of RAST screening in clinical allergy: a cost-effective approach to patient care. Ear, Nose, Throat, J., 107, 15.
4. Nalebuff, D.J. and Fadal, R.G. (1979). The modified RAST assay: an aid in the diagnosis and management of allergenic disorders. Contin. Educ. Family Phys., 10, 64.
5. Hoffman, D.R. (1981). Comparison of methods of performing the radioallergosorbent test: Phadebas. Fadel/Nalebuff, and Hoffman Protocols. Ann. Allergy, 45, 343.
6. De Filippi, I., Yman, L. and Schroder, H. (1981). Clinical accuracy of updated version of the Phadebas RAST test. Ann. Allergy, 46, 249.
7. King, W.P. (1982). Efficacy of a screening radioallergosorbent test. Arch. Otolaryngol., 108, 81.
8. Perelmutter, L. (1984). IgG4: non-IgE mediated atopic disease. Ann. Allergy, 52, 64.
9. Zimmerman, E.M., Yunginger, J.W. and Gleich, G.J. (1980). Interference in ragweed pollen and honeybee venom radioallergosorbent test. J. Allergy Clin. Immunol., 66, 386.
10. Reid, M.J., Cheung, H.V. and Lewiston, N.J. (1981). Microtiter solid phase radioimmunoassay for specific immunoglobulin E. J. Allergy Clin. Immunol., 67, 263.

Liquid-phase *in vitro* Allergen-specific IgE Assay with *in situ* Immobilization

A. Said El Shami and Olusola Alaba

Diagnostic Products Corporation, Los Angeles, CA 90045, USA

ABSTRACT

A new test system (AlaSTAT) has been developed for the *in vitro* measurement of circulating allergen-specific IgE — a system which allows the reaction between allergen molecules and IgE antibodies to proceed in liquid phase. The system is based on a novel method for attaching individual allergens to a soluble polymer matrix labeled with an identifiable ligand. The system also invites automation through its use of liquid reagent formats. The AlaSTAT procedure is as follows. A ligand is covalently linked to a liquid matrix which in turn attaches to NH_2-, OH-, COOH- or SH- groups on the allergen. A tube coated with the same ligand serves as the vessel for the initial reaction between allergen-specific IgE in the patient sample and the allergen-ligand-labeled matrix; this first reaction proceeds with liquid-phase kinetics. After one hour, a multivalent anti-ligand is added, and reacted for another hour to link the allergen-ligand-labeled matrix/specific IgE complexes to the ligand-coated tube. The tube is then washed and reacted for one hour with a blend of two horseradish peroxidase-labeled monoclonal anti-IgE antibodies. Finally, substrate is added for 15 minutes. The resulting color is directly proportional to the amount of allergen-specific IgE in the patient sample.

KEYWORDS

AlaSTAT; liquid matrix; ligand; anti-ligand; biotin; avidin; RAST.

INTRODUCTION

After the discovery by Ishizaka *et al.* [1] that reaginic activity mediating the passive transfer of Prausnitz and Kustner test is gamma E, and following the elucidation of its biochemical structure by Johansson and Bennich [2], the ideal of using an *in vitro* method for allergy testing finally became attainable. Shortly thereafter, Wide *et al.* [3] developed the radioallergosorbent test (RAST) for *in vitro* measurement of allergen-specific IgE in serum. The RAST assay is now widely used as an adjunct to, or even sometimes

Printed in Great Britain

as a replacement for, the classical skin-testing procedures used in allergy diagnosis. Several modifications of this classical method have been introduced. All are based on attaching allergens to one type of solid phase or another (paper disc, cellulose thread, microtiter well or polystyrene tube).

In the RAST system the patient sample is reacted with individual allergens immobilized on a solid phase. This format has some inherent disadvantages which ultimately compromise the reliability of the RAST approach. The most significant of these disadvantages is a tendency towards non-specific binding of IgE, especially when the total serum IgE level is high [4-9]. This problem is still evident with all allergosorbent-based assays, despite the various modifications that have been introduced over the years. Relying on solid-phase kinetics, allergosorbent test systems require long incubation times; moreover, because amino groups are required for cyanogen bromide coupling to cellulose allergosorbents, allergens that contain significant amounts of carbohydrates will not be represented. Accordingly, an assay format where allergen and patient sample are reacted in the liquid phase would improve the reliability of *in vitro* measurements of allergen-specific IgE.

In the past, several successful attempts have been made to label allergens with haptens such as dinitrophenol (DNP) or trinitrophenol (TNP) [10-15]. The problem with these attempts is the possibility that some allergens may not be labeled, due to either the lack of accessible amino groups for haptenization of the allergen molecule, or to the absence of any such groups if the allergen molecule is a carbohydrate. In case these groups are available and do get labeled, specific IgE-binding sites may nevertheless become altered. Furthermore, this approach would require affinity purification of the labeled allergen preparation in order to separate labeled from unlabeled allergens and ensure high specific activity for the labeled allergen molecules.

The AlaSTAT system contrasts with RAST and this direct method of labeling allergens in that the allergen molecules are chemically attached to a soluble polymer in the AlaSTAT system. It likewise avoids the problems, mentioned above, that occur in labeling allergen molecules directly. In the AlaSTAT system an identifiable ligand, biotin, is covalently attached to a backbone (the soluble polymer) to which the allergen molecules are also attached.

MATERIALS AND METHODS

Allergenic Extracts

The allergenic extracts used in the AlaSTAT system are obtained from Greer Laboratories (Lenoir, NC). Each lot of allergenic extract is assessed by RAST-inhibition using Pharmacia disks (Piscataway, NJ) and a pool of atopic sera from individuals allergic to the allergen. Only extracts that totally inhibit IgE binding to Pharmacia disks are used.

Preparation of Allergen-soluble Matrix

Two soluble carbohydrates are utilized in the following reaction sequence:

(i) soluble dextran with molecular mass varying from 6,000 to 80,000 daltons (Serva, Westbury, NY), and

(ii) soluble Ficoll with molecular mass varying from 70,000 to 400,000 daltons (Sigma, St. Louis, MO).

These two carbohydrates are referred below as Matrix-OH.

The amino acid copolymers used are polylysine, alanine, and polyglutamic, glutamic ester (Sigma, St. Louis, MO).

Reaction Sequence 1

1,1'-carbonyldiimidazole (CDI) activation and coupling with amino acid copolymer and allergen

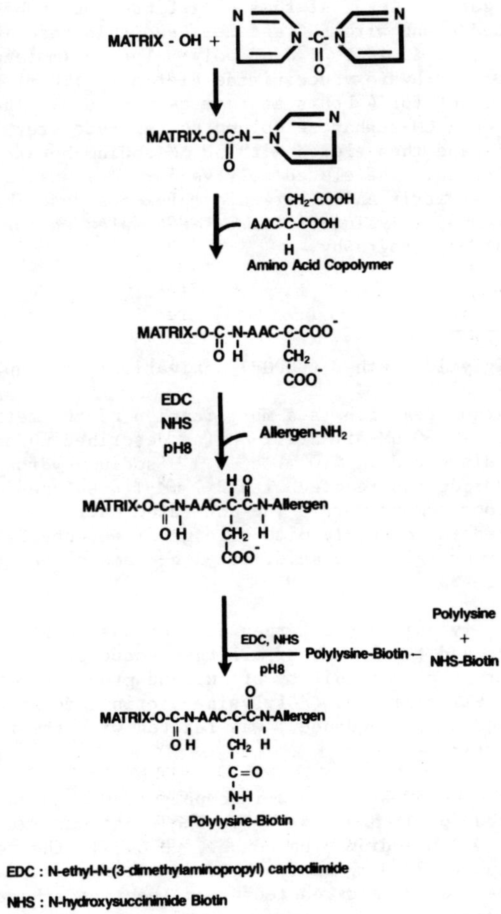

Reaction Sequence 1

This CDI reaction is a modification of the method of Bethel *et al.* [16]. Matrix-OH (dextran or Ficoll) 124 nM was dissolved in 2.0 ml of dimethylsulfoxide containing 10.0 mg of 1,1'-carbonyldiimidazole (CDI) and reacted at room temperature for 30 minutes, with stirring. Amino acid copolymer (eg. glutamic acid, ethyl glutamate, 770 nM) in 2.0 ml of dimethylsulfoxide was added and the reaction continued for 24 hours at room temperature. Following dialysis against 0.2 M sodium bicarbonate, pH 8, the Matrix-AAC reaction product was separated on Sephacryl-300 from free reactants. To the purified Matrix-AAC, 20.0 mg of lyophilized allergen extract was added, followed by the addition of 0.15 M N-ethyl-N'-(2-dimethylaminopropyl)carbodiimide (EDC) and 0.15 M N-hydroxysuccinimide (NHS), after which the mixture was allowed to react for 24 hours at 4°C, while stirring. The reaction was maintained at pH 8 with 1.0 M sodium carbonate. Reaction products were chromatographed on Sephacryl-300 to separate free, unreacted materials from Matrix-AAC-Allergen complex. At this stage, previously biotinylated poly-L-lysine was added along with EDC and NHS, and this reaction was allowed to continue for 16 hours at 4°C. 1.8 μM poly-L-lysine (molecular mass 3,800 daltons) and 180 μM M-hydroxysuccinimide biotin in 1.0 ml dimethylformamide were allowed to react for 4 hours at room temperature. The product was chromatographed on a CM-sepharose column to separate free reactants from polylysine-biotin and then eluted with 50 mM sodium borate, pH 9.0 containing 2.0 M sodium chloride. The eluted polylysine-biotin was adjusted to pH 8.1 and coupled to the Matrix-AAC-allergen complex (as described above) to form Matrix-AAC-allergen-polylysine-biotin. The complex so formed was purified by gel exclusion chromatography.

Reaction Sequence 2a

1,4-butanedioldiglycidyl ether (BDDGE) activation and coupling

The BDDGE activation reaction is a modification of the method of Sundberg *et al.* [17]. Briefly, 150 nM of Matrix-OH (as described under reaction sequence 1) was dissolved in 4.0 ml of 0.3 M sodium hydroxide containing 0.26 mM sodium borohydride and reacted with 2.7 mM 1,4-butanedioldiglycidyl ether for 4 hours at room temperature. The activated Matrix was then adjusted to pH 8.5 and coupled (a) directly with allergen (sequence 2a) or (b) with amino acid copolymer (glutamic acid, ethyl glutamate) and then with the allergen (sequence 2b).

With sequence 2a, lyophilized allergen extract was coupled directly to the activated Matrix, and the amount of allergen added to the reaction was adjusted to ensure the availability of oxirane groups for the subsequent coupling of polylysine-biotin. Polylysine-biotin made according to the procedure outlined under sequence 1 was reacted with the residual oxirane groups on the Matrix.

With sequence 2b, 600 nM of amino acid copolymer was added to the activated matrix and reacted for 16 hours at 4°C. The reaction mixture was then dialyzed against 0.1 M sodium bicarbonate, pH 8. To the resultant Matrix-amino acid copolymer, 20.0 mg of lyophilized allergen extract, 0.15 M EDC and 0.15 M NHS were added and reacted as in reaction scheme 1. Following a Sephacryl-300 chromatography to separate unreacted components from Matrix-ACC-Allergen, polylysine-biotin (synthesized as described above) was reacted with 0.15 M EDC and 0.15 M NHS. Finally, the Matrix-allergen-polylysine-biotin and Matrix-amino acid copolymer-allergen-polylysine-biotin complexes were purified using gel exclusion chromatography.

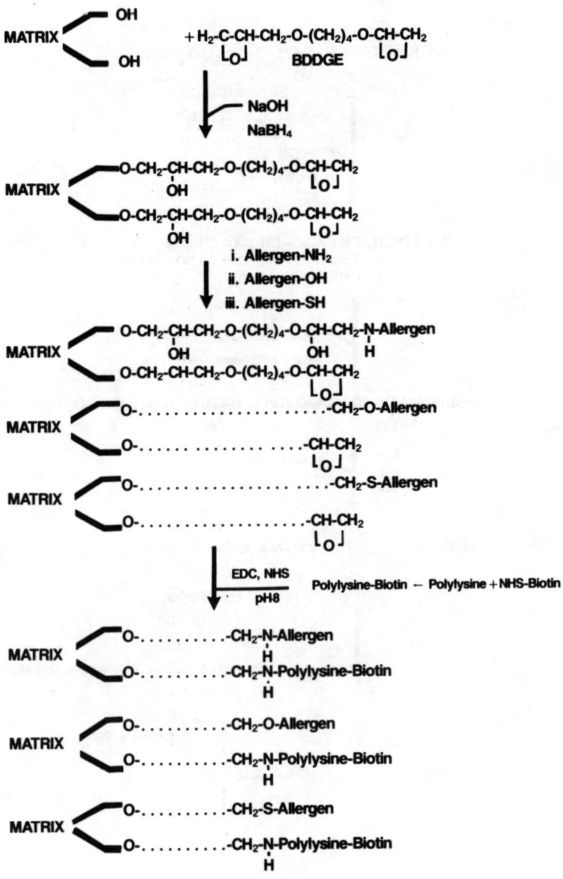

Reaction Sequence 2a

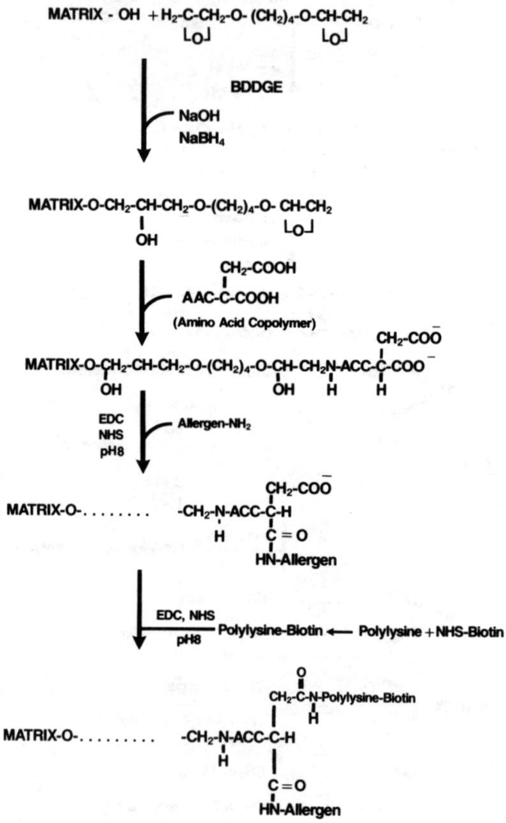

Reaction Sequence 2b

Other reaction sequences are explained in detail elsewhere (European Patent 0 245 926, and other patents pending).

Ligand-coated Tubes

The ligand-coated tubes were prepared by immobilizing biotinylated protein on 12 x 75 mm polystyrene tubes using a modification of the method of Engvall and Perlmann [18].

Preparation of Monoclonal Anti-IgE Antibodies and Horseradish Peroxidase-anti IgE Conjugate

Two different monoclonal antibodies (1C5 and 2G4) produced against human IgE, and shown to recognize two different epitopes on the epsilon chain of the IgE molecule, were each purified on DEAE-sepharose CL-6B, after ammonium sulfate precipitation. Each was then conjugated to horseradish peroxidase enzyme, using the periodate method of Nakane *et al.* [19], and diluted to a working solution in 0.05 M phosphate buffer, pH 7.0, containing 0.15 M sodium chloride, 0.01% thimerosal and 0.1% human serum albumin. The two monoclonal anti-IgE-horseradish peroxidase conjugates were then mixed in a previously determined ratio.

Preparation of Enzyme Substrate

The substrate components consisted of o-phenylenediamine (OPD), formulated into a tablet, and a solution of hydrogen peroxide (0.025%).

Reference Allergen and Atopic Calibrator Materials

Timothy grass allergen extract attached to Ficoll through polyglutamic, glutamate ester (as described under sequence 1 above) was used as the reference allergen in the AlaSTAT system. A pool of Class IV atopic serum for Timothy grass (G6), as classified by Phadebas-RAST kit from Pharmacia (Piscataway, NJ), was diluted in a non-atopic serum to simulate Phadebas-RAST atopic Classes III, II and I.

Determination of Allergen-specific IgE using the AlaSTAT Method

In the performance of the AlaSTAT test [20], 50 μl of the patient serum sample is first incubated in a ligand-coated tube for one hour with 100 μl of an individual allergen (prepared as described above). During this liquid-phase reaction, IgE specific for the given allergen, if present, binds to the allergen on the liquid matrix. The allergen-IgE complex thus formed in the liquid phase is then reacted for one hour with 50 μl of avidin which, during this second incubation period, binds the biotin-Matrix-allergen-IgE complex and also the biotin on the coated-tube, thereby forming a bridge between the solid phase and the reacted IgE. IgE antibodies not specifically bound to the biotin-coated tube are removed by washing the test tube. The amount of specific IgE that is immobilized on the tube is quantitated by adding 200 μl of horseradish peroxidase-labeled monoclonal anti-IgE antibodies (described above). This reaction is allowed to proceed for one hour, after which excess enzyme is washed away. 500 μl of enzyme substrate is now added and allowed to react with the bound enzyme for fifteen minutes, producing a color directly proportional to the amount of endogenous IgE specific for the allergen. A semi-quantitative result in terms of a Class number is assigned to the sample with respect to the chosen allergen by comparing the color generated in the sample tube with the color produced in a series of standards (designated A, B, C and D) processed in the same assay against the reference allergen, Timothy grass.

RESULTS

Direct Labeling, Versus the AlaSTAT Method

Table 1 summarizes the results obtained when three different allergen extracts were subjected to direct labeling with biotin using NHS-biotin [14]. The table also shows the results obtained with the indirect labeling method employed in the AlaSTAT system. The AlaSTAT method, as is evident from the results shown in Table 1, is superior to the direct method by at least ten orders of magnitude in terms of its ability to bind allergen-specific IgE.

Table 1

Comparison of direct labeling of allergens vs. AlaSTAT method

	Direct labeling	AlaSTAT method
	Absorbance at 490nm	
Bermuda grass	0.190	1.984
Meadow grass	0.955	>2.500
D. pteronyssinus	0.516	>2.500

AlaSTAT Reference Allergen and Calibration

Timothy grass was selected as the AlaSTAT reference allergen. A pool of atopic serum to Timothy grass, classified by the Pharmacia's Phadebas-RAST kit as Class IV, was diluted in a non-atopic serum 1:5, 1:25 and 1:50 (representing Classes III, II and I respectively) and assayed with the AlaSTAT system as described above for Timothy grass. These dilutions are arbitrary and are only intended to simulate the calibration already established for the Phadebas-RAST. The results, which are summarized in Table 2, demonstrate the linearity of the AlaSTAT calibration curve.

Table 2

AlaSTAT Scoring and Calibration Values with the reference allergen, Timothy grass

Calibrator	Class	Absorbance (490nm)
D	1	0.077
C	2	0.121
B	3	0.470
A	4	1.925

DISCUSSION

The methods and results presented in this study are representative of our efforts to develop an allergen-specific IgE assay system that circumvents the problems associated with traditional methods of allergy testing, such as skin testing and RAST-type assays in which allergens are attached to a solid phase. Several studies have shown that allergosorbent assays have a tendency for non-specific IgE-binding which affects their reliability [4-9]. This observation has been confirmed in a separate study presented elsewhere in this publication (Alaba and El Shami).

Because of the heterogeneous nature of allergen extracts, there are technical difficulties in producing solid-phase allergens with a high degree of consistency from batch to batch. In addition to problems of non-specific IgE binding, RAST and RAST-like methods require long reaction times and are not easily automated. A logical alternative to this is the development of a liquid-phase assay system where the reaction between allergen and serum IgE proceeds in liquid phase. The key, however, is to devise a method to effectively label all allergen molecules and to do so without destroying their specific IgE-binding sites. Given this, the advantages of liquid-phase kinetics would be achieved, and allergen molecules would still have conformations close to their native states, thus making them maximally available to bind to specific IgE antibodies.

Recently, Aalberse *et al.* reported the use of hapten-modified allergens as an alternative to solid-phase-coupled allergens in a RAST-type assay [15]. In that study allergens were labeled directly with a hapten, trinitrophenol (TNP). We have shown that such direct labeling of allergens is not adequate for the level of sensitivity and specificity required for allergen-specific IgE assays (Table 1). The problem of sensitivity becomes critical when a large number of allergen molecules fail to get labeled and must compete with labeled allergens for the binding of specific IgE antibodies. There is, therefore, a problem of maintaining the batch-to-batch consistency of labeled allergens. The method presented in the present study was designed to circumvent these problems and to allow the presentation of allergen molecules to serum specific IgE antibodies in conformation close to their native states.

The AlaSTAT specific IgE assay is calibrated to the Phadebas-RAST and employs a similar classification system. These assays are semi-quantitative. We find that this classification system is adequate so long as it does not deviate from linearity. Results by the AlaSTAT system correlate very well with those by Phadebas-RAST except in a few cases where the discrepant results are attributable to non-specific IgE binding in the RAST, as confirmed by Western blot analysis (Alaba and El Shami, this publication).

The AlaSTAT technology offers the following significant advantages over other *in vitro* allergen-specific IgE assay systems:

1. It is a liquid-phase assay that uses a soluble backbone or polymer for linking allergen molecules, such that their spatial orientation closely approximates that of the physiological situation. Furthermore, the identifiable ligand (biotin) is not present on allergen molecules but on the soluble backbone, thus helping to ensure that IgE-binding sites on allergen molecules are preserved.

2. Because of the use of a soluble backbone and amino acid copolymer, an amplification system is achieved that helps to drive the kinetics

of the liquid-phase reaction and increase the possible number of immunocomplexes that could be immobilized on a solid support. The overall result of this amplification is that the AlaSTAT assay is complete in 3½ hours or less without loss of sensitivity.

3. The liquid format of the AlaSTAT makes it convenient for either manual performance or complete automation.

REFERENCES

1. Ishizaka, K., Ishizaka, T. and Hornbrook, M.M. (1966). Physico-chemical properties of human reaginic antibody. IV. Presence of a unique immunoglobulin as a carrier of reaginic activity. J. Immunol., 97, 75.
2. Johansson, S.G.O. and Bennich, H. (1967). Studies on a new class of human immunoglobulins. I. Immunological properties. In: Nobel Symposium 3 (ed. J. Killander), Stockholm, Almqvist & Wiksell, p.193.
3. Wide, C., Bennich, H. and Johansson, S.G.O. (1967). Diagnosis of allergy by an *in vitro* test for allergen antibodies. Lancet, ii, 1105.
4. Adkinson, N.F. (1981). The radioallergosorbent test in 1981 - limitations and refinements. J. Allergy Clin. Immunol., 67, 87.
5. Moroz, L.A. (1982). RAST for diagnosis of IgE-mediated disease. J. Allergy Clin. Immunol., 70, 315.
6. Adkinson, N.F. (1982). J. Allergy Clin. Immunol., 70, 316.
7. Platts-Mills, T.A.E. (1984). Diagnostic tests: laboratory techniques in immediate hypersensitivity. In: Allergy (ed. M.H. Lessof), New York, John Wiley & Sons, p.107.
8. van Toorerenbergen, A.W., Gerth van Wijk, R., Dieges, P.H. and Leijnse, B. (1987). Irrelevant IgE binding to cellulose disks in RAST. Int. Arch. Allergy Appl. Immunol., 83, 436.
9. Hamilton, R.G. (1987). The fluorescent allergosorbent test (FAST): performance in comparison to the RAST. J. Clin. Immunoassay, 10, 93.
10. Takatsu, K. and Ishizaka, K. (1975). Reaginic antibody formation in the mouse. VI. Suppression of IgE and IgG antibody response to ovalbumin following the administration of high dose urea-denatured antigen. Cell. Immunol., 20, 276.
11. Kishimoto, T. and Ishizaka, K. (1973). Regulation of antibody response *in vitro*. V. Effect of carrier-specific helper cells on generation of hapten-specific memory cells of different immunoglobulin classes. J. Immunol., 111, 1.
12. Kishimoto, T. and Ishizaka, K. (1974). Regulation of antibody response *in vitro*. VIII. Multiplicity of soluble factors released from carrier-specific cells. J. Immunol., 112, 1685.
13. Okudaira, H. and Ishizaka, K. (1973). Reaginic antibody formation in the mouse. III. Collaboration between hapten-specific memory cells and carrier-specific helper cells for secondary anti-hapten antibody formation. J. Immunol., 111, 1420.
14. Ishizaka, I. and Adachi, T. (1976). Generation of specific helper cells and suppressor cells *in vitro* for the IgE and IgG antibody responses. J. Immunol., 117, 40.
15. Aalberse, R.C., van Zoonen, M., Clemens, J.G.J. and Winkel, I. (1986). The use of hapten-modified antigens instead of solid-phase-coupled antigens in a RAST-type assay. J. Immunol. Methods, 87, 51.
16. Bethell, G.S., Ayers, J.S., Hancock, W.S. and Hearn, M.T.W. (1979). A novel method of activation of cross-linked agaroses with 1,1'-carbonyl diimidazole which gives a matrix for affinity chromatography devoid of additional charged groups. J. Biol. Chem., 254, 2572.
17. Sundberg, L. and Porath, J. (1974). Preparation of adsorbents for bio-

specific affinity chromatography. 1. Attachment of group-containing ligands to insoluble polymers by means of bifunctional oxiranes. J. Chromatography, 90, 87.

18. Engvall, E. and Perlmann, P. (1971). Enzyme-linked immunosorbent assay (ELISA): quantitative assay of immunoglobulin G. Immunochemistry, 8, 871.
19. Nakane, P.K. and Kawaoi, A. (1974). Peroxidase-labeled antibody: a new method of conjugation. J. Histochem. Cytochem., 22, 1084.
20. Diagnostic Products Corporation (1988). AlaSTAT Allergen-specific IgE (product insert). Los Angeles, CA.

Evaluation of Non-specific IgE Binding: Comparison of two *in vitro* Allergen-specific IgE Assays

Olusola Alaba and A. Said El Shami
Diagnostic Products Corporation, Los Angeles, CA 90045, USA

ABSTRACT

The likely incidence of non-specific IgE interference during the *in vitro* determination of allergen-specific IgE was investigated using the RAST, a solid-phase based assay format, and AlaSTAT, a new liquid-phase test system. Atopic sera with total IgE values varying between 1,220 IU/ml and 30,000 IU/ml were tested for a variety of allergens. The overall agreement within one class difference with Pharmacia RAST was 90%. One serum was found to be positive (class 3) in the RAST but negative in the AlaSTAT to dog dander and two other sera tested positive (classes 2 and 3) for *Aspergillus fumigatus* in the RAST but showed no binding of specific IgE in the AlaSTAT. When these sera were tested with the Western blot technique we were not able to demonstrate the presence of specific IgE to either dog dander in the first serum or *Aspergillus fumigatus* in the other AlaSTAT-negative sera. The AlaSTAT system was further evaluated for non-specific IgE binding by using human serum albumin as an allergen according to the AlaSTAT chemistry. Sera with total IgE of up to 30,000 IU/ml were tested and found to be non-reactive to human serum albumin.

KEYWORDS

AlaSTAT; RAST; allergen-specific IgE; non-specific binding; competitive RAST-inhibition; Western blotting.

INTRODUCTION

The *in vitro* determination of IgE antibodies specific for one of hundreds of known allergens has been shown to be a valuable diagnostic tool for human allergic disease [1-5]. In the past, the trend has been towards the development of solid phase format assay systems where the allergenic molecules are attached directly to a solid support such as paper disk, microtiter plate or cellulose thread. One common feature with all these assay systems is the obligatory reaction of serum IgE with allergen on the solid phase. Several reports have, however, shown evidence of non-specific IgE sticking to these solid phases especially when the serum total IgE is over 1,000 IU/ml [6-11]. This problem can become very severe with serum samples from patients with atopic eczema or helminthic infections where the total serum IgE can be very high. The problem is further compounded by the unsuitability for skin testing of such patients for fear of anaphylactic shock or the state of skin condition. These factors underscore the importance of having a very reliable *in vitro* assay alternative. It has been generally accepted that the contribution of total IgE to non-specific binding in RAST assays is largely a function of the allergosorbent [8,10,11]. Recently, we developed a new assay system, AlaSTAT (described elsewhere in this publication), where the allergenic molecules, presented in a liquid matrix, are reacted with a serum sample in the liquid phase. Since the decision about immunotherapy may have to depend largely on the result of *in vitro* diagnostic tests for specific IgE we decided to assess the level of non-specific IgE interference in the AlaSTAT system. In the present study the performance of the AlaSTAT is contrasted to the allergosorbent test with regard to the influence of serum total IgE.

MATERIALS AND METHODS

AlaSTAT Specific IgE Test

AlaSTAT allergen-specific IgE test system (Diagnostic Products Corporation, Los Angeles, CA, 90045) [12] is a liquid-phase amplified immunoenzymometric assay designed for the semi-quantitative determination of allergen-specific IgE in serum. Unlike other commercially available allergen-specific IgE tests the AlaSTAT employs a unique concept of covalently attaching allergen molecules to a liquid matrix to which an identifiable ligand is also attached. In the performance of the AlaSTAT test, 50 µl of the patient serum sample is first reacted, in a ligand-coated tube, for one hour with 100 µl of specific modified allergen in the liquid phase. During this incubation period IgE specific for the given allergen, if present, binds to the modified allergen. The allergen-IgE complex thus formed in the liquid phase is then reacted for one hour with 50 µl of a multivalent antiligand directed at the ligand attached to the liquid matrix to which the allergen is also covalently attached. During this second incubation, the antiligand binds to the ligand-labeled matrix-allergen-IgE complex and also to the ligand-coated tube thereby forming a bridge between the solid phase and the reacted IgE. Following a washing step to remove any IgE not specifically bound to the allergen in the liquid matrix, 200 µl of Horse radish peroxidase-labeled monoclonal anti-IgE (mixture of two DPC monoclonal antibodies, 2G4 and 1C5) is added to the *in situ* immobilized specific IgE now on the solid phase. This reaction is allowed to proceed for one hour and excess enzyme is washed away. 500 µl enzyme substrate is now added and allowed to react with the bound enzyme for fifteen minutes, producing a color directly proportional to the amount of endogenous IgE specific for the allergen. A semi-quantitative result in terms of a class number is assigned to the sample with respect to the chosen allergen by comparing the final color with the color produced in a series of standards processed in the same assay against a reference allergen.

AlaSTAT Total IgE

Total IgE was determined according to the AlaSTAT protocol (Diagnostic Products Corporation, Los Angeles, CA 90045) [13]. AlaSTAT total IgE is an amplified immunoenzymometric assay based on ligand-coated tubes and three liquid monoclonal anti-IgE antibodies (1C5, 2G4 and 1F1, two are enzyme-labeled, and the third is ligand-coupled). IgE in the patient sample is captured between the enzyme-labeled and the ligand-labeled monoclonals. Separation is achieved by bridging the ligand-coated tube and the ligand-coupled anti-IgE-IgE complex upon the addition of an anti-ligand. Enzyme substrate is then added and the colored product is directly related to the IgE concentration in the patient sample.

Allergen Extracts

The allergenic extracts used in the present studies are obtained from Greer Laboratories, Lenoir, North Carolina, 28645. Each lot of allergenic extract is assessed by RAST-inhibition using Pharmacia disks and a pool of atopic sera from individuals allergic to the allergen. Only extracts that totally inhibited IgE binding to Pharmacia disks were used.

RAST-inhibition Assays

Competitive RAST-inhibition assays were performed using Pharmacia allergosorbent disks and reagents (Pharmacia, Diagnostics Division, Piscataway, NJ 08854). Briefly, 25 μl of serially diluted allergen extract was incubated with 25 μl of specific atopic serum pool in a microtiter well. After one hour on a shaker at room temperature one allergosorbent disk was added to each well and incubation continued overnight at 4°C. Following three washes with 200 μl of wash buffer, 50 μl of anti-IgE-enzyme conjugate was added and the reaction continued for two hours at room temperature. Excess anti-IgE-enzyme conjugate was removed by washing three times with wash buffer. Substrate was added and the colored product generated was measured at 420 nm after addition of the stop solution. The intensity of the color generated would be inversely proportional to the degree of similarity between the components on the RAST disk and the extract tested.

SDS-PAGE, Electrotransfer and Immunochemical Detection

SDS-polyacrylamide gel electrophoresis (12.5%) was performed according to the method of Laemmli [14]. Allergenic extracts (20-30 μg) were boiled in 0.0625 M Tris, 2% SDS, 5% 2-mercaptonethanol, 10% glycerol, pH 6.8. The resolved components were electrotransfered for 16 hours, 4°C (30 volts, 0.13 amps.) to nitrocellulose sheet (0.45 μ, Schleicher and Shuell, Keene, NH, 03431). Transfer buffer consisted of 25 mM Tris, 192 mM glycine, 20% methanol. After transfer to nitrocellulose, the membrane was blocked by incubation with 3% Blotto (3 g non-fat milk in 100 ml of PBS) for one hour at room temperature. The membrane was washed three times with PBS and then 4 mm wide strips of the membrane were probed with undiluted pools of specific atopic sera (minimum RAST class 3) overnight, rocking, at 4°C. Following three times washing with PBS, the membrane strips were incubated for 2 hours at room temperature with a mixture of two monoclonal anti-IgE antibodies (1C5 and 2G4 produced at DPC and used in the AlaSTAT total IgE assay) diluted in 0.5% Blotto, 0.05% thimerosal in PBS. The strips were then washed and further reacted with goat anti-mouse IgG antibodies conjugated with alkaline phosphatase (Bio-Rad, Richmond, California) for one hour at room temperature. Resolved allergenic components in the extract were visualized by the addition of enzyme substrate (1.5%, 5-Bromo-4-Chloro-3-indoyl Phosphate and 3% Nitro Blue Tetrazolium in 0.1 M sodium bicarbonate, pH 8.1).

RESULTS

Comparison of AlaSTAT and RAST

In the initial phase of development of AlaSTAT the Pharmacia RAST system was used as a reference to screen the atopic sera in our collection. The RAST class obtained for a serum sample would later be checked against the AlaSTAT class with respect to a specific allergen. With at least 200 different sera and 104 different allergens comprising of grasses, trees, weeds, epidermals, molds, mites and foods we observed an overall concordance with Pharmacia RAST of 88% and overall agreement within one class was 90%. In two of these serum samples we observed class 3 for RAST but class 0 for AlaSTAT and with another sample a RAST class 2 and AlaSTAT class 0. We therefore decided to investigate these sera further using an independent technique such as Western blotting. Table 1 summarizes the comparison results of RAST and AlaSTAT with respect to three patients and dog dander allergen. These sera (1009, 394 and 1010) were used to probe the PAGE-SDS-resolved components of dog dander extract. The dog dander extract used for this Western blot experiment was from the same lot that was used in the AlaSTAT kit. The results are presented in Fig. 1. We could not demonstrate the presence of IgE specific for dog dander in serum 1009 whereas the specific IgE for dog dander could be demonstrated in serum samples 394 and 1010).

Table 1. AlaSTAT and RAST values for dog dander

Serum	Total serum IgE (IU/ml)	AlaSTAT class	RAST class
1009	5,090	0	3
394	8,800	3	4
1010	9,020	3	4

To determine if the dog dander allergen extract components used for the Western blot experiment were similar to the components present on the RAST dog dander disk we performed RAST inhibition tests using a pool of atopic sera from skin test positive individuals to dog dander. The results of the competitive inhibition test presented in Fig. 2 show that the dog dander extract was able to inhibit binding of dog dander-specific IgE to the RAST disk (E5), which suggests that the extract contained components similar to those present on the allergosorbent disk. From the results presented in Fig. 2 we concluded that the dog dander extract used in the AlaSTAT assay was not inferior to the components coupled to the allergosorbent disk. We then repeated the RAST inhibition assay using the same dog dander extract but substituted serum 1009 for the pool of atopic sera used in the above experiment. The result is also summarized in Fig. 2. No inhibition of IgE binding could be demonstrated with the dog dander extract using this serum 1009. It is evident from these RAST inhibition and the Western blot results that serum 1009 does not contain specific IgE to dog dander.

Similar RAST inhibition experiments were carried out with another patient serum (serum 415) which gave a RAST class 3 but was class 0 in the AlaSTAT

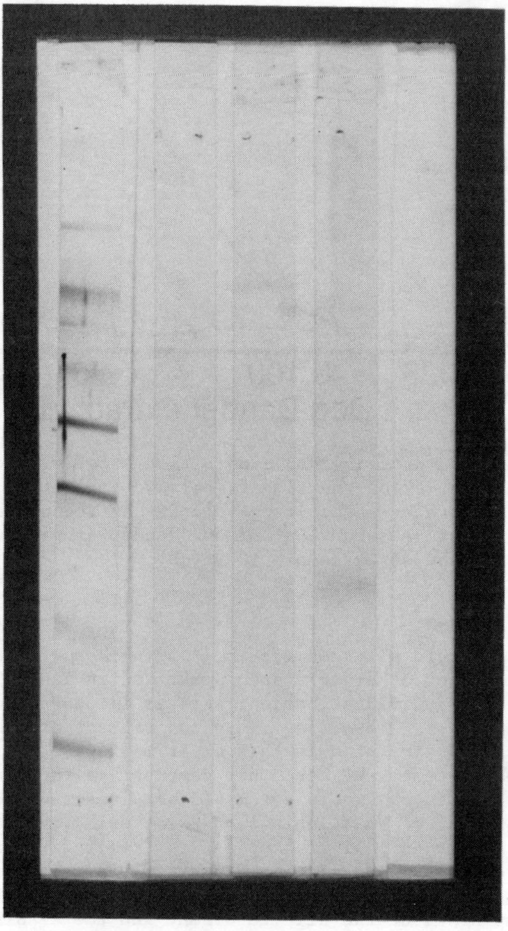

Fig. 1. Nitrocellulose electroblotting of dog dander extract separated by SDS-PAGE (12.5%) and immuno-probed with serum samples 1009 (lane 2), 394 (lane 3) and 1010 (lane 4). Lane 1 is SDS-PAGE of molecular weight markers stained with Auro dye protein stain (116 kd, 97.4 kd, 66 kd, 45 kd, 36 kd, 29 kd, 20.1 kd and 14.2 kd).

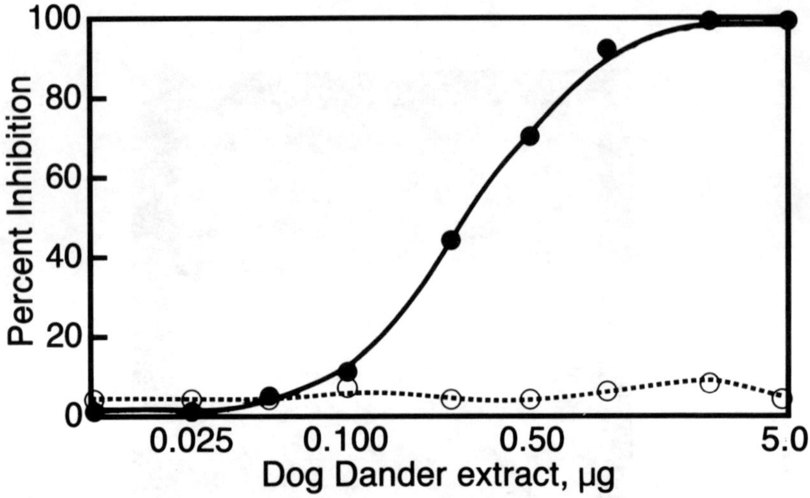

Fig. 2. Competitive RAST-inhibition assay with dog dander allergosorbent disk (E5) and inhibition of dog dander-specific IgE binding by dog dander extract using serum pool of dog dander-atopic subjects (●) and serum sample 1009 (O).

with regard to the mold allergen *Aspergillus fumigatus*. The RAST results for serum 415 and four other sera are summarized in Table 2. The results of Western blots using individual patient sera as immuno-probes of SDS-PAGE-resolved components of *Aspergillus fumigatus* extract are shown in Fig. 3. The *Aspergillus fumigatus* used in this experiment was the same lot of extract used in the AlaSTAT test. We could not demonstrate the presence of specific IgE for *Aspergillus fumigatus* in serum 415 nor in serum 5163. Figure 4 summarizes the result of RAST inhibition assay using the same *Aspergillus fumigatus* extract with a pool of atopic sera from individuals that were skin test positive for this mold. The mold extract was able to inhibit the binding of IgE to RAST *Aspergillus fumigatus* disk (M3) suggesting similarity between the components on the allergosorbent and the AlaSTAT mold extract. Contrary to the results obtained when a pool of atopic sera was used, no inhibition of IgE binding was seen with the 415 and 5163 serum samples in the presence of the mold extract (Fig. 4).

Reaction of Sera 1009, 5163 and 415 with Other Allergens

To ascertain that the IgE antibodies from these three individuals were recognized in the AlaSTAT assay we tested for other allergens to which these patients had skin tested positive using both AlaSTAT and RAST (Table 3) and confirmed the results with Western blots (Fig. 5). It was clear from these experiments that the IgE antibodies from these patients were reconized by the monoclonal anti-IgE antibodies used in AlaSTAT and the studies presented above. The lack of evidence for dog dander specific IgE in 1009 serum and *Aspergillus fumigatus*-specific IgE in sera 415 and 5163 would therefore imply

Table 2. AlaSTAT and RAST values for *Aspergillus fumigatus*

Serum	Total serum IgE (IU/ml)	AlaSTAT class	RAST class
415	2,800	0	3
1214	10,640	2	3
5163	1,220	0	2
1367	8,970	1	2
394	8,800	1	2

Table 3. AlaSTAT and RAST for *D. pteronyssinus* and Timothy grass

Allergen	Serum	Total serum IgE (IU/ml)	AlaSTAT class	RAST class
D. pteronyssinus	1009	5,090	4	4
	415	2,800	4	4
Timothy grass	5163	1,220	4	4

that those specific IgE antibodies are simply absent in these sera.

Determination of Non-specific IgE Binding in AlaSTAT Allergen-specific IgE Assay

From the results of the studies presented above we could demonstrate that non-specific IgE binding was possible in RAST with certain serum samples. We investigated the possibility of non-specific IgE binding in the AlaSTAT system by using human serum albumin as an allergen in the AlaSTAT format and testing it with a variety of serum samples with total IgE values of up to 30,000 IU/ml. With each of the serum tested in these experiments we have included allergens to which the patient had tested positive by skin tests as positive controls. These serum samples were similarly tested for other allergens against which the patients showed no positive skin reaction. The results summarized in Table 4 clearly indicate that serum samples with total IgE of up to 30,000 IU/ml do not interfere in the AlaSTAT assay.

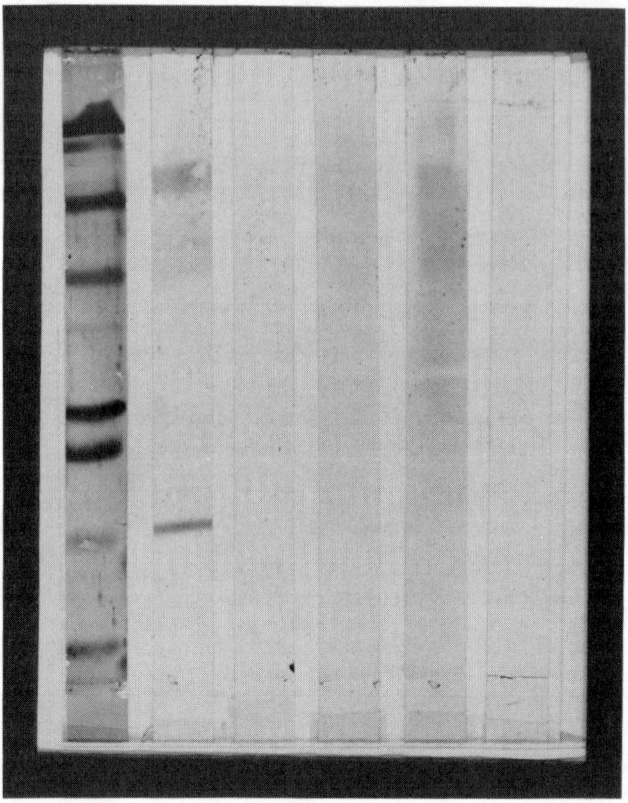

Fig. 3. Nitrocellulose electroblotting of *A. fumigatus* mold extract separated by SDS-PAGE (12.5%) and immuno-probed with serum samples 1214 (lane 2), 415 (lane 3), 394 (lane 4), 1367 (lane 5) and 5163 (lane 6). Lane 1 is SDS-PAGE of molecular weight markers stained with Auro dye protein stain (same as in Fig. 1).

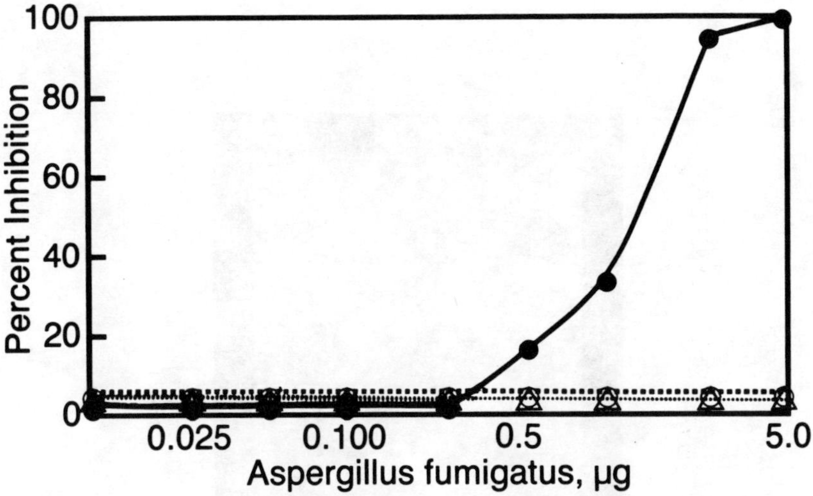

Fig. 4. Competitive RAST-inhibition assay with *A. fumigatus* allergosorbent disk (M3) and inhibition of *A. fumigatus*-specific IgE by *A. fumigatus* mold extract using serum pool of *A. fumigatus*-atopic subjects (●), serum sample 415 (O) and serum sample 5163 (Δ).

Discussion

Several studies [6-11,13] have been published which suggest that non-specific IgE interference with the allergosorbent test method can be a problem, especially where high total serum IgE values are encountered. The studies that we have presented above have examined the problem of non-specific IgE interference using two commercial allergen-specific IgE assay systems which utilize different technologies to measure a similar serum component. In the AlaSTAT system the allergen is presented to serum IgE in a liquid format. After the allergen-IgE complex has formed an anti-ligand is added to effect a separation of the IgE bound to the liquid allergen on to a ligand-coated tube from the unreacted IgE. This is crucial but by no means the only aspect of the reaction that helps to prevent non-specific IgE binding in the AlaSTAT system. With all allergosorbent techniques serum IgE antibodies, both specific and non-specific IgE, must contact the allergen directly on a solid matrix. From the results presented above non-specific binding of IgE can occur with some serum samples when allergens are presented in a solid-phase format. With most of the serum samples tested the agreement between the AlaSTAT and Pharmacia RAST within one class difference is about 90%. As reported above, however, certain serum samples give higher than one class difference between the AlaSTAT and the RAST. The evidence presented would seem to clearly indicate that in the three cases studied the discordant results can be explained by non-specific binding of IgE. The RAST inhibition studies coupled with the results of the Western blots are obvious proofs that in the cases presented there could be no doubt that non-specific IgE binding to the allergosorbent was a significant source of error. In studies

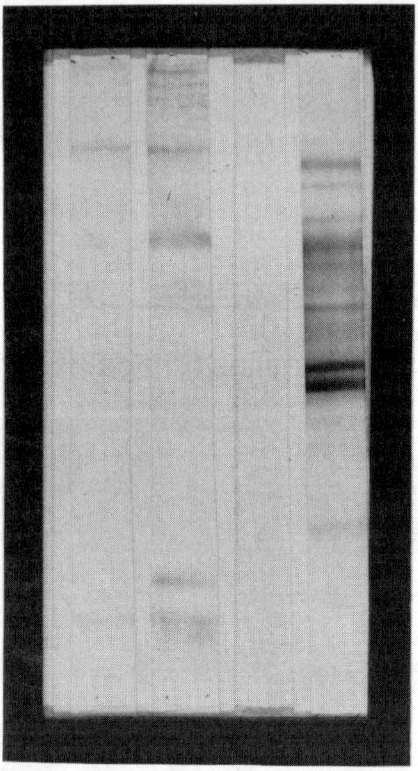

Fig. 5. Nitrocellulose electroblotting of mite *D. pteronyssinus* (lanes 1 and 2) and Timothy grass (last lane) extracts separated by SDS-PAGE (12.5%) and immuno-probed with serum sample 1010 (lane 1), 415 (lane 2) and 5163 (lane 3).

Table 4. Determination of non-specific IgE binding in AlaSTAT allergen specific IgE assay

Serum	Total serum IgE (IU/ml)	AlaSTAT allergen	IgE binding (absorbance)
1423	1,800	Human serum albumin	0.006
		Common ragweed	1.09
		Pecan	0.004
		Acacia	0.58
1010	9,020	Human serum albumin	-0.004
		Walnut	1.26
		D. pteronyssinus	2.50
		Codfish	2.38
		Lamb	0.001
1214	10,640	Human serum albumin	-0.001
		Shrimp	0.61
		Cat epithelium	-0.003
		A. fumigatus	0.43
136	30,000	Human serum albumin	0.008
		Milk	0.86
		Timothy grass	2.45

Absorbance = A(490 nm) of atopic serum - A(490 nm) of normal, nonatopic serum.
The cut-off absorbance value for the D calibrator (class 1) in this experiment was A(490) = 0.074.

of this type where two systems are compared it is important to be certain that the source of discrepancy was not the difference in the allergenic extracts used in the two different systems. We have shown that the AlaSTAT extracts were able to completely prevent specific IgE binding to the RAST disk when pools of sera from atopic patients were used in competitive RAST inhibition assays. These results suggest to us that the components present in the extracts were at least similar to those coupled to the allergosorbent. Competitive RAST inhibition assay continues to be a very useful method of comparing the components present in different allergenic extracts [16-18].

Certain investigators have attributed the reason for the non-specific binding of IgE that they observed with the RAST to be due to possible cross-reactivity of specific IgE with the cellulose structure of the allergosorbent [10,15]. Although we no not disagree that with certain serum samples this may well be true, it would not, however, explain a similar observation when a microtiter plate is the allergosorbent [11]. We believe that non-specific binding of IgE would be a feature of all solid-phase based assays. This feature becomes rather pronounced as the total serum IgE value increases beyond 2,000 - 5,000 IU/ml depending on the type of allergosorbent, which would make the quantitation of specific IgE rather difficult in patients with atopic eczema or helminthic infection. Furthermore, we have often observed that increasing the sample size of those serum samples with high total serum IgE also would increase the amount of non-specific IgE binding when tested in the RAST system. The AlaSTAT system does not suffer from this limitation of non-specific IgE binding. We have studied serum samples with total IgE levels as high as 30,000 International Units/ml and no non-specific IgE binding could be demonstrated. We attribute this lack of non-specific binding of IgE in the AlaSTAT to the advantages inherent in the liquid-phase based reaction format, and also to the fact that the binding of allergen-complexed IgE and free IgE to the ligand-coated tube (solid phase) is prevented until the addition of a specific multivalent anti-ligand.

REFERENCES

1. Stenius, B., Wide, L., Seymour, W.M., *et al.* (1971). Clinical significance of specific IgE to common allergens. Clin. Allergy, 1, 37.
2. Aas, K. and Johansson, S.G.O. (1971). The radioallergosorbent test in the *in vitro* diagnosis of reaginic allergy. A comparison of diagnostic approaches. J. Allergy Clin. Immunol., 48, 134.
3. Norman, P.S., Lichtenstein, L.M. and Ishizaka, K. (1973). Diagnostic tests in ragweed hay fever: a comparison of skin tests, IgE antibody measurements and basophil histamine release. J. Allergy Clin. Immunol., 52, 210.
4. Berg, T., Bennich, H. and Johansson, S.G.O. (1971). *In vitro* diagnosis of atopic allergy. I. A comparison between provocation tests and the radioallergosorbent test. Int. Arch. Allergy Appl. Immunol., 40, 770.
5. Evans, R., Reisman, R.E., Wypych, J.I. and Arbesman, C.E. (1972). An immunologic evaluation of ragweed-sensitive patients by newer techniques. J. Allergy, 49, 285.
6. Adkinson, Jr., N.F. (1981). The radioallergosorbent test in 1981 - limitations and refinements. J. Allergy Clin. Immunol., 67, 87.
7. Moroz, L.A. (1982). RAST for diagnosis of IgE-mediated disease. J. Allergy Clin. Immunol., 70, 315.
8. Adkinson, Jr., N.F. (1982). J. Allergy Clin. Immunol., 70, 316.
9. Platts-Mills, T.A.E. (1984). Diagnostic tests: laboratory techniques in immediate hypersensitivity. In: Allergy (Ed. M.H. Lessof), John Wiley & Sons Ltd, pp.107-125.
10. van Toorenenbergen, A.W., Gerth van Wijk, R., Dieges, P.H. and Leijnse, B. (1987). Irrelevant IgE binding to cellulose disks in RAST. Int. Archs. Allergy Appl. Immunol., 83, 436.
11. Hamilton, R.G. (1987). The fluorescent allergosorbent test (FAST): performance in comparison to the RAST. J. Clin. Immunoassay, 10, 93.
12. AlaSTAT Allergen-specific IgE product insert, Diagnostic Products Corporation, April 1988.
13. AlaSTAT Total IgE product insert, Diagnostic Products Corporation, April 1988.
14. Zeiss, C.R., Grammer, L.C. and Levitz, D. (1981). Comparison of the radioallergosorbent test and a quantitative solid-phase radioimmunoassay for the detection of ragweed-specific immunoglobulin E antibody in patients undergoing immunotherapy. J. Allergy Clin. Immunol., 67, 105.
15. Aalberse, R.C., Koshte, V. and Clemens, J.G.J. (1981). Immunoglobulin E antibodies that crossreact with vegetable foods, pollens, and hymenoptera venom. J. Allergy Clin. Immunol., 68, 356.
16. Gleich, G.J., Larson, J.B., Jones, R.T. and Baer, H. (1974). Measurement of the potency of allergy extracts by their inhibitory capacities in the radioallergosorbent test. J. Allergy Clin. Immunol., 53, 158.
17. Schroder, H. and Yman, L. (1980). Standardization of the RAST inhibition assay. Allergy, 35, 234.
18. Baldo, B.A. (1983). Standardization of allergens. Allergy, 38, 535.

Discussion: Session Four

Chairman: Terry Merrett

1. Questions to Bruce Mitchell

Sola Alaba. What could be the possible reason for the artefactual specific IgE reactivity to ovalbumin and β-lactoglobulin in these subjects?

Bruce Mitchell. I should emphasize that because there was a good correlation between the skin tests and IgE-RAST to these foods I believe IgE antibodies were really present. My point is that in the presence of high total serum IgE levels, such as frequently occurs in cases of atopic dermatitis, the antigen binding assay produces artefactually high results. Our explanation revolves around the specific IgG:IgE ratio being so high for foods, much higher than for inhalants, that it is difficult to maintain antigen excess without diluting the test serum to a level where the specific IgE is no longer detectable.

Alec Sehon. It has long been recognized that IgG antibodies are produced in response to allergenic stimuli. In the 1950's, well before the discovery of IgE, chromatography was used by Rosa Augustin, Julius Gordon, Louis Perelmutter and several other people to demonstrate these "blocking IgG antibodies". When the RAST procedure was developed, people forgot all about measuring IgG antibodies in these patients. Now had you used as an immunosorbent monoclonal antibodies to IgE linked to particles, don't you think you would have had fewer complications and no interference from total IgE?

Bruce Mitchell. To overcome the problems we saw in this relationship between immunoglobulins E and G there are many assay systems we could have pursued, and had we decided to do so your suggestion is certainly one which would have merited attention.

Farouk Skakib. I think there is one snag which we should bear in mind when applying antigen binding assays, and that is to be wary of monovalent antibodies. For example, your assay will underestimate the concentration of IgG4.

Bruce Mitchell. Yes, we were aware that our assay might have been underestimating IgG4 concentrations because of valency considerations. However, this made no difference to our conclusions that IgG4 antibodies against food

Advances in the Biosciences Vol. 74

Printed in Great Britain

antigens constituted a very much greater proportion of total IgG specific antibody than did IgG4 antibodies against inhalants. These differences are presumably due to the amount and rate of exposure as well as the route of administration.

Sheldon Spector. I was intrigued by the disproportionate amount of IgG4 with food allergens compared to the inhalants, and the likelihood that it was due to the route of administration. It would be interesting to know if this hypothesis has already been tested by the various clinical studies of oral immunotherapy; perhaps the ingestion of inhalants in these studies resulted in a disproportionate amount of inhalant specific IgG4 antibody? Another point is that an insect venom sting also results in a disproportionate IgG4 level, and this is given by yet another route. Have you looked at any of these patients?

Bruce Mitchell. No, we have restricted ourselves to patients with atopic dermatitis.

Terry Merrett. I should just like to comment that in a population study as long ago as 1982 Mike Burr and I showed that IgG4 antibodies to normal every day foods were as commonly found among non-atopics as among atopics. The reverse was true for inhalants, i.e., inhalant specific IgG4 antibodies were commonly found among atopics but not non-atopics. A short while afterwards Barry Kay sent us sera from asthmatics who were divided into groups which in bronchial provocations exhibited (a) an immediate response, (b) a dual reaction and (c) an isolated late response - in no instance was IgG4 antibody to the provoking allergen found without there also being IgE antibody, although the reverse happened. These results point to the irrelevance of IgG4 antibody measurements as an aid to the management of allergic patients, and so it is rather disturbing that several diagnostic companies are geared to market IgG4 diagnostic kits. Sheldon you may be interested to know that together with Gordon Reeves we found positive insulin-specific IgG4 to be present in the circulation of diabetics within three months of commencing insulin treatment.

2. Questions to Louis Perelmutter

John Warner. It is worth reviewing the "gold" standards for allergy diagnoses. Obviously provocation testing and clinical history assessment deserve consideration, but you also talked about the skin-prick test being the gold standard in several studies. However, the question a clinician asks is: "What is the meaning of a positive skin test, when there are 15% of the population who have positive skin tests without any disease?" Then what do we make of *in vitro* tests? If indeed skin tests were a gold standard then why do any *in vitro* tests at all. To exemplify the problem Joan Longbottom and I undertook to compare the *in vitro* test results obtained with MAST and RAST in a group of children with various problems. Interestingly, most specificities and sensitivities compared rather well, but cat and dog danders proved the exception because MAST produced fewer positives. However, when we reviewed the childrens' skin tests and clinical histories for animal danders, the MAST seemed to be more accurate than RAST. The irony was that the UK distributors of MAST were so concerned by the disagreement with RAST that they changed the composition of these animal allergosorbents so as to ensure that the two test systems produced equivalent results. It is my contention that clinical history and provocation responses are the reference tests.

Louis Perelmutter. I agree with your comments. However, when you raise the question of re-evaluating the skin test, you really have to challenge the whole system of the diagnosis of allergic disease.

Robert Hamburger. There is a little technical problem that bothers me Louis; it concerns your labelled allergen procedure. Some years ago we played with a similar test, but found so many problems with the "stickiness" of the solid phase for labelled allergens that we gave it up as a bad job. Did you have problems with positivities which you could not explain?

Louis Perelmutter. No, I didn't have these non-specific binding problems. One critical factor seems to be the amount of enzyme (horse radish peroxidase) attached to allergens, and another concerns the optimization of the amount of labelled allergen used in the test.

3. Questions to Said El Shami

Sami Bahna. Are there any clinical comparisons of this new method with other established procedures?

Said El Shami. Yes, a great deal of clinical comparison work has been done, but we have not done it ourselves. Dr Sheldon Spector has carried out a large comparison study of AlaSTAT versus skin-prick testing.

Terry Merrett. Sheldon has previously intimated to me that if called upon to do so he could present his data. While he gathers his thoughts could I perhaps ask you, Sami, what kind of clinical comparison would you most like to see? But please let us all bear in mind that there is no established "gold standard" for allergy diagnosis, since all procedures have their critics.

Sami Bahna. I would prefer to go for food allergy testing.

Terry Merrett. Where would you carry out the tests? Remember that some investigators claim that to test in the clinic does not necessarily reproduce the home environment where the symptoms occur - perhaps because they result as a mixture of food plus environmental antigens.

Sami Bahna. I am confident that the tests can be performed in the clinic.

Said El Shami. Dr Bahna, as newcomers in this field we have had no choice but to standardize our tests against the Pharmacia RAST. This is not to say that the RAST is properly standardized for all allergens, and I think Dr Alaba will show us at least one food allergen he has standardized.

Robert Hamburger. I just wanted to ask Said the same question I asked of Louis Perelmutter. Did you have some sera which gave you falsely positive results because of "stickiness"?

Said El Shami. We have examined correlation data for AlaSTAT and RAST which have been provided by Dr Abrahamson (who has done a lot of work in Sweden) and by Dr Spector, and there are many specimens which read at least one class lower in one of the tests - sometimes the lowest result is recorded by AlaSTAT and for other cases it is RAST. Sometimes even one test is negative while the other is positive, and because of the lack of a diagnostic gold standard we are usually unable to decide which test is providing the correct result. Dr Alaba will talk about this problem.

Terry Merrett. We should remember that AlaSTAT is a very different procedure from RAST, using as it does ligand labelled allergens in solution instead of allergosorbents as a solid phase. With two such different approaches it is reasonable to anticipate that they will produce different results on occasions. I see Dr Spector has his slides ready, which should show us just how well AlaSTAT performs in his clinic.

Sheldon Spector. This preliminary data shows first of all the concordance rate between AlaSTAT and RAST in approximately 5000 patient-allergen evaluations. When comparing first of all the same classification score the concordance was 81%, and this increased to 91% if we allowed one class difference in either direction. When comparing AlaSTAT with skin tests we found an overall concordance of 85%. This is a good concordance rate, but it did differ according to allergen; I think grasses showed the largest discrepancy followed by *D. farinae*. It is not exactly clear why this should have been so, unless it relates to differences in the source of allergen; we used Greer allergens for our skin testing, which I imagine to be different from Pharmacia's source.

Herbert Mansmann. What is the commercial source of allergen used by DPC?

Said El Shami. Greer Laboratories. Bob Esch does a good job of preparing allergen extracts for us which have been purified according to our requirements. These special batches are different from those which his Company provides for skin testing.

Herbert Mansmann. It is absolutely critical that the same allergens be used for the skin tests and for the *in vitro* system.

Terry Merrett. On a purely scientific basis of course you are correct, but in the real world we usually have to compare "apples with pears". Even if we could obtain extracts from the same manufacturer you will find differences from batch to batch, as Joan Longbottom showed this morning occurs with moulds. Furthermore the molecular cloning of allergens has clearly demonstrated that the allergenic content of pollen can vary considerably from season to season. Perhaps yours is not a practical proposition?

Herbert Mansmann. I would suggest that it is.

Terry Merrett. In that case, can I ask you Sheldon if you used the same specially processed Greer extracts which DPC have labelled with ligand?

Sheldon Spector. No, we used the commercial skin test extracts which are freely available. My objective was to see how well AlaSTAT performed compared with my routine skin testing procedure.

Herbert Mansmann. If you are going to do it that way, then I suggest that in order to be fair to Pharmacia, you ought to use the allergens which they are using on their discs. However, I admit that it might not be possible to obtain them.

Martin Chapman. I would just like to say a word in defence of the RAST procedure. We prepare the reagents in our laboratory from a variety of commercial sources. Our routine RAST for five allergens is not the Phadebas RAST although it uses cyanogen bromide activated discs. Incidentally, although much has been circulated to say that cyanogen bromide is a terrible thing for preparing leak-proof immunosorbents, we have found it to be very satisfactory. I can understand the need to look for alternative procedures for the measurement of IgG antibodies, but the IgE-RAST procedure works well in our hands and appears not to be dependent on using the same source of allergen as Pharmacia. Okay, it is labour intensive, so we just hire someone to do it.

Terry Merrett. I agree with your sentiments for the most part Martin, but must admit that Pharmacia supply some paper disc allergosorbents which we have been unable to mimic, presumably because our allergenic source material is inappropriate. Perhaps I could deflect the questions back to skin testing. John Warner earlier raised the point about positive skin tests in asymptomatic patients, and if I might term these "clinical false-positives", just to bait Jack Pepys, might we not suppose that T-cell receptors in the skin might be recognising different allergen epitopes to the T-cell receptors in the respiratory tract?

Jack Pepys. I wonder if I could just comment on that. The value for positive prick tests with common allergens in the asymptomatic population is anything between 30-40%. These skin tests are positive. Kaliner in studying such subjects has found that in association with a positive skin test reaction there are altered pharmacological responses that decrease adrenergic activity, increase alpha and increase cholinergic activities.

John Warner. On one of your slides I think you had either antigen or antibody attached to your matrix. Does this mean you are going to use the procedure to measure antigen concentrations?

Said El Shami. Yes we are, I got those slides from our patent application.

4. Questions to Sola Alaba

Terry Merrett. I was interested in the data you showed for some sera which were RAST positive but AlaSTAT negative to *Aspergillus fumigatus*. The obvious assumption to be made was that this was due to differences in the extracts used in the two tests. Indeed your extract was unable to demonstrate any IgE activity in the Western blot, but contrary to expectations it was able to inhibit the positive RAST. Have you any answers to this anomaly?

Sola Alaba. We were not able to demonstrate with our extract any IgE-binding activity in the serum in question but the same extract was shown to be quite good when a pool of *Aspergillus fumigatus*-positive serum was used. This suggests to us a lack of *A. fumigatus* specific IgE in the Western blot-negative serum.

Jack Pepys. First of all a question, then a comment. In the two sera that had apparent positives on the RAST for *Aspergillus fumigatus* and were negative in your test, what were their skin test reactions?

Sola Alaba. They were negative.

Jack Pepys. Good. Now to the point which may explain this anomaly. High total serum IgE levels go hand-in-hand with falsely positive RAST results - there is no doubt about that and I think that is what is happening here. The other point is purely anecdotal, but as it concerns the WHO International IgE Reference it may be worth relating. The original reference standard had been supplied by Gunnar Johansson from Ethiopian school children, but when it was found to contain hepatitis antigen another source was required. To cut a long story short we supplied the Medical Research Council at very short notice with 4 litres of serum taken from our patients with allergic aspergillosis - the pool had a total IgE level in excess of 10,000 kU/l and may very well contain a substantial amount of *Aspergillus*-specific IgE antibody.

Sami Bahna. Most of us in this room have been involved in allergy diagnosis and all recognize the limitations of current skin testing and *in vitro* techniques. While I welcome your new test, I worry about the confusion it will bring. It will be compared with those tests with which we are already dissatisfied. Will it be better for some allergens than other commercial kits, and will it be worse for others? Will it be used in laboratories already overloaded with tests such that the assessment of quality control will be made even more difficult than it is now?

Sola Alaba. We have already discussed the lack of a "gold standard" for allergy testing, so I need not dwell further, other than to say inevitably the AlaSTAT will be compared with RAST. However, I think we can offer the overburdened technician some relief: AlaSTAT is a liquid phase assay system and can be much more easily automated than RAST.

Louis Perelmutter. I agree that discrepant results can be due to non-specific serum responses due to high total IgE levels; nonetheless, in practise only rarely do we see these very high levels. Another potentially troublesome interference is IgG antibody and I am wondering how you intend to cope with this problem.

Sola Alaba. The presence of allergen-specific IgG antibody in serum from patients being hyposensitized loomed large in our mind when we were developing the test. The cascade reaction described by Said El Shami was designed to load an excess of allergen on the ligand back-bone, so much of an excess that even small amounts of IgE could be specifically detected in the presence of high concentrations of IgG antibody. The direct method of doing the same thing is obvious and we are looking into that right now.

Louis Perelmutter. By purposefully adding IgG antibodies to the test sera?

Sola Alaba. Exactly so.

Terry Merrett. Actually, we have never been very worried by the interference in RAST of IgG antibodies, because the allergen seems to be present in sufficient excess on the paper discs. More of a concern to us has been the evaluation of different methods for interference caused by the choice of test specimen. For example, haemolysed serum and plasma can cause artefactually high results, and I imagine you have checked these possible sources of interference.

Sola Alaba. Serum and plasma produce identical results.

Session Five

PAEDIATRIC ALLERGY AND EPIDEMIOLOGY

Chairman: S. Bahna

Development of Pediatric Food Allergy

Sami L. Bahna

Department of Allergy and Immunology, Cleveland Clinic Foundation, 9500 Euclid Avenue, Cleveland, OH 44195-5124, USA

ABSTRACT

Development of food allergy in infants is influenced by several factors. Genetic predisposition for atopy probably provides a base for sensitization to occur following adequate exposure to the food allergen. Both sensitization and precipitation of symptoms are often enhanced by certain contributory factors, particularly in the gut.

KEYWORDS

Allergy; atopy; diet; feeding; food; infants; pediatrics.

INTRODUCTION

Development of food allergy, as well as of atopy in general, seems to depend on three main groups of factors (Table 1). Genetic factors seem to provide an infrastructure for sensitization with specific allergens. Certain contributory factors facilitate the appearance of clinical manifestations. Probably both genetic and environmental factors are involved in determining the age of onset of clinical manifestations, the target organ, the offending allergens, and the course of the disease.

This manuscript provides a concise discussion of how some intrinsic and extrinsic factors influence the development of food allergy in young children.

GENETICS

The relationship between heredity and atopy has been well known. Allergy occurs with a higher frequency and appears at an earlier age when a positive history of allergy is present in both parents than when only one or neither of the parents is allergic (Orgel *et al.*, 1975; Kaufman and Frick, 1976; Croner *et al.*, 1983). In one study, the incidence of atopic disease by 18

Advances in the Biosciences Vol. 74

Printed in Great Britain

Table 1. Factors that determine development of pediatric food allergy

A. Genetics
Atopic constitution.

B. Allergen exposure
1. Infant's diet:
a. Bottle feeding.
b. Early introduction of solid foods.
2. Maternal diet during pregnancy and lactation.

C. Contributing factors
1. Gastrointestinal maturation.
2. Immunologic defect.
3. Gastrointestinal disease.

months of age was 23% when atopy was present in both parents, 12% when it was in one parent only, and 5.3% when there was no family history of allergy (Croner *et al.*, 1983).

The literature provides sufficient data to indicate that the IgE response is under genetic control (Meyers and Marsh, 1981). A high incidence of allergy to certain foods has been noted in occasional families (Gerrard, 1966; Langeland, 1983; Bahna, 1987). Figure 1 illustrates data on a family where cow's milk caused allergy symptoms in all three siblings, their mother, an uncle, and two cousins. Nevertheless, there are no data to indicate that sensitization to specific allergens is genetically determined. In studying a large series of twins, we noticed no evidence that heredity determines the sensitivity to specific allergens or the development of specific allergy manifestations (Prather and Bahna, 1981; Bahna, 1983).

INFANT'S DIET

The advantages of breast feeding in reducing the incidence of allergy in infants, particularly to foods, have been documented by numerous studies (Bahna and Heiner, 1980; Bahna, 1986). The magnitude of prevention or delay of occurence of allergy is particularly obvious when breast feeding is used exclusively for six months or more and in the offspring of atopic families (Saarinen *et al.*, 1979; Saarinen and Kajosaari, 1980; Businco *et al.*, 1983a). Lactating mothers also tend to avoid early introduction of solid foods (Sleigh and Ounsted, 1975), which further contributes to a lower incidence of allergy in their infants (Saarinen and Kajosaari, 1980; Fergusson *et al.*, 1981). Contradictory conclusions of some studies on the role of breast feeding in protecting against allergy might be due to methodological differences in study design or data analysis (Kramer, 1988).

Infants who are not breast fed are commonly fed a cow's milk formula, hence bovine milk is the most common allergen in this age group, affecting perhaps 3-5% (Bahna, 1987). Crossed immunoelectrophoresis studies of bovine milk proteins revealed that whey contains at least 37 individual antigenic components and casein contains at least seven (Gjesing *et al.*, 1986). Several milk-substitute formulas for infant feeding have been available and their allergenicity is generally less than the regular cow's milk formulas.

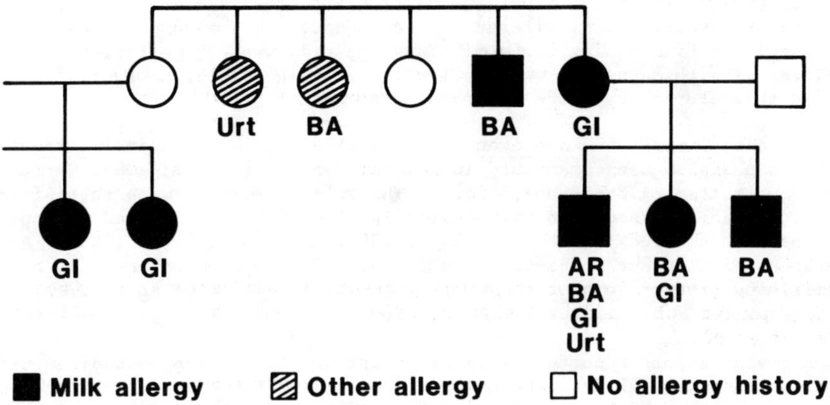

Fig. 1. Familial incidence of milk allergy. Most subjects, particularly those with multiple manifestations, had multiple allergens. AR: allergic rhinitis; BA: bronchial asthma; GI: gastrointestinal; Urt: urticaria.

A lower allergenicity of soybean formula than of cow's milk formula has been observed in some studies (Businco *et al.*, 1983a) but not in others (Kjellman and Johansson, 1979).

Heat treatment of bovine milk denatures mainly its serum albumin and immunoglobulin. These protein fractions, however, are much less allergenic than beta-lactoglobulin and casein (Bahna and Heiner, 1980).

Formulas made of enzymatically-hydrolysed bovine casein (eg. Nutramigen by Mead Johnson) and recently whey hydrolysate (LHA by Nestle) have shown promising hypoallergenic qualities (Kahn *et al.*, 1985; Kahn *et al.*, 1987). It seems that hypoallergenicity of proteins can be more readily attained by enzymatic hydrolysis than by heat treatment. Hydrolysis denatures both the conformational and sequential types of epitopes whereas heat denatures mainly the conformational type (Kilshaw *et al.*, 1982; Pahud *et al.*, 1985).

Elemental diet formula made of synthesized amino acids (Vivonex by Norwich Eaton) has also been used in the diagnosis and management of food allergy (Dockhorn and Smith, 1981; Hill and Lynch, 1982). It would be especially useful in patients who had severe allergy symptoms to multiple foods and thus need to be on an especially restricted elimination diet.

MATERNAL DIET DURING PREGNANCY AND LACTATION

Several investigators have described infants who were born already sensitized to foods (Kaufman, 1971; Kurome *et al.*, 1976; Dannaeus *et al.*, 1978; Gerrard and Perelmutter, 1986). Detection of specific IgE antibodies to food in cord blood of newborns whose mothers had no such antibodies supports an intrauterine sensitization (Michel *et al.*, 1980; Businco *et al.*, 1983b). Such information suggests that in families at high risk of atopy, limitation of the common allergenic foods in the pregnant woman's diet might lessen the chance of development of allergy in the infant. This assumption could not be confirmed by avoiding cow's milk and egg from week 28 gestation to delivery (Fälth-Magnusson *et al.*, 1987). It is possible, however, that the third trimester is too late for such intervention.

Sensitization to foods, and even manifest allergy, in exclusively breast-fed infants has also been described in several reports (Warner, 1980; Gerrard and Perelmutter, 1986; Bahna, 1987). The relationship between the infant's symptoms and the detection in breast milk of maternally ingested food protein has been documented (Jakobsson *et al.*, 1985; Axelsson *et al.*, 1986). Allergy in breast-fed infants to maternally ingested food may be enhanced by a diminished production by the mammary gland of specific IgA antibodies to the offending food (Machtinger and Moss, 1986).

Some investigators reported striking exceptions to the long-standing rule that an increased allergen exposure enhances sensitization (Firer *et al.*, 1981; Gerrard and Shenassa, 1983; Bahna, 1987). It is quite possible that infants who become sensitized by such minute quantities of antigen received across the placenta or the mammary gland are genetically determined to be exquisitely sensitive.

GASTROINTESTINAL MATURATION

The increased incidence of food allergy during infancy is attributed at least partly to the immaturity of the gastrointestinal tract; histologically, enzymatically, and immunologically. The intestinal mucosa of infants exhibits increased permeability and is not adequately protected by secretory IgA (Udall *et al.*, 1981). This is most remarkable in premature infants. Concentrations of circulating beta-lactoglobulin in preterm neonates were higher than in term neonates given similar amounts of cow's milk formula (Robertson *et al.*, 1982). The influx of large quantities of protein macromolecules enhances the induction of abnormal immunologic responses, both locally in the gut as well as systemic.

IMMUNOLOGIC DEFECT

An increased incidence of food allergy in infants has been observed in association with increased IgE production (Michel *et al.*, 1980; Businco *et al.*, 1983b; Croner *et al.*, 1983), IgA deficiency (Taylor *et al.*, 1973), transient hypogammaglobulinemia of infancy (Fineman *et al.*, 1979), and T cell defect (Juto, 1980).

GASTROINTESTINAL DISEASE

It is conceivable that an injury to the histologic or immunologic integrity of the gut mucosa by any of a wide variety of disorders might enhance the development of food allergy (Walker, 1975; Aiuti and Paganelli, 1983). Incomplete digestion of food protein and increased intestinal mucosal permeability during diarrheal diseases facilitate the absorption of excessive quantities of macromolecules that may enhance sensitization or exacerbation of pre-existing food allergy. Control of the associated gastrointestinal disease is often followed by increased tolerance to an offending food.

PROGNOSIS

The prognosis of food allergy is generally much better in children than in adults. The duration of clinical sensitivity to food tends to be shorter in infants than in older children; in patients with single rather than multiple symptoms or multiple food allergens; when the symptoms are gastrointestinal rather than in other systems; when the food allergen is cow's milk, soybean, cereals, or fruit rather than nuts, seafood, or fish; and when the degree of sensitivity (total IgE; skin reactivity; RAST score) is low (Bock, 1982; Ford and Taylor, 1982).

In patients who frequently take limited quantities of the offending food without apparent symptoms, the intestinal mucosa may continue to have histologic and immunologic abnormalities (Shiner *et al.*, 1975; Harris *et al.*, 1977) that might perpetuate the state of sensitivity for prolonged durations.

The child who once had food allergy often continues to be allergy-prone for years, during which time the allergy symptoms and/or the offending allergens may change.

REFERENCES

Aiuti, F. and Paganelii, R. (1983). Food allergy and gastrointestinal disease. Ann. Allergy, 51, 275-280.

Axelsson, I., Jakobsson, I., Lindberg, T. and Benediktsson, B. (1986). Bovine β-lactoglobulin in the human milk. Acta Pediatr. Scan., 75, 702-707.

Bahna, S.L. and Heiner, D.C. (1980). Allergies to Milk. Grune and Stratton, New York.

Bahna, S.L. (1983). Concodances in atopic twins. J. Allergy Clin. Immunol., 71 (No. 1 part II), 100.

Bahna, S.L. (1986). Is breast the best? Am. J. Reprod. Immunol. Microbiol., 10, 127-131.

Bahna, S.L. (1987). Milk allergy in infancy. Ann. Allergy, 59, 131-136.

Bock, A.S. (1982). The natural history of food sensitivity. J. Allergy. Clin. Immunol., 69, 173-177.

Businco, L., Marchetti, F., Pellegrini, G., Cantani, A. and Perlini, R. (1983a). Prevention of atopic disease in "at-risk newborns" by prolonged breast-feeding. Ann. Allergy, 51, 296-299.

Businco, L., Marchetti, F., Pellegrini, G. and Perlini, R. (1983b). Predictive value of cord blood IgE levels in at risk newborn babies and influence of type of feeding. Clin. Allergy, 13, 503-508.

Croner, S., Kjellman, N.I.M., Eriksson, B. and Roth, A. (1983). IgE screening in 1701 newborn infants and the development of atopic disease during infancy. Arch. Dis. Child., 57, 364-368.

Dannaeus, A., Johansson, G. and Foucard, T. (1978). Clinical and immunological aspects of food allergy in childhood. II. Development of allergic symptoms and humoral immune response to foods in infants of atopic mothers during the first 24 months of life. Acta Pediatri. Scand., 67, 497-504.

Dockhorn, R.J. and Smith, T.C. (1981). Use of a chemically defined hypoallergenic diet (Vivonex) in the management of patients with suspected food allergy/intolerance. Ann. Allergy, 47, 264-266.

Fälth-Magnusson, K. and Kjellman, N.I.M. (1987). Development of atopic disease in babies whose mothers were receiving exclusion diet during pregnancy - A randomized study. J. Allergy Clin. Immunol., 80, 868-875.

Fergusson, D.M., Horwood, L.J., Beautrais, A.L., Shannon, F.T. and Taylor, B. (1981). Eczema and infant diet. Clin. Allergy, 11, 325-331.

Fineman, S.M., Rosen, F.S. and Geha, R.S. (1979). Transient hypogammaglobulinemia, elevated immunoglobulin E level, and food allergy. J. Allergy Clin. Immunol., 64, 216-222.

Firer, M.A., Hosking, C.S. and Hill, D.J. (1981). Effect of antigen load on development of milk antibodies in infants allergic to milk. Br. Med. J., 283, 693-696.

Ford, R.P.K. and Taylor, B. (1982). Natural history of egg hypersensitivity. Arch. Dis. Child., 57, 649-652.

Gerrard. J.W. (1966). Familial recurrent rhinorrhea and bronchitis due to cow's milk. JAMA, 198, 605-607.

Gerrard, J.W. and Shenassa, M. (1983). Food allergy: two common types as seen in breast-fed and formula fed babies. Ann. Allergy, 50, 375-378.

Gerrard, J.W. and Perelmutter, L. (1986). IgE-mediated allergy to peanut, cow's milk, and egg in children with special reference to maternal diet. Ann. Allergy, 56, 351-354.

Gjesing, B., Osterballe, O., Schwartz, B., Wahn, U. and Lowenstein, H. (1986) Allergen-specific IgE antibodies against antigenic components in cow milk and milk substitutes. Allergy, 41, 51-56.

Harris, M.J., Petts, V. and Penny, R. (1977). Cow's milk allergy as a cause of infantile colic: Immunofluorescent studies of jejunal mucosa. Aust. Paediatr. J., 13, 276-281.

Hill, D.J. and Lynch, B.C. (1982). Elemental diet in the management of severe eczema in childhood. Clin. Allergy, 12, 313-315.

Jakobsson, I., Lindberg, T., Benediktsson, B. and Hansson, B.G. (1985). Dietary bovine β-lactoglobulin is transferred to human milk. Acta Pediatr. Scan., 74, 342-345.

Juto, P. (1980). Elevated serum immunoglobulin E in T cell deficient infants fed cow's milk. J. Allergy Clin. Immunol., 66, 402-407.

Kahn, A., Mozin, M.J., Casimir, G., Montauk, L. and Blum, D. (1985). Insomnia and cow's milk allergy in infants. Pediatrics, 76, 880-884.

Kahn, A., Rebuffat, E., Blum, D., Casimir, G., Duchateau, J., Mozin, M.J. and Jost, R. (1987). Difficulty in initiating and maintaining sleep associated with cow's milk allergy in infants. Sleep, 10, 116-121.

Kaufman, H.S. (1971). Allergy in the newborn: skin tests reactions confirmed by Prausnitz-Kustner test at birth. Clin. Allergy, 1, 363-367.

Kaufman, H.S. and Frick, O.L. (1976). The development of allergy in infants of allergic parents: A prospective study concerning the role of heredity. Ann. Allergy, 37, 410-415.

Kilshaw, P.J., Heppell, L.M.J. and Ford, J.E. (1982). Effects of heat treatment of cow's milk and whey on the nutritional quality and antigenic properties. Arch. Dis. Child., 57, 842-847.

Kjellman, N.I.M. and Johansson, S.G.O. (1979). Soy versus cow's milk in infants with a biparental history of atopic disease: Development of atopic disease and immunoglobulins from birth to four years of age. Clin. Allergy, 9, 347-358.

Kramer, M.S. (1988). Does breast feeding help protect against atopic disease? Biology, methodology, and a golden jubilee of controversy. J. Pediatrics, 112, 181-190.

Kuroume, T., Oguri, M., Matusumara, T., Iwasaki, I., Kanbe, Y., Yamada, T., Kawabe, S. and Negishi, K. (1976). Milk sensitivity and soybean sensitivity in the production of eczematous manifestations and breast-fed infants with particular reference to intra-uterine sensitization. Ann. Allergy, 37, 41-46.

Langeland, T. (1983). A clinical and immunological study of allergy to hen's eggwhite. I. A clinical study of egg allergy. Clin. Allergy, 13, 371-382.

Machtinger, S. abd Moss, R. (1986). Cow's milk allergy in breast-fed infants: the role of allergen and maternal secretory IgA antibody. J. Allergy Clin. Immunol., 77, 341-347.

Meyers, D.A. abd Marsh, D.G. (1981). Committee report: Report on a National Institute of Allergy and Infectious Diseases - sponsored workshop on the genetics of total immunoglobulin E levels in humans. J. Allergy Clin. Immunol., 67, 167-170.

Michel, F.B., Bousquet, J., Greillier, P., Robinet-Levy, M. and Coulomb, Y. (1980). Comparison of cord blood immunoglobulin E concentrations and maternal allergy for the prediction of atopic diseases in infancy. J. Allergy Clin. Immunol., 65, 422-430.

Orgel, H.A., Hamburger, R.N., Bazaral, M., Gorin, H., Groshong, T., Lenoir, M., Miller, J.R. and Wallace, W.M. (1975). Development of IgE and allergy in infancy. J. Allergy Clin. Immunol., 56, 296-307.

Pahud, J.J., Monti, J.C. and Jost, R. (1985). Allergenicity of whey protein: its modification by tryptic *in vitro* hydrolysis of the protein. J. Ped. Gastro. Nutr., 4, 408-413.

Prather, B. and Bahna, S.L. (1981). Does heredity determine the allergic manifestation or the sensitivity to a specific allergy? J. Allergy Clin. Immunol., 67 (January suppl.), 61.

Robertson, D.M., Paganelli, R., Dinwiddie, R. and Levinsky, R.J. (1982). Antigen absorption in the preterm and term neonate. Arch. Dis. Child., 57, 369-372.

Saarinen, U.M., Kajosaari, M., Backman, A. and Siimes, M.A. (1979). Prolonged breast-feeding as prophylaxis of atopic disease. Lancet, 2, 163-168.

Saarinen, U.M. and Kajosaari, M. (1980). Does dietary elimination in infancy prevent or only postpone a food allergy?: A study of fish and citrus allergy in 375 children. Lancet, 1, 166-167.

Shiner, M., Brook, C.G.D., Ballard, J. and Herman, S. (1975). Intestinal biopsy in the diagnosis of cow's milk protein intolerance without acute symptoms. Lancet, 2, 1060-1063.

Sleigh, G. and Ounsted, M. (1975). Present day practice in infant feeding. Lancet, I, 753.

Taylor, B., Norman, A.P., Orgel, H.A., Stokes, C.R., Turner, M.W. and Soothill, J.F. (1973). Transient IgA deficiency and pathogenesis of infantile atopy. Lancet, 2, 111-113.

Walker, W.A. (1975). Antigen absorption from the small intestine and gastro-intestinal disease. Pediatr. Clin. North Am., 22, 731-746.

Warner, J.O. (1980). Food allergy in fully breast-fed infants. Clin. Allergy, 10, 133-136.

Udall, J.N., Pang, K., Fritze, L., Kleinman, R.E. and Walker, W.A. (1981). Development of gastrointestinal mucosal barrier. I. The effect of age on intestinal permeability of macromolecules. Pediatr. Res., 15, 241-244.

Does the Heisenberg Principle Apply to the Predictive Value of IgE in Atopic Allergy?

Robert N. Hamburger

Pediatric Immunology and Allergy Division, Pediatric Department M-009-D, School of Medicine, University of California San Diego, La Jolla, CA 92093, USA

ABSTRACT

The relationship of the level of immunoglobulin E (IgE) in the serum of an individual and the likelihood that that person will have an atopic allergic disease is explored. Factors contributing to a lack of predictability include: (a) problems of reproducibility in the laboratory measurement of serum IgE levels; (b) partial independence of the genetic regulation of total serum IgE versus allergen-specific IgE levels; (c) stability of serum IgE level in an individual over time; (d) admixture of maternal blood with the infant's cord blood producing misleadingly high levels of cord serum IgE; (e) environmental variation in allergenic stimuli; (f) unsuspected helminthiasis; and (g) other intercurrent illnesses.

Despite all of the foregoing, population studies have confirmed the high predictive value of serum IgE levels in unselected groups of newborns, infants, children and adults. In studying the individual, with exceptions in both high and low levels of serum IgE, it is necessary to combine the laboratory data with history, physical examination and environmental study to make reliable diagnostic predictions.

KEYWORDS

Immunoglobulin E (IgE); human serum (cord, infants, children, adults); allergy prediction; atopy; genetic regulation.

As proposed in 1927 by Werner Heisenberg, the "Uncertainty Principle" or "Indeterminacy Principle" states that the position and the velocity of an object cannot both be measured exactly, at the same time, even in theory. In fact, the two together (velocity and position) have no meaning in nature. The complete rule states that the product of the uncertainties equals (or is greater than) 10^{-34} joule-second and may be expressed by the formula: h/(2pi) where h is Planck's constant (6.626196×10^{-34}). This is only significant for very small particles, such as an electron. The principle applies to

Advances in the Biosciences Vol. 74

other pairs such as energy and time; and in recent years has been used to examine large phenomena, which some claim behave in a similar manner but, of course, expressed by different formulae. The idea has come full circle in that, seemingly random events that appear to be totally unrelated and unpredictable, actually do behave in a predictable manner.

Our question today is: "What is the relationship, if any, between the level of immunoglobulin E (IgE) in the serum of an individual and the likelihood that that person has or will have an allergic disease?" We, and several other researchers, have been addressing that issue for the past twenty years. In this presentation I will summarize what we know, what we believe, and what we guess (hypothesize) is the relationship of IgE to atopic allergic diseases.

First let me state that the inability to measure velocity and position at the exact same time has nothing to do with inadequacies in the measuring instruments, the techniques, or the observer. Similarly, in discussing the predictive value of IgE levels, I'm going to ignore the inadequacies of the methods and the immunology laboratories' ability to accurately and reproducibly measure serum IgE levels. Additionally, I will assume that we all agree on the definition and diagnosis of major atopic allergic diseases. While all three (accuracy, reproducibility and diagnosis) are subject to marked differences in opinions of experts in the field, for this analysis we will assume that technical and technological problems have been overcome in answering the question of the value of total serum IgE level in predicting allergic disease.

As can be seen in the references accompanying this paper, we have written over fifty abstracts and articles on the measurement of total and specific IgE in the diagnosis, treatment and prophylaxis of allergic disease. The *In Vitro* Tests Committee of the American Academy of Allergy has been studying the problems of PRIST and RAST (total and specific IgE measurements) for many years and has raised serious questions about the reliability of most commercial laboratories and several research laboratories as well. The one conclusion with which most agree, is that the results between laboratories are not comparable; and only in a few selected laboratories, with samples run at the same time using the same lot-number reagents, could consistent reproducibility be demonstrated. Fortunately for this thesis, our laboratory was one of the top ones in the USA for both accuracy and reproducibility [38,41].

At the time when we began our work in this field, there were two myths extant which required data to dispel. The first was that IgE levels in the serum varied from hour to hour, or at best from day to day, so widely that genetic, regulation and prediction studies would not be possible. The second piece of misinformation was that the cord serum contained large amounts of IgE; on average about 40 units per ml. By 1971 we had corrected both of these misconceptions [1-5] and had demonstrated clearly that the serum IgE level is under Mendelian genetic control [8-14], that the normal newborn [2] has less than 0.5 I.U. of IgE per ml of serum (<1.0 ng/ml), that 1.0 I.U. equalled 2.42 ng [7], and that the reason that the human newborn would not make an atopic allergic reaction is that there was no detectable IgE on the mast cells or basophils at birth and for some weeks after birth [11,21,24, 43].

With all of the foregoing information available for the past fifteen years, why is there still debate over the value of using the serum IgE level to help make a diagnosis of allergic disease or a prediction of the appearance of allergic disease in the future? The answer probably lies in the exceptions. That is, the number of individuals with a low serum IgE level with positive RAST and positive skin tests and clear-cut, obvious atopic allergic disease [33,34,36]; and the others with elevated (greater than two standard deviations above the geometric mean) serum IgE levels, with no signs or symptoms of allergy [40].

Let me review with you Kjellman and Croner's data [49] to help to clarify the problem. Table 1 is adapted from their data on the relationship of cord serum IgE level to the subsequent development of atopic disease in 1,651 children. Based on their data, two standard deviations above the mean cord serum IgE is 0.9 I.U./ml.

Table 1. Cumulative incidence of atopic disease from birth to 6 years

	Cord blood IgE I.U./ml	
	<0.9 (n=1443)	>0.9 (n=208)
18 months	29 (2%)	108 (52%)
3 years	274 (19%)	144 (69%)
6 years	433 (30%)	171 (82%)
Continuous 1-6 years	5 (0.3%)	58 (28%)

Our data [46], in a study of the value of prophylaxis measures in the prevention of allergy, utilizing greater or less than 0.5 I.U./ml as the dividing line, found almost twice as much (21% vs 38%) allergic disease by three years of age, in those infants who had 0.5 units or more of IgE in their cord serum. We test for admixture of maternal and infant blood in any elevated IgE cord serum specimen to avoid misclassification into a high risk category [48].

Taken together, our two prophylaxis studies [19,20,30,46,50,55] provide an appreciation of the effects of environmental modification on the genetic endowment. Our original twin studies [8,12] in children indicated that 84% of the control of serum IgE levels was due to heritable (genetic) factors and only 16% was influenced by the environment. In adult twins the analysis of variance indicated that by middle age, the environmental influences will have increased to over 30%.

By combining our data with data published by John Gerrard, Max Kjellman, Gunnar Johansson, Bengt Bjorksten and others, and applying some mathematical corrections based upon observed incidence derived from family history studies, we developed predictive tables for major allergic disease based solely upon serum IgE level [15,42]. It is interesting that our earliest approximations have stood the test of time and have been confirmed by investigators in Western Europe [37,44,45,47,51], Canada [53] and the USA [28] (Table 2).

Data is not yet available for other areas of the world, but it is clear that basal levels are much higher in some ethnic groups, such as Filipinos and

Table 2. Prediction of atopic allergic disease based upon serum IgE level

Serum IgE level (I.U./ml)		Percent of Population	Percent Atopic
Infants	Children/Adults		
<6	<60	67	5
6 - 15	61 - 210	21	20
16 - 60	211 - 450	9	35
>60	>450	<3	~100*

*In the absence of helminthiasis or Job-Buckley Syndrome

Modified from reference #42.

Japanese. In a study [13] comparing caucasian children with age, sex, and socio-economically matched Filipino children (born in the USA of bilateral Filipino parents), the Filipino children had over three times higher mean serum IgE levels (227 vs 69 I.U./ml; $p<0.01$) and almost twice as much allergic disease (48% vs 25%).

In our first study of the possible prophylaxis or prevention of the development of allergic disease in the infants of highly allergic parents, significant effect was seen primarily correlated with the length and exclusivity of breast feeding, and mother and infant's diet [15,19,20,30]. A large, more complex, double blind and controlled collaborative study was undertaken with preliminary data [46,55,60] tending to confirm the earlier work. With the development of a laboratory test to measure compliance with the dietary restrictions of the protocol [50,58] an important variable was discovered: both unwitting and deliberate feeding of cow's milk and/or eggs to the study infants, usually by the grandmother or the baby-sitter who felt it necessary to surreptitiously feed these foods to the infants. Surprise home-visits confirmed the problem of significant non-compliance as shown in Table 3.

Table 3. Effect of educational home visits on parental performance

Environmental factor (n)	Results observed on follow-up		
	Improved	Unchanged	Worse
Pets in home (169)	56	99	14
Excessive dust (169)	78	86	5
Mold (168)	71	67	30
Mean of all factors (%)	68 (40%)	84 (50%)	16 (10%)

Modified from reference #55.

It therefore appears that to measure the true effectiveness of a prophylaxis regimen [47,51,53,56] it will be necessary to re-analyze the data in those families who really adhered to the rather rigorous protocols. In addition, a comparison of the value of each of the parameters of the prophylaxis study [52], that is, dietary restrictions [47], home environmental restrictions [55] and exclusive breast feeding [53,57] can be independently examined in the light of the laboratory and home-visit compliance data.

In my opinion, the use of the allergy/immunology laboratory [54,59] in the diagnosis of atopic allergic disease is more influenced by economic considerations than by objective data. It is reminiscent of the failure of acceptance of my work on the peptide inhibition of allergic reactions [16,29, 31] despite published confirmation by Stanworth and successful clinical use in over one thousand allergic patients (data on file with the FDA from Immunetech Pharmaceuticals, 11045 Roselle St., San Diego, CA 92121). One can only wait for truth to win over ignorance, fear and avarice.

In conclusion, there is little doubt of the validity and value of measuring both total and specific IgE in the diagnosis and prediction of the likelihood of developing atopic allergic disease in the study of populations. However, in studying the individual, with exceptions in both high and low serum IgE levels, it continues to be important to combine the laboratory data with environmental study and a thorough and complete history and physical examination to make the most reliable diagnosis and predictions possible.

The "uncertainty" of the ability to measure accurately the serum IgE level, combined with the individual exceptions in both directions for the relationship between level and allergic disease and the "indeterminacy" of establishing the diagnosis of IgE mediated atopic allergic disease, makes it appear to some that the two parameters are intrinsically impossible to measure at the same time. My thesis is that there is no inherent or theoretical reason why we should not continue to strive to perfect our diagnostic and prognostic abilities so that we can focus our prophylactic and therapeutic efforts on those with the greatest likelihood of success.

REFERENCES

1. Bazaral, M., Orgel, H.A. and Hamburger, R.N. (1971). Absence of serum immunoglobulin E in newborns: Heterogeneity in time of synthesis in infants. Ped. Res., 5, 334 (Abstr.).
2. Bazaral, M., Orgel, H.A. and Hamburger, R.N. (1971). IgE levels in normal infants and mothers and an inheritance hypothesis. J. Immunol., 107, 794.
3. Orgel, H.A., Reid, R.T., Stevenson, D.D. and Hamburger, R.N. (1971). Immunoglobulin E in breast and bottle-fed infants. Clin. Res., 19, 212 (Abstr.).
4. Stevenson, D.D., Orgel, H.A., Hamburger, R.N. and Reid, R.T. (1971). Development of IgE in newborn infants. J. Allerg. Clin. Immunol., 48, 61.
5. Hamburger, R.N. and Mendelson, L.M. (1971). Current status of immunoglobulin E (IgE). Calif. Med., 115, 46 (Epitome).
6. Hamburger, R.N. and Bazaral, M. (1972). IgE levels in asthmatics. Brit. Med. J., 1, 178 (Letter).
7. Bazaral, M. and Hamburger, R.N. (1972). Standardization and stability of immunoglobulin E (IgE). J. Aller. Clin. Immunol., 49, 189.
8. Hamburger, R.N. and Bazaral, M. (1972). IgE levels in twins confirms genetic control in human beings. J. Allerg. Clin. Immunol., 49, 91

(Abstr.).
9. Hamburger, R.N., Orgel, H.A. and Bazaral, M. (1973). Genetics of human serum IgE levels. In: Mechanisms in Allergy: Reagin-mediated Hypersensitivity. Eds. Goodfriend, Sehon and Orange; Marcel Dekker Inc., New York.
10. Hamburger, R.N. (1973). Niveles de inmunoglobulina E., herencia a la alergia e inmunidad a los helmintos. Bol. Med. Hosp. Infantil, 30, 309.
11. Miller, J.R., Lenoir, M.A., Groshong, T., Wallace, W.W., Bazaral, M., Orgel, H.A. and Hamburger, R.N. (1973). Development of serum IgE during the first year of life. Ped. Res., 7, 368 (Abstr.).
12. Bazaral, M., Orgel, H.A. and Hamburger, R.N. (1974). Genetics of IgE and allergy: Serum IgE levels in twins. J. Allerg. Clin. Immunol., 54, 288.
13. Orgel, H.A., Lenoir, M.A. and Bazaral, M. (1974). Serum IgG, IgA, IgM, and IgE levels and allergy in Filipino children in the United States. J. Allergy Clin. Immunol., 53, 213.
14. Hamburger, R.N. (1974). Development and inheritance of IgE and allergy. Allergol. et Immunopath., II, 101.
15. Orgel, H.A., Hamburger, R.N., Bazaral, M., Gorrin, H., Groshong, T., Lenoir, M., Miller, J.R. and Wallace, W.W. (1975). Development of IgE and allergy in infancy. J. Allerg. Clin. Immunol., 56, 296.
16. Hamburger, R.N. (1975). Peptide inhibition of the Prausnitz-Kustner reaction. Science, 189, 389.
17. Hamburger, R.N. (1976). Alergy and the immune response. Amer. Sci., 64, 157.
18. Hamburger, R.N. (1976). IgE and allergy. In: Advances in Pediatrics, ed. L. Barness; Year Book Medical Publishers, 23: 95.
19. Hamburger, R.N. and Orgel, H.A. (1976). The prophylaxis of allergy in infants debate. Ped. Res., 10, 387 (Abstr.).
20. Hamburger, R.N. and Orgel, H.A. (1976). Prophylaxis of allergy. West. J. of Med., 125, 215 (Epitome).
21. Mellon, M.H., Orgel, H.A. and Hamburger, R.N. (1977). The fate of IgE in infants following exchange transfusion. Clin. Res., 25, 182A (Abstr.).
22. Orgel, H.A., Kemp, J.P., Meltzer, E.O. and Hamburger, R.N. (1977). Atopy and IgE in a pediatric practice. Annals of Allergy, 39, 161.
23. O'Connor, R., Prenner, B. and Hamburger, R.N. (1977). Tissue mast cell numbers in allergic and non-allergic individuals. Annals of Allergy, 39, 76 (Abstr.).
24. Mellon, M.H., Tomaszewski, M., Orgel, H.A. and Hamburger, R.N. (1977). Serum levels and skin fixation of transfused IgE in infants. Ped. Res., 11, 490 (Abstr.).
25. Hamburger, R.N., Mellon, M.H., Tomaszewski, M. and Orgel, H.A. (1977). The hazard of IgE mediated allergic reaction after blood transfusion. The West. J. of Med., 127, 408 (Abstr.).
26. Wilcsek, R. and Hamburger, R.N. (1978). Variation in the measurement of serum IgE levels. Aller. et Immunopath., 6, 129.
27. Hamburger, R.N. (1978). The measurement of IgE. West. J. of Med., 129, 227 (Epitome).
28. Lichtenstein, L.M. and Hamburger, R.N. (1978). IgE and atopic disease. In: Immunological Diseases. Third Edition, Vol.II, ed. Mas Samter; Little Brown and Co., Inc. Boston, Mass.
29. Hamburger, R.N. (1979). Cell and IgE receptor response to pentapeptide. Clin. Res., 27, 109A (Abstr.).
30. Ziering, R.W., O'Connor, R.D., Mellon, M., Cook, D.A., Tomaszewski, M., Street, D.H. and Hamburger, R.N. (1979). University of California San Diego prophylaxis in infancy study: An interim report. J. Allerg. Clin. Immunol., 63, 199 (Abstr.).
31. Hamburger, R.N. (1979). Peptide inhibition of IgE binding to cell

receptors. Ped. Res., 13, 449 (Abstr.).
32. Hamburger, R.N. (1979). The measurement of IgE. Conn. Med., 43, 488.
33. Hamburger, R.N., Melvin, D. and Street, D.H. (1980). Radioallergosorbent test (RAST) and skin tests in low-IgE individuals. J. Allerg. Clin. Immunol., 65, 232 (Abstr.).
34. Hamburger, R.N., Melvin, D. and Street, D.H. (1980). Development of antigen specific IgE antibody in low total IgE atopic individuals. Clin. Res., 28, 108A (Abstr.).
35. Hamburger, R.N. (1980). IgE in cord serum. Annals of Allergy, 44, 44 (Letter).
36. Hamburger, R.N. (1980). The genetic implications of positive RAST in low IgE individuals. Immunol. Allergy. Pract., 2, 182.
37. Michel, F.B., Bousquet, J., Greillier, P., Robinet-Levy, M. and Coulomb, J.R. (1980). Comparison of cord blood immunoglobulin E concentration and maternal allergy for the prediction of atopic diseases in infancy. J. Allerg. Clin. Immunol, 65, 422.
38. Evans, R. (1981). For the *In Vitro* Tests Committee, R.N. Hamburger, Member. A US reference for human immunoglobulin E. J. Allerg. Clin. Immunol., 68, 79.
39. Hamburger, R.N. (1981). Development of atopic allergy in children. In: International Allergy Symposium, Uppsala Sept. 24-26, 1980, ed. S.G.O. Johansson, Excerpta Medica, Amsterdam-Oxford and Princeton.
40. Hamburger, R.N. (1981). A cautious view of the use of RAST in clinical allergy. Immunol. Allerg. Pract., 5, 151.
41. Evans, R. (1982). For the *In Vitro* Tests Committee, R.N. Hamburger, Member. Variability in the measurement of specific immunoglobulin E antibody by the RAST procedure. J. Allerg. Clin. Immunol., 69, 245.
42. Hamburger, R.N. (1982). The immunogenetics of IgE provides predictive value for the development of allergy. Annals of Allergy, 49, 9.
43. Hamburger, R.N. and Orgel, H.A. (1982). Transfused immunoglobulin E in newborns. J. Allerg. Clin. Immunol., 70, 143.
44. Kjellman, N.I. (1982). Prediction and prevention of atopic allergy. Allergy, 37, 463.
45. Pantin, C.F. and Merrett, T.G. (1982). Allergy screening using a microcomputer. Br. Med. J. (Clin. Res.), 285, 483.
46. Hamburger, R.N., Heller, S., Mellon, M.H., O'Connor, R.D. and Zeiger, R.S. (1983). Current status of the clinical and immunological consequences of a prototype allergic disease prevention program. Annals of Allergy, 51, 281.
47. Businco, L., Marchetti, F., Pellegrini, G. and Perlini, R. (1983). Predictive value of cord blood IgE levels in 'at-risk' newborn babies and influence of type of feeding. Clin. Allergy, 13, 503.
48. O'Connor, R.D., Zeiger, R.S., Mellon, M.H., Heller, S. and Hamburger, R.N. (1984). The occurrence of elevated cord sera IgE levels due to maternal blood admixture. New Engl. and Regiona Allergy Proc., 5, 149.
49. Kjellman, N.I. and Croner, S. (1984). Cord blood IgE determination for allergy prediction - a follow-up to seven years of age in 1,651 children. Annals of Allergy, 53, 167.
50. Hamburger, R.N. (1984). Diagnosis of food allergies and intolerances in the study of prophylaxis and control groups in infants. Annals of Allergy, 53, 673.
51. Bjorksten, B. (1984). IgE and lymphocyte function in small children as allergy predictors (Ger). Z. Hautkr., 59, 817.
52. Asser, S. and Hamburger, R.N. (1985). Atopic sensitization of the fetus and newborn in relation to the development of the human immune system. In: Advances in Perinatal Medicine, ed. A. Milunsky, E.A. Friedman and L. Gluck; 4, 299-321.
53. Chandra, R.K., Puri, S. and Cheema, P.S. (1985). Predictive value of

cord blood IgE in the development of atopic disease and role of breast-feeding in its prevention. Clin. Allergy, 15, 517.

54. Asser, S. and Hamburger, R.N. (1986). The use of the laboratory in the diagnosis of allergic disease. Schumpert Med. Q., 4, 272.

55. Zeiger, R.S., Heller, S., Mellon, M., O'Connor, R.D. and Hamburger, R.N. (1986). Effectiveness of dietary manipulation in the prevention of food allergy in infants. J. Allerg. Clin. Immunol., 78, 224.

56. Michel, F.B., Bousquet, J., Dannaeus, A., Hamburger, R.N., Bellanti, J.A. Businco, M.L. and Soothill, J. (1986). Preventive measures in early childhood allergy. J. Allerg. Clin. Immunol., 78, 1022.

57. Johnson, R., Casillas, R., Mellon, M. and Hamburger, R.N. (1987). The relationship between cow's milk (cm) ingestion in breast milk fed infants and serum IgE levels. Annals of Allergy, 58, 283 (Abstr.).

58. Hamburger, R.N., Casillas, R., Johnson, R., Mellon, M., O'Connor, R.D. and Zeiger, R. (1987). Long-term studies in prevention of food allergy: patterns of IgG anti-cow's milk antibody responses. Annals of Allergy, 59, 175.

59. Casillas, R., Johnson, R., Tubelis, D. and Hamburger, R.N. (1987). Comparison of RIA, EIA and FIA methods for the quantitation of total and specific IgE antibody (Ab). Annals of Allergy, 58, 290 (Abstr.).

60. Hamburger, R.N. and Asser, S.M. (1987). Prevention of atopy: Neonatal aspects. In: Immunology of the Neonate, eds. G.R. Burgio, L.A. Hanson and A.G. Ugazio. Springer-Verlag, Berlin, Heidelberg, New York, pp153-160.

A Prospective Study of Allergy in High-risk Children

M. L. Burr*, T. G. Merrett†, J. Merrett†, B. K. Butland*, and F. G. Miskelly‡

*MRC Epidemiology Unit, Cardiff;
†RAST Allergy Unit, Benenden Hospital, Cranbrook, Kent;
‡Department of Paediatrics, Merthyr General Hospital, Merthyr Tydfil, UK

ABSTRACT

A cohort of 497 infants, all with a high risk of atopic disease, were identified at birth and followed up for periods of up to 4 years. IgE and IgG4 antibodies were measured at birth, at 3 months and (in a subset) at 12 months. From birth to 12 months total IgE levels rose and total IgG4 declined; IgE titres tended to be higher in boys while IgG4 levels at 3 months were higher in girls. The incidence of wheeze and eczema was recorded annually, and boys had a higher incidence of wheeze than girls during each year. A high IgE titre in cord blood and at 3 months was associated with a higher risk of wheeze during the fourth year of life; total and egg-specific IgE antibodies at 3 months were associated with eczema during the first year of life. Egg-specific IgG4 antibodies in cord blood were associated with a reduced incidence of wheeze during the first year, while total IgG4 at three months tended to be associated with a lower incidence of wheeze during each year. Infants who received any breast milk had a reduced incidence of wheeze, especially during their first two years.

These findings support suggestions that IgE antibodies predict atopic disease, that IgG4 antibodies are protective against it, that egg is an important allergen in infancy, and that breast milk protects against wheezing, either by an anti-infective or by an anti-allergic action.

KEYWORDS

Allergy; eczema; breast feeding.

INTRODUCTION

Allergic disease is an important cause of illness in childhood, and its incidence appears to be increasing (Burr, 1987). It is therefore important to seek clues about its causation and the mechanisms involved. Prospective studies of high-risk infants can reveal associations between exposure to various environmental factors and the subsequent development of symptoms.

Advances in the Biosciences Vol. 74

Such studies also permit observations of immunological mechanisms which appear to be implicated in allergic disease. The IgE class of antibodies is known to be associated with atopic conditions, while IgG antibodies, and particularly the IgG4 subclass, may be protective (Dannaeus *et al.*, 1978; Merrett *et al.*, 1984). This paper presents data from a cohort of infants at high risk of atopic disease, supplying evidence of the predictive effects of total and specific IgE and IgG4 antibodies on the development of symptoms during the first four years of life.

SUBJECTS AND METHODS

Women attending antenatal clinics at two hospitals in South Wales were asked about allergic disease in themselves and their families. Those who gave a history of asthma, hay fever or eczema in themselves, their husbands or their children were invited to participate in a study designed to investigate the causes of allergic disease. Those who agreed to take part were admitted to a randomized controlled trial of the effects of withholding cow's milk protein in the prevention of allergy; details of this trial have already been published (Miskelly *et al.*, 1988). Cord blood was taken from the infants at birth for assay of IgE and IgG4 antibodies (Nye *et al.*, 1975; Merrett, 1986). A further sample of blood was taken from the infants when they were 3 months old, and again, from a small subset, at 12 months. Skin tests against a range of common allergens were performed at 6 months and 12 months. At 3, 6 and 12 months the infants were examined by a physician for the presence of eczema, and the mother was asked whether the child had wheezed since the previous visit. Thereafter the mothers have been sent a short postal questionnaire at each birthday asking about eczema and wheezing during the previous year. The local ethical committee consented to the study.

This paper describes the results of the blood and skin tests in relation to the incidence of eczema and wheezing up to the age of 4 years.

RESULTS

There were 497 infants who entered the study; information about symptoms was obtained for 483 for the first year. Thereafter a few moved out of the area and could not be traced, so that the 2-year, 3-year and 4-year data relate to 469, 460 and 403 children respectively. It is anticipated that some further information will be obtained, since the 4-year follow-up is not quite complete, so the following results are provisional only.

Cord blood was obtained from 394 babies. In eight cases there was evidence of contamination with maternal blood: total IgE concentration was over 5 kU/l and the same inhalent-specific IgE antibodies were found in the cord and the maternal blood. These eight values were therefore excluded from the results.

Table 1 shows the serum IgE and IgG4 levels at birth, 3 months and 12 months. Blood was taken from only 66 infants at 12 months because it was found that the mothers became distressed by the procedure when their children were at this age, and it was feared that participation in the study would be jeopardized. Although these 66 infants are not strictly a random sample of the whole cohort, there is no reason to suppose that they differ from the others in any material way. Because the number of 12 month results is small, these values have not been used in relation to the symptoms of the children.

Table 1. IgE and IgG4 at birth, 3 months and 12 months

	No.	Median	Upper quartile
IgE (kU/l)			
Newborn boys	200	0.3	0.88
Newborn girls	186	<0.3	0.60
3 month boys	243	1.8	4.0
3 month girls	215	1.2	2.4
12 month boys	33	14.0	48.0
12 month girls	33	6.0	20.5
IgG4 (mg/ml)			
Newborn boys	199	0.31	0.60
Newborn girls	170	0.40	0.72
3 month boys	244	<0.05	0.08
3 month girls	214	0.05	0.08
12 month boys	33	<0.05	0.09
12 month girls	32	<0.05	<0.05

The distributions of serum IgE and IgG4 values were very skewed, many being below the limit of measurability. Means and standard deviations were therefore inappropriate, so medians and upper quartiles are shown here. Serum IgE rose steadily from birth to 12 months, and at each stage the boys tended to have higher values than the girls.

The IgG4 values were highest in the cord blood and declined thereafter; at 12 months the majority were below the limits of the assay. At 3 months the proportion of girls with IgG4 >0.05 mg/ml (101/214) was significantly higher than the proportion of boys (88/244, $p<0.05$).

Table 2 shows the relationship between serum IgE (at birth and 3 months) and skin tests to egg, performed at 6 months. Cord IgE was unrelated to the skin test results; at 3 months, however, there was a highly significant association, which was even stronger for egg IgE. The cut-off points (0.3 and 5.0 kU/l for total IgE and 0.17 Phadebas RAST units (PRU)/ml for egg IgE) were chosen because they were the values above which a measurement could always be made. The lower limit of sensitivity for the assays was not the same for all specimens because some had to be diluted to a greater degree, owing to the small quantity available. It seemed reasonable to accept a higher cut-off level for IgE at 3 months than at birth because of the rise which occurred in that time.

Figure 1 shows the risk of symptoms in boys compared with that in girls.

Table 2. Relationship between IgE and positive egg skin tests at 6 months

	Total No.	No. (%) with positive egg skin test	χ^2 with Yates' correction
Cord IgE			
>0.3 kU/l	153	13 (9%)	0.6 (NS)
⩽0.3 kU/l	221	13 (6%)	
3 month IgE			
>5.0 kU/l	72	13 (18%)	13.3 (p<0.001)
⩽5.0 kU/l	373	19 (5%)	
3 month egg IgE			
>0.17 PRU/ml	23	12 (52%)	67.4 (p<0.0001)
⩽0.17 PRU/ml	412	19 (5%)	

"Wheeze" meant that the mother reported having heard a whistling sound coming from the child's chest during the previous year; "eczema" meant that the child had been treated by a doctor for eczema during the previous year. Odds ratios were calculated and the 95% confidence intervals (shown by the vertical bars) computed by Woolf's method (Breslow & Day, 1980). A bar which does not cross the line implies a risk significantly different from unity at the 5% level. The incidence of wheeze was significantly higher in boys than in girls during each of the first four years of life. The incidence of eczema during the first two years also tended to be higher in boys, but the excess risk did not quite achieve statistical significance and disappeared at 3 years of age.

Figures 2-8 show the relationships between certain IgE and IgG4 measurements and the incidence of symptoms during the first four years of life; the eight specimens suspected of contamination are excluded. The cut-off point for each immunoglobulin is the level above which a definite value could always be assigned, except for cord IgG4, for which the median value was selected. Cord IgE was not predictive of eczema, nor of wheeze during the first three years of life, but during the fourth year the risk of wheeze was substantially higher in children whose cord IgE had exceeded 0.3 kU/l in comparison with other children (Fig. 2). Figures 3 and 4 show that the incidence of eczema during the first year of life was related to a high serum IgE at 3 months and even more strongly to a high egg IgE at that time. At 2 years these relationships were less marked, and were not apparent in the third or fourth years. The risk of wheeze was somewhat raised (non-significantly) at one year, and significantly raised at 4 years, in children with a high 3-month IgE, but was unrelated to egg-specific IgE. When the data from boys and girls were analysed separately, the relationships between IgE and wheeze

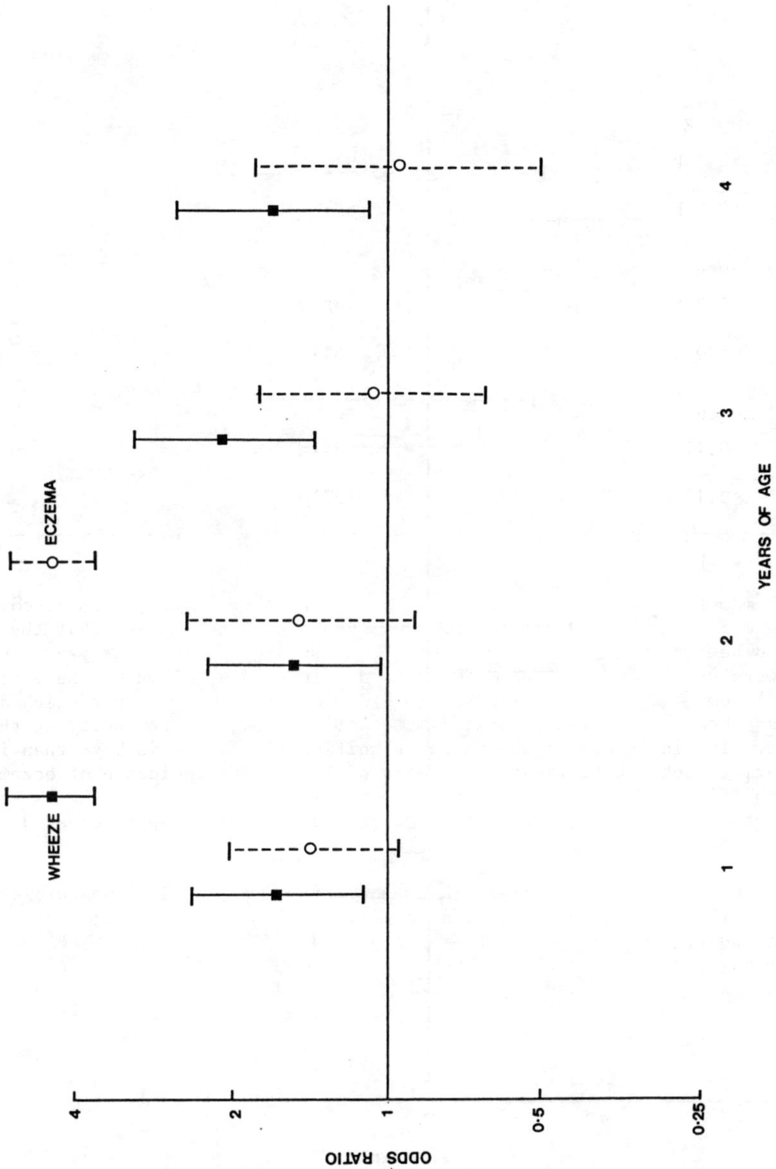

Fig. 1. Risk of symptoms in boys relative to girls.

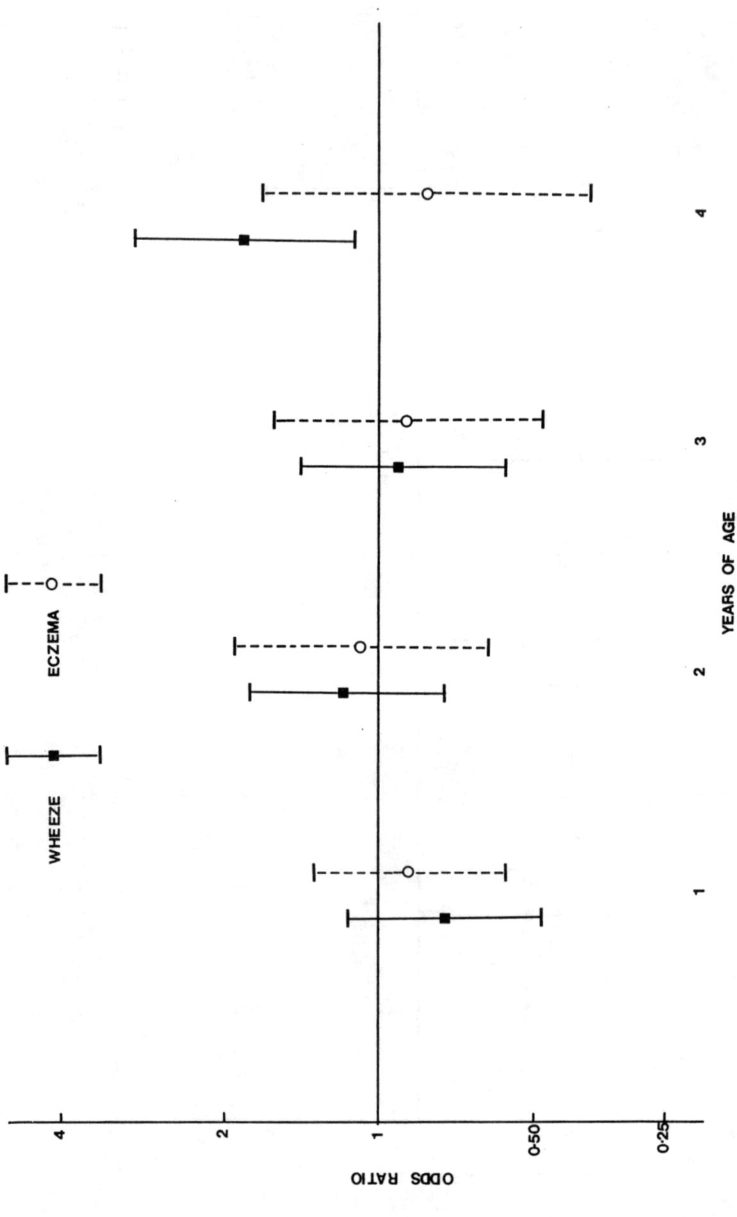

Fig. 2. Risk of symptoms for cord IgE >0.3 relative to ⩽0.3 kU/1.

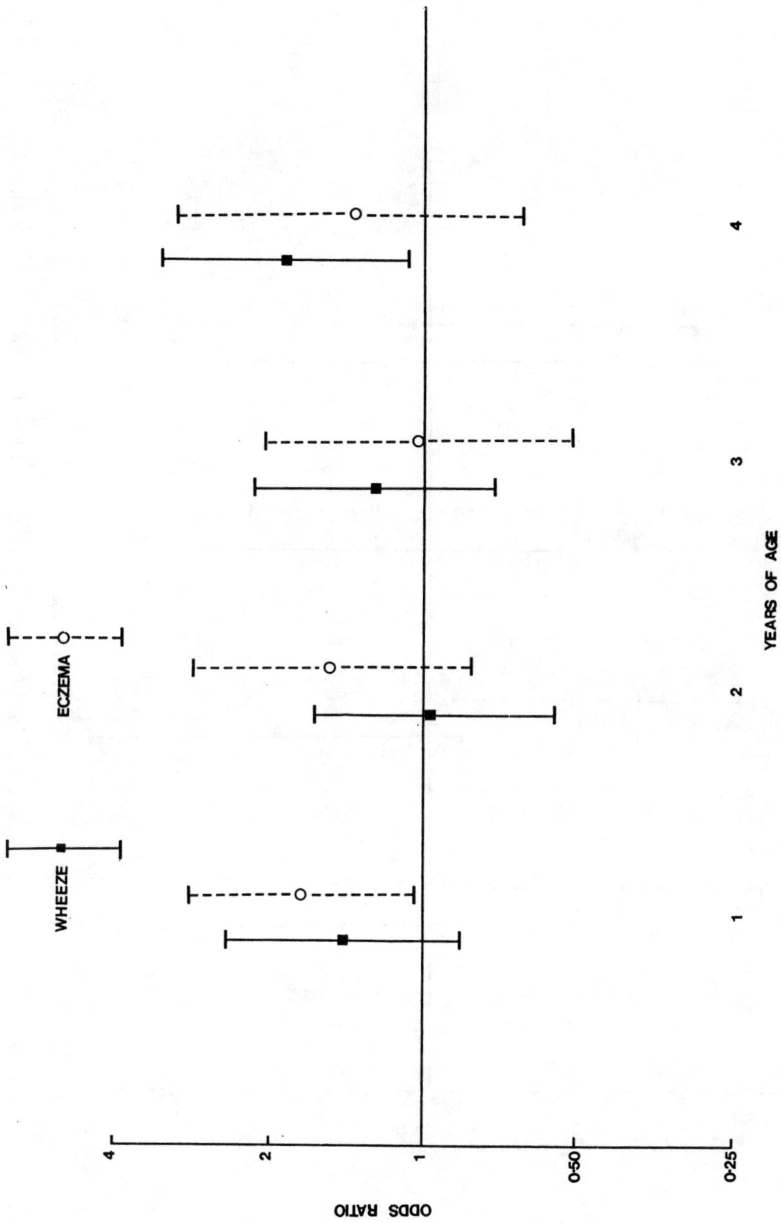

Fig. 3. Risk of symptoms for 3 month IgE >5 relative to ⩽5 kU/1.

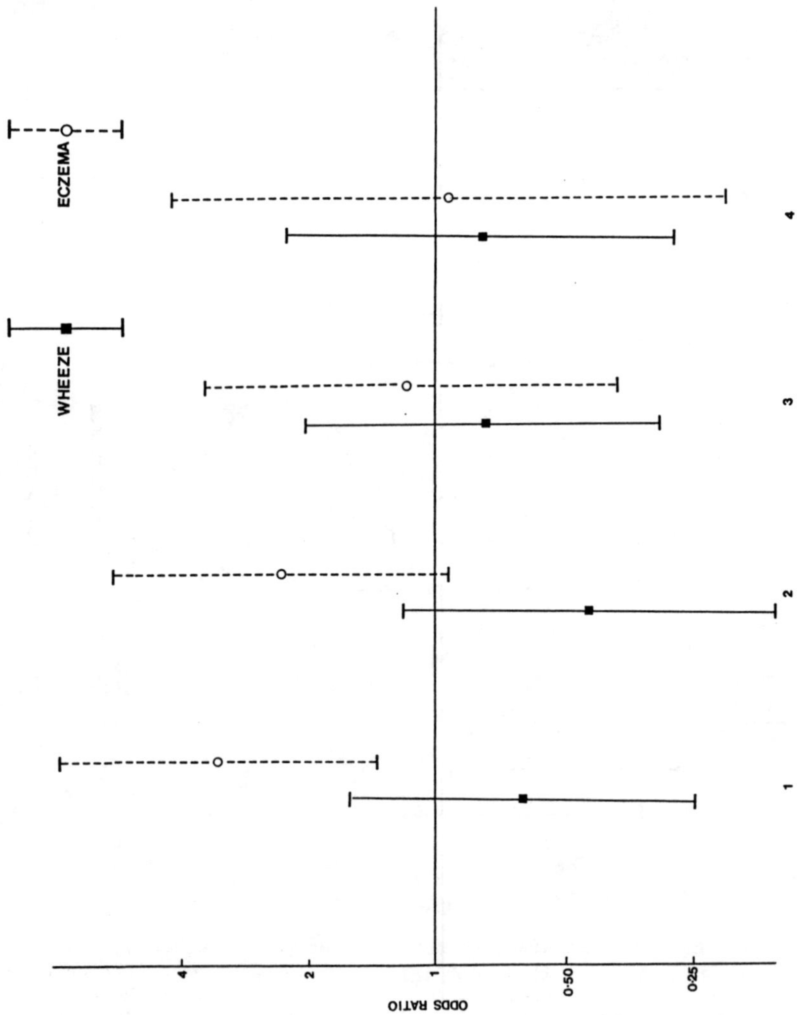

Fig. 4. Risk of symptoms for 3 month egg IgE >0.17 relative to ⩽0.17 PRU/ml.

at 4 years persisted, showing that they did not arise simply because the boys had a higher IgE and a greater tendency to wheeze than the girls.

Cord IgG4 was not clearly predictive of symptoms at any time (Fig. 5). A high IgG4 level at 3 months was associated with a reduced risk of wheeze during all four years, although the relationship never achieved statistical significance (Fig. 6). A high titre of egg-specific IgG4 in cord blood predicted a lower incidence of wheeze during the first year of life (Fig. 7); the 3-month egg IgG4 was less clearly predictive of wheeze (Fig. 8), while eczema was not predicted by any of the IgG4 indices. A separate analysis of data in boys and girls suggested that these relationships were not simply due to the sex-differences in IgG4 and symptoms.

Figure 9 shows the predictive effect of any breast feeding on symptoms. The breast fed infants are those who received any breast milk at all; nearly half were breast fed for 4 weeks or less, and exclusive breast feeding was rare. Nevertheless the breast fed group had a lower incidence of wheeze, the effect being particularly strong during the first two years of life. During the first year of life wheeze was also associated with maternal smoking, number of siblings, month of birth (August - November), and low social class. A multiple logistic regression analysis showed that the negative association between breast feeding and wheeze was independent of all these factors. Three month IgE levels were similar in the breast fed and the non-breast fed babies.

DISCUSSION

This paper presents preliminary results of a four-year follow-up of a cohort of infants at high risk of allergy. The randomized controlled trial showed that withholding cow's milk in early infancy did not reduce the incidence of eczema or wheezing during the first year of life (Miskelly *et al.*, 1988). Breast feeding seemed to confer protection against wheeze during the first 2 years and possibly to a lesser extent thereafter. This appeared to be due to a positive effect of breast milk rather than to the withholding of other foods, since hardly any children were exclusively breast fed, and was all the more remarkable in view of the short duration of breast feeding for many babies. The incidence of eczema was not affected by breast feeding in this study. Other studies relating infant feeding to symptoms have produced conflicting results (Burr, 1983).

Many investigators have reported that cord IgE predicts atopic disease (Dannaeus *et al.*, 1978; Michel *et al.*, 1980; Croner *et al.*, 1982; Businco *et al.*, 1983; Chandra *et al.*, 1985). Most of these studies have shown a predictive effect within the first two years of life, so it is somewhat surprising that in the present cohort it was not until 4 years that cord IgE (and 3 month IgE) predicted wheeze.

It must be recognised, of course, that wheeze does not necessarily have an allergic cause. In early childhood it is often caused by viral bronchiolitis. The valleys of South Wales, where these children live, have an unusually high incidence of respiratory disease (Colley & Read, 1970; Dean *et al.*, 1978), probably because of the cold damp climate, and it is possible that an excess of respiratory infection in early infancy masks any trend in allergy-related wheezing. By the age of 4 years atopic asthma is beginning to emerge as the predominant cause of wheezing. The protective effect of breast feeding may, in part at least, be attributable to some immunological defence against infection, and not only to an anti-allergic effect, particu-

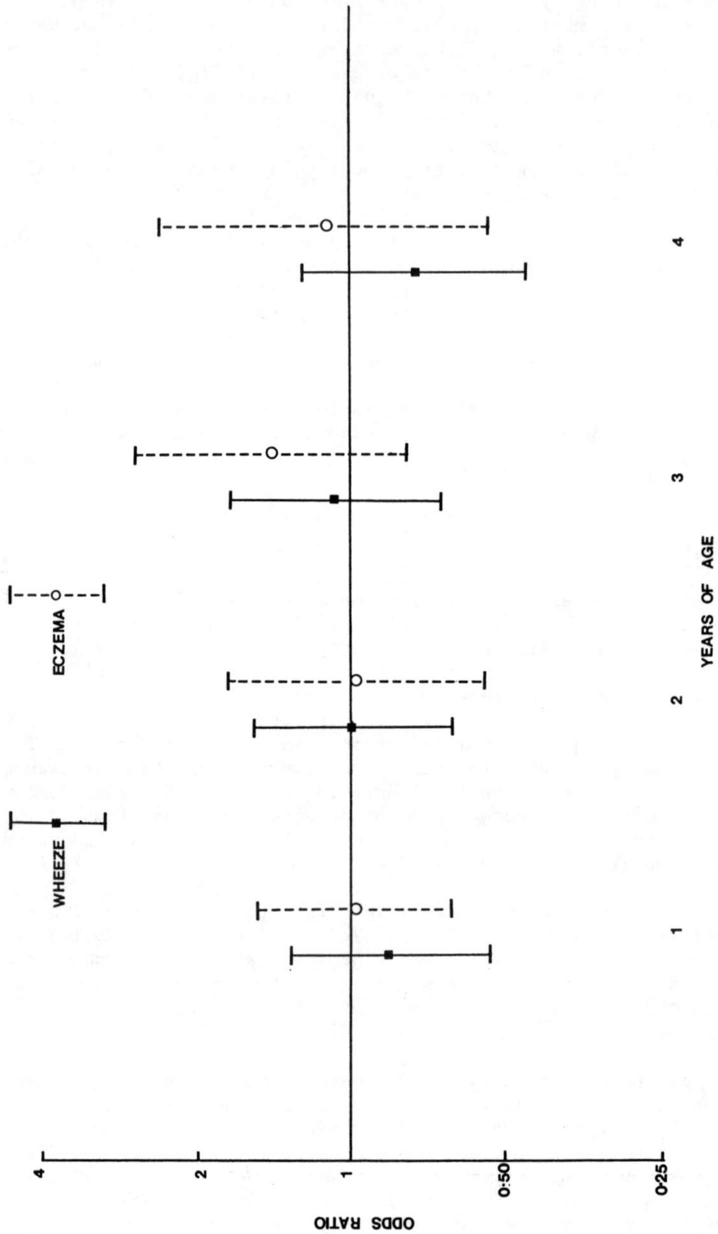

Fig. 5. Risk of symptoms for cord IgG4 >0.35 relative to ≤0.35 mg/ml.

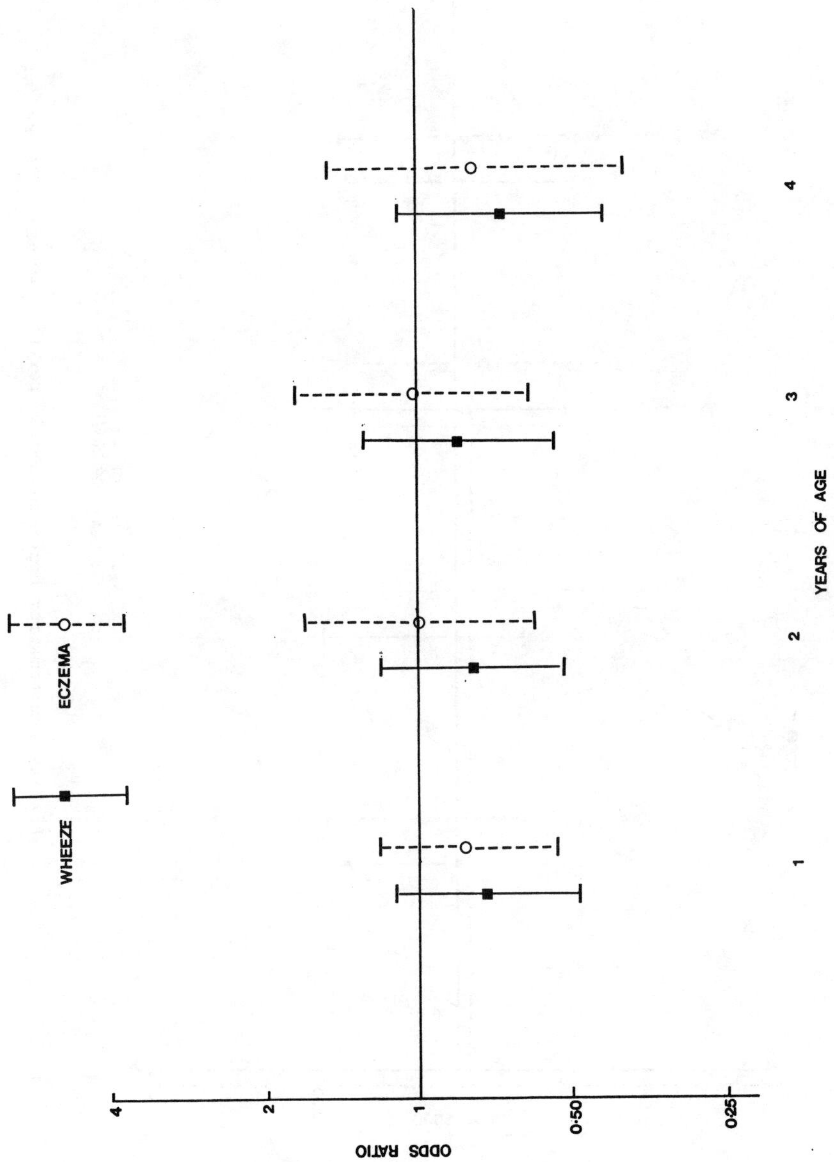

Fig. 6. Risk of symptoms for 3 month IgG4 >0.05 relative to ⩽0.05 mg/ml.

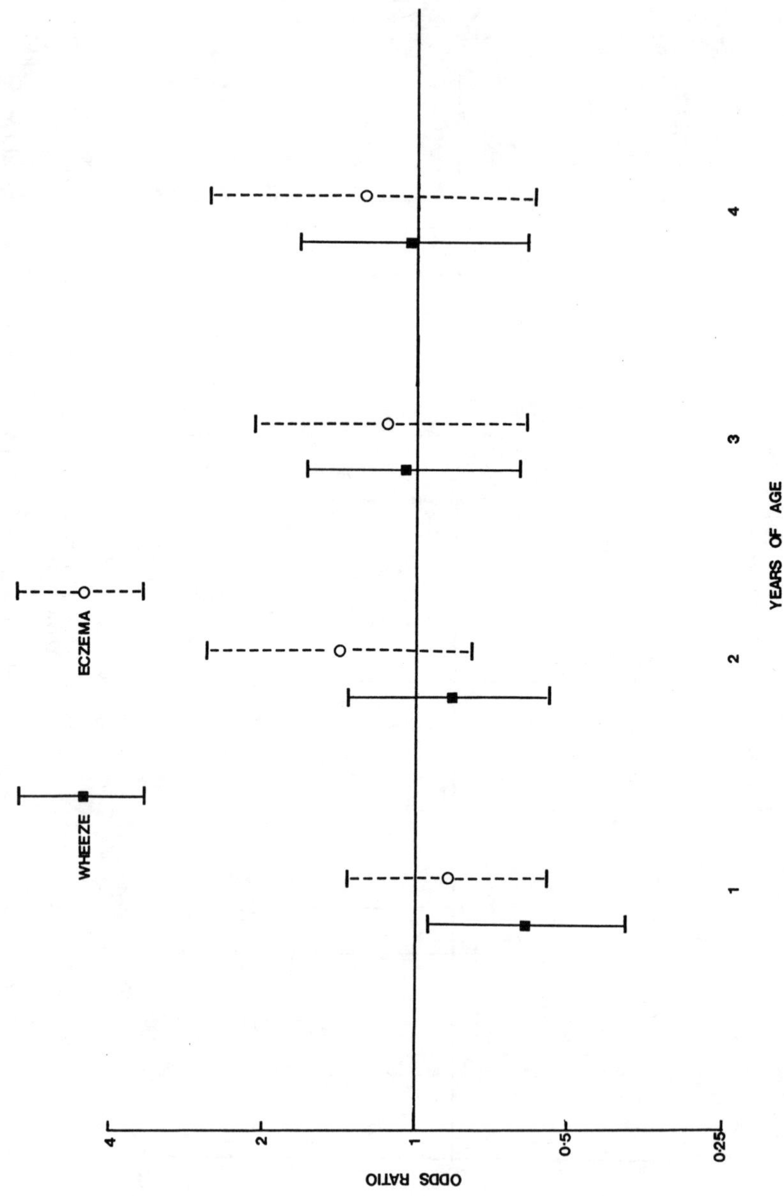

Fig. 7. Risk of symptoms for cord egg IgG4 >31 relative to ≤31 PRU/ml.

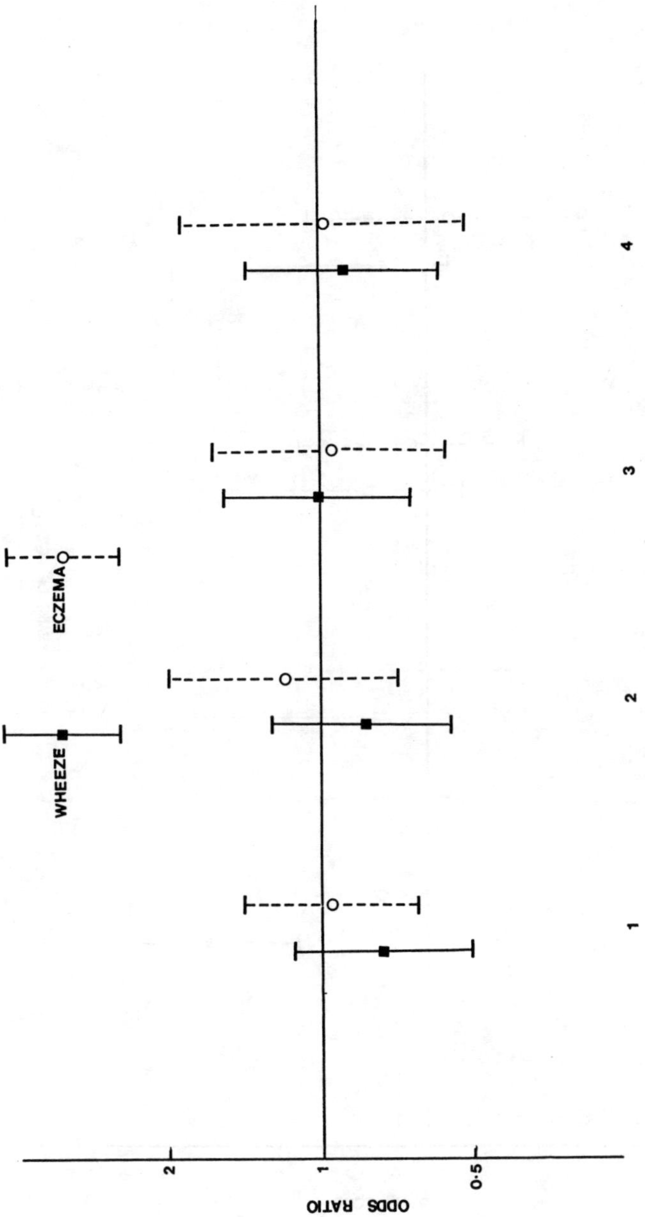

Fig. 8. Risk of symptoms for 3 month egg IgG4 >6.2 relative to ⩽6.2 PRU/ml.

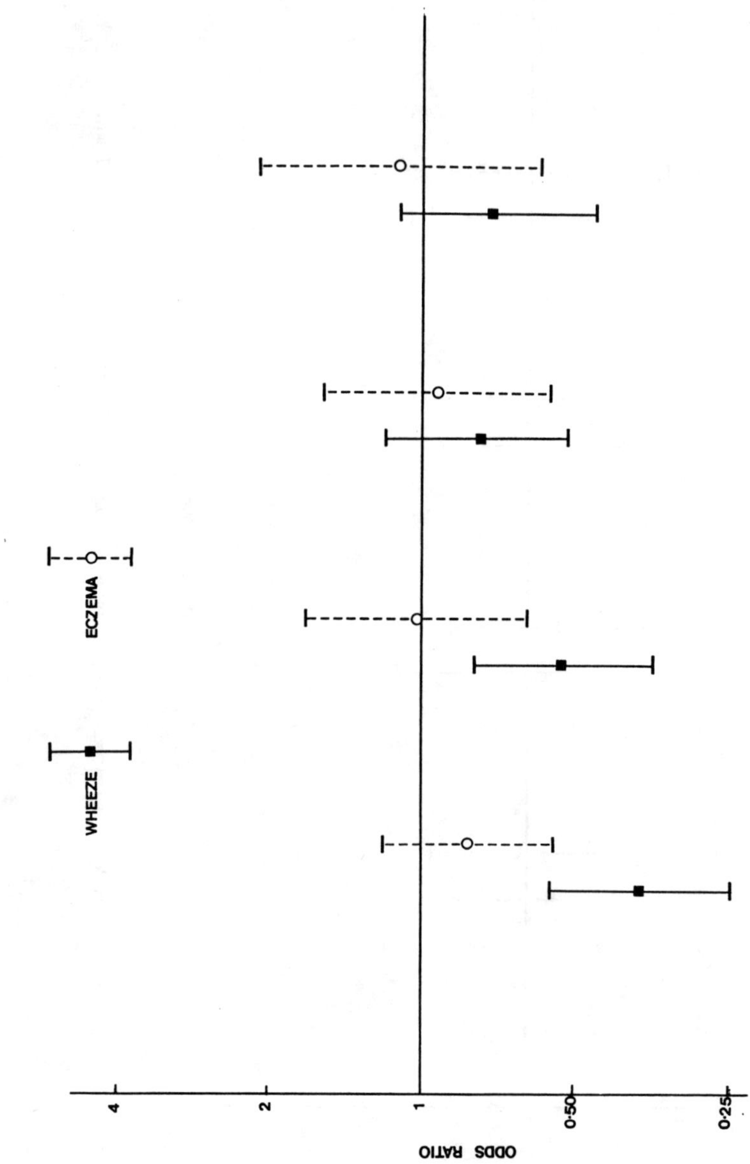

Fig. 9. Risk of symptoms for breast fed relative to never breast fed infants.

larly since IgE levels were no lower in the breast-fed than in those who were never breast-fed. Perhaps the geographical variation in risk of respiratory disease for infants explains the lack of agreement between different studies of the effect of breast feeding on symptoms. The higher incidence of wheeze in boys compared with girls at each age is striking, and reflects the higher prevalence of asthma (Dawson *et al.*, 1969; Burr *et al.*, 1974; Hamman *et al.*, 1975) and of bronchial hyperreactivity (Sears *et al.*, 1986) in boys compared to girls later in childhood. The tendency for boys to have higher IgE levels than girls presumably reflects the same excess male risk of allergic disease. Eczema was not predicted by cord IgE, although in the first year it was associated with a high IgE at 3 months, as others have reported (Dannaeus *et al.*, 1978; Gordon *et al.*, 1982). The association between eczema in the first year and egg IgE antibodies suggests that egg allergy contributes to eczema in infancy as it does in later childhood (Atherton *et al.*, 1978). Positive skin tests to egg occurred more frequently in infants with a high 3 month IgE and were very strongly associated with egg specific IgE antibodies, corroborating this hypothesis.

IgG4 antibodies were measured because of the possibility that they constitute a natural defence mechanism against allergic disease (Merrett *et al.*, 1984). Unlike IgE antibodies they cross the placental barrier, and the serum levels then decline during early infancy until the child synthesizes its own IgG4. A protective function of IgG4 is supported by the lower risk of wheeze in the first year among infants with a high cord egg IgG4 titre, and also by the tendency for a lower incidence of wheeze to occur throughout the first four years of life among children with a high 3 month IgG4 titre. It was noteworthy that a high 3 month IgG4 titre occurred more commonly in girls than in boys, suggesting that it may contribute to the reduction in risk of allergy in girls.

It must be conceded that relatively few significant associations were found, given the number of relationships examined. It is unlikely that the associations detected were fortuitous, since they all were in the directions expected and showed overall consistency. The predictive value of immunological measurements probably varies in different areas according to the relative importance of allergic, infective, irritant and other factors in the causation of symptoms.

The follow-up of these children is continuing and it is intended that they shall all be examined at the age of 7 years, when the predictive effects of immunoglobulin assays and other factors will be further examined.

The following conclusions can be drawn from this study:

1. IgE antibodies in early infancy are associated with an increased risk of allergic disease;
2. IgG4 antibodies are probably protective against allergy;
3. Egg appears to be an important allergen in infancy;
4. Breast milk, even in small amounts, is protective against respiratory symptoms in infancy;
5. The predictive effects of these indices may vary in different areas according to the relative importance of allergic, infective and other factors in causing symptoms.

ACKNOWLEDGEMENTS

We gratefully acknowledge a financial grant from the Welsh Scheme for the development of Health & Social Research.

REFERENCES

Atherton, D.J., Sewell, M., Soothill, J.F., Wells, R.S. and Chilvers, C.E.D. (1978). A double-blind controlled crossover trial of an antigen-avoidance diet in atopic eczema. Lancet, i, 401-403.

Breslow, N.E. and Day, N.E. (1980). Statistical methods in cancer research, Vol.1: the analysis of case-control studies. IARC scientific pulications No.32, International Agency for Research on Cancer, Lyon; p.134.

Burr, M.L. (1983. Does infant feeding affect the risk of allergy? Arch. Dis. Child., 58, 561-565.

Burr, M.L. (1987). Is asthma increasing? J. Epidemiol. Community Health, 41, 185-189.

Burr, M.L., Eldridge, B.A. and Borysiewicz, L.K. (1974). Peak expiratory flow rates before and after exercise in schoolchildren. Arch. Dis. Child. 49, 923-926.

Businco, L., Marchetti, F., Pellegrini, G. and Perlini, R. (1983). Predictive value of cord blood IgE levels in 'at risk' newborn babies and influence of type of feeding. Clin. Allergy, 13, 503-508.

Chandra, R.K., Puri, S. and Cheema, P.S. (1985). Predictive value of cord blood IgE in the development of atopic disease and role of breast-feeding in its prevention. Clin. Allergy, 15, 517-522.

Colley, J.R.T. and Reid, D.D. (1970). Urban and social origins of childhood bronchitis in England and Wales. Br. Med. J., 2, 213-217.

Croner, S., Kjellman, N-I.M., Eriksson, B. and Roth, A. (1982). IgE screening in 1701 newborn infants and the development of atopic disease during infancy. Arch. Dis. Child., 57, 364-368.

Dannaeus, A., Johansson, S.G.O. and Foucard, T. (1978). Clinical and immunological aspects of food allergy in childhood. Acta Paediatr. Scand. 67, 497-504.

Dawson, B., Horobin, G., Illsley, R. and Mitchell, R. (1969). A survey of childhood asthma in Aberdeen. Lancet, i, 827-830.

Dean, G., Lee, P.N., Todd, G.F., Wicken, A.J. and Sparks, D.N. (1978). Factors related to respiratory and cardiovascular symptoms in the United Kingdom. J. Epidemiol. Community Health, 32, 86-96.

Gordon, R.R., Noble, D.A., Ward, A.M. and Allen, R. (1982). Immunoglobulin E and the eczema-asthma syndrome in early childhood. Lancet, i, 72-74.

Hamman, R.F., Halil, T. and Holland, W.W. (1975). Asthma in schoolchildren: demographic associations and peak expiratory flow rates compared in children with bronchitis. Brit. J. prev. soc. Med., 29, 228-238.

Merrett, T.G. (1986). Quantification of IgE and IgG4. In: Applications of Immunological Methods in Biomedical Sciences, 4th Ed. (ed. D.M. Weir), Blackwell Scientific Publications, Oxford, pp.130.1-130.23.

Merrett, J., Barnetson, R.St.C., Burr, M.L. and Merrett, T.G. (1984). Total and specific IgG4 antibody levels in atopic eczema. Clin. Exp. Immunol. 56, 645-652.

Michel, F.B., Bousquet, J., Greillier, P., Robinet-Levy, M. and Coulomb, Y. (1980). Comparison of cord blood IgE concentrations and maternal allergy for the prediction of atopic disease in infancy. J. Allergy Clin. Immunol., 65, 422-430.

Miskelly, F.G., Burr, M.L., Vaughan-Williams, E., Fehily, A.M., Butland, B.K.

and Merrett, T.G. (1988). Infant feeding and allergy. Arch. Dis. Child. 63, 388-393.

Nye, L., Merrett, T.G., Landon, J. and White, R. (1975). A detailed investigation of circulatory IgE levels in a normal population. Clin. Allergy, 4, 13-24.

Sears, M.R., Jones, D.T., Holdaway, M.D., Hewitt, C.J., Flannery, E.M., Herbison, G.P. and Silva, P.A. (1986). Prevalence of bronchial reactivity to inhaled methacholine in New Zealand children. Thorax, 41, 283-289.

Allergen Exposure in Infancy

J. O. Warner and J. A. Price

Cardiothoracic Unit, Brompton Hospital, Fulham Road, London SW3 6HP, UK

ABSTRACT

The increased prevalence of allergy is reviewed briefly in relation to predisposing states and environmental trigger factors. The protective effects of breast feeding are discussed, but data is also presented to show that human milk can be more cause for concern than the highly modified cows' milk formulae nowadays available. Clinical studies of airborne allergens have been performed during the month after the child's month of birth. In contrast to other studies which utilized high volume air samplers, using low volume samplers we have found detectable amounts of aerosolysed house dust mite and cat antigen in bedrooms. The amounts of antigen were quantified by an ELISA technique which used specific monoclonal antibodies, and could be correlated in a highly significant manner with the degree of skin test positivity to the sampled antigen. These results substantiate the postulate that an infant can inhale quantities of allergen which influence the subsequent development of allergy.

KEYWORDS

Allergy; environmental factors; breast feeding; airborne allergens.

INTRODUCTION

There is strong evidence that there has been an increase in the incidence of allergic disorders in our population [1]. As there has been poor concordance for asthma in identical twins [2] and there has been no change in the genetic stock to account for the increased incidence of allergic disease, it is likely that environmental factors have increased the chances of genetically predisposed individuals manifesting their genotype. Recent studies from developing countries would suggest that increasing exposure to house dust mite is a significant factor in the increased prevalence of asthma [3]. The same could be true in western countries.

Advances in the Biosciences Vol. 74

Printed in Great Britain

PREDISPOSING FACTORS

A number of predisposing states have been identified which appear to increase susceptibility to the development of allergy. These states include family history of atopy [4], opsonization defects [5], immunoglobulin A deficiency [6], second component of complement deficiency [7], diminished suppressor T-lymphocyte numbers [8] and possibly possessing the CF gene [9]. Of these factors only a family history and a high cord blood IgE have a reasonable degree of sensitivity, though specificity is low [10]. In the predisposed infant the type of allergy and the likelihood that it will develop have been associated with maternal diet in pregnancy and during breast feeding [11], month of birth [12], infant feeding regimes [13,14], parental smoking [15], infections in infancy with particular reference to gastro-enteritis [16].

The relative contribution of each factor to the development of allergic disease is unknown and before any sensible preventive regimes can be recommended in susceptible families, this must be delineated. The nature of sensitization to either high concentration exposure or repeated low dose exposure requires characterization.

DIET

Most studies to date are focused on diet in the evolution of eczema. This has occurred because it is easier to monitor diet and to document the presence or absence of eczema. Furthermore, this condition is likely to evolve within the first two years of life. On the whole studies have suggested that breast feeding is protective by comparison with cow's milk formula based bottle feeding [13,14]. However, with the advent of the highly modified, and less allergenic cow's milk formula feeds [17], more recent studies have failed to show such dramatically protective effects [18]. Indeed, it has been suggested that the environment is now having a greater adverse influence on human milk compared with cow's milk to account for the loss of protective effect of natural breast feeding [19].

BREAST MILK

It is widely accepted that potentially allergenic macromolecules are absorbed by the normal adult gut and can pass from mother to infant in the breast milk [20]. Several decades ago physicians realized that such substances could produce adverse reactions in susceptible infants [21]. Unfortunately, most reports have been anecdotal and have relied on unsubstantiated information. We have conducted a controlled study in an infant with egg allergy whilst being fully breastfed. Double blind challenges of the mother with egg resulted in relapse in the infant's eczema during periods when the mother was eating eggs. Furthermore, breast milk collected from the mother and skin tested in the baby only produced positive reactions from milk collected whilst the mother had egg in her diet [22]. These observations have led to more recent prospective studies which have shown that high risk babies of mothers who avoid egg, cow's milk and fish during the first three months of lactation are less likely to develop allergic disease than breastfed infants of mothers on a normal diet [23]. This confirms the importance of antigen in breast milk as a route of sensitization.

AERO-ALLERGENS

Far less information is available on aeroallergen exposure in the development of allergic disease. Indirect evidence that this is important comes from month of birth studies, particularly in relation to birch pollen allergy [24]. Our own observations have shown significant differences in month of birth of house mite sensitive asthmatics, compared with asthmatics without this sensitivity [12].

Our month of birth observations in relation to house mite allergy are very difficult to reconcile with studies of the Platts-Mills' Group utilizing air sampling techniques. They found that in an undisturbed bedroom, airborne mite allergen cannot be detected when using sampling rates of 17 litres per minute for up to eight hours [25]. Most of the particles of house mite allergen found in the air during vigorous domestic disturbances were on particles of greater than 10 microns [25]. Thus, it would be unlikely that such particles would penetrate into airways. Furthermore, a study attempting to estimate the likely exposure of infants to house mite allergen using an infant respiratory pump only detected very minute quantities of the allergen during bed making when the apparatus was held extremely close to the blankets. In other situations no allergen was detected [26].

STUDY

We have therefore conducted a study using a low volume Casella personal sampler. This samples at a rate of 2 litres per minute which happens to be the approximate tidal minute volume of a newborn baby. Contrary to previous observations we have been able to detect significant quantities of the principle allergen of house mite (*Der p I*) following three hours of sampling at two litres per minute. The samples were taken in the patients' homes whilst normal activities were in progress in the sampling area. The house mite antigen was detected by an ELISA technique utilizing a monospecific antiserum to *Der p I*. In addition, the principle allergen of cat fur (*Fel d I*) was analysed by an ELISA technique utilizing a monoclonal antibody provided by Dr T. Platts-Mills. Samples were eluted from glass fibre filters used in the Casella sampler. All the sampling was done in the homes of children with asthma, in the month after the child's month of birth provided that there had been no changes to the living quarters since the child's birth.

RESULTS

There was an excellent linear correlation between the quantity of *Der p I* detected in dust collected from the bedroom carpet and the bedroom air ($R = 0.95$). There was a less satisfactory correlation between bedroom air and mattress dust ($R = 0.5$). In other words the level of *Der p I* detected by the low volume air sampler was valid as it correlated well with levels found in the reservoir from which it arises. The observations, furthermore, suggest that carpets are a more important source of aerolysed *Der p I* than mattresses, at least during the daytime.

The asthmatic children with positive skin tests to *Dermatophagoides pteronyssinus* were infinitely more likely to have detectable quantities of *Der p I* in bedroom air from the low volume air sampler than those who were skin test negative ($p<0.01$). Similarly, children who were skin test positive to cat fur were significantly more likely to have detectable levels of *Fel d I* in the bedroom air than those who were skin test negative. This is in contrast to our observations in a previous study that there was no association between the degree of sensitivity to house mite, either from skin testing or IgE antibody measurement and the presence of *Der p I* in the mattress dust sampled at any time of the year in randomly selected homes of asthmatic children.

CONCLUSIONS

Thus, low volume air sampling is a satisfactory way of monitoring true aeroallergen exposure and our observations would suggest that early aeroallergen exposure is important and validates the hypothesis that an infant is able to inhale quantities of allergen which influences the subsequent development of allergy.

REFERENCES

1. Fleming, D.M. and Crombie, D.L. (1987). Prevalence of asthma and hay fever in England and Wales. Br. Med. J., 294, 279-283.
2. Edfors Lubs, M.L. (1971). Allergy in 7,000 twin pairs. Acta Allergologica, 26, 249.
3. Woolcock, A.J., Dowse, G., Temple, K., Stanley, H., Alpers, M.P. and Turner, K.J. (1983). Prevalence of asthma in the South Fore people of Papua-New Guinea. A method for field studies of bronchial reactivity. Eur. J. Respir. Dis., 64, 571-581.
4. Wittig, H.J., McLaughlin, E.T., Leifer, K.L. and Bellert, J.D. (1987). Risk factors for the development of allergic disease: analysis of 2190 patient records. Ann. Allergy, 41, 4-8.
5. Richardson, V.F., Larcher, V.F. and Price, J.F. (1983). A common congenital immunodeficiency predisposing to infection and atopy in infancy. Arch. Dis. Child., 58, 799.
6. Taylor, B., Norman, A.P., Urgel, H.A., Stokes, C.R., Turner, M.W. and Soothill, J.F. (1973). Transient IgA deficiency and pathogenesis of infantile atopy. Lancet, 2, 111-113.
7. Hutchings, P. and Walker-Smith, J.A. (1982). The gastrointestinal system. Clinics in Immunology and Allergy, 2, 43-76.
8. Chandra, R.K. and Baker, M. (1983). Numerical and functional deficiency of suppressor T cells precedes development of atopic eczema. Lancet, 1, 193.
9. Warner, J.O., Norman, A.P. and Soothill, J.F. (1976). Cystic fibrosis heterozygosity in the pathogenesis of allergy. Lancet, 1, 990-991.
10. Chandra, R.K., Puri, S. and Cheema, P.J. (1985). Predictive value of cord blood IgE in the development of atopic disease and role of breast feeding in its prevention. Clin. Allergy, 15, 517-522.
11. Chandra, R.K., Puri, S., Suraiya, C. and Cheema, P.S. (1986). Influence of maternal food antigen avoidance during pregnancy and lactation on incidence of atopic eczema in infants. Clin. Allergy, 16, 563-569.
12. Warner, J.O. and Price, J.F. (1978). House mite sensitivity in childhood asthma. Arch. Dis. Child., 53, 710-713.
13. Glaser, J. and Johnstone, D.E. (1953). Prophylaxis for allergic disease in the newborn. JAMA, 153, 620-622.
14. Matthew, D., Taylor, B., Norman, A.P., Turner, M.W. and Soothill, J.F. (1977). Prevention of eczema. Lancet, 1, 321.
15. Taylor, B. and Wadsworth, J. (1987). Maternal smoking during pregnancy and lower respiratory tract illness in early life. Arch. Dis. Child., 62, 76-791.

16. Manuel, P.D. and Walker-Smith, J.A. (1981). A comparison of three infant feeding formulae for the prevention of delayed recovery after infantile gastroenteritis. Acta Paediatr. Belg., 34, 13-20.
17. McLaughlin, P., Anderson, K.J., Widdowson, E.M. and Coombs, R.A. (1981). Effect of heat on the anaphylactic-sensitizing capacity of cow's milk, goat's milk and various infant formulae fed to guinea pigs. Arch. Dis. Child., 56, 165-171.
18. Warner, J.O. (1985). Allergies in childhood. Progress in child health. (ed. J.A. Macfarlane), 2, 63-76.
19. Taylor, B., Wadsworth, J., Wadsworth, M. and Peckham, C. (1984). Changes in the reported prevalence of childhood eczema since the 1939-45 war. Lancet.
20. Matsumara, T., Kuroume, T., Oguri, M., Iwasaki, I., Karobe, Y., Yamadu, J., Kawabe, S. and Negishi, K. (1975). Egg sensitivity and eczematous manifestations in breast fed newborns with particular reference to intra-uterine sensitization. Ann. Allergy, 35, 221.
21. Talbut, F.B. (1918). Eczema in childhood. Med. Clin. N. Amer., 1, 95.
22. Warner, J.O. (1980). Food allergy in fully breast fed infants. Clin. Allergy, 10, 133-136.
23. Bjorkstein, B. Personal communication.
24. Bjorkstein, F., Suoniemi, I. and Koski, V. (1980). Neonatal birth pollen contact and subsequent allergy to birch pollen. Clin. Allergy, 10, 581.
25. Tovey, E.R., Chapman, M.D. and Platts-Mills, T.A.E. (1981). The distribution of dust mite allergen in the houses of patients with asthma. Am. Rev. Respir. Dis., 124, 630-635.
26. Carswell, F., Clark, J., Robinson, D.W. and Platts-Mills, T.A.E. (1982). Bristol house dust mites. Modern Problems in Paediatrics, 21, 86-93.

Discussion: Session Five

Chairman: Sami Bahna

Jack Pepys. I would like to add one extra piece of information which may assist John Warner in his interesting studies. In the early 1950's Kate Maunsell was surveying the houses of asthmatics in order to see if there was a common environmental factor. On a map of London she pinned flags which identified the homes where the asthma began. The result was that the flags followed underground streams. This was a very important finding because it identified houses which are asthma houses, simply because they are located over water.

John Warner. Thank you for making the point, I omitted to say that we have been monitoring both the wall moisture and relative humidity in all the homes, but as yet the results still have to be analysed.

Jack Pepys. You still ought to know what is happening under the homes.

John Warner. Wall moisture should be higher where there is an underground stream, and it does seem that the mite content in mattress dust correlates with the moisture readings. As a result I think we should be able to substantiate that earlier observation.

Sheldon Spector. I was also impressed with Dr Warner's study and I am wondering if you plan to incorporate into it any measurements relating to cockroach sensitivity. This does seem to be an increasingly recognized cause of allergy, at least in the USA, and I wonder if you would comment on that. I would like the panellists to summarize what recommendations we should make to an allergic family in order to try and prevent allergy among their children; for example, should we recommend not only breast feeding but also a very rigid diet for the mother?

Sami Bahna. Can we deal first with the question about cockroach hypersensitivity?

John Warner. I have no experience of this at all, other than to say that we certainly have cockroaches in our hospital. Perhaps we should try and identify whether or not they are present in our patients' homes.

Advances in the Biosciences Vol. 74

Robert Hamburger. We certainly see an enormous amount of cockroach positivity in San Diego, particularly in certain sections of the city. I recall that because people get upset if you insinuate their homes are infested with cockroaches, we had to euphemistically call our cockroach extract a test for "household insects".

Sami Bahna. Perhaps we can now address the question of preventing allergy. First, may I add here that to control the environment will not be enough; we will be faced with parents who have rigorously followed all our precautionary measures, yet still have very allergic children, because they are genetically predisposed to allergy.

Robert Hamburger. If there is sufficient allergy in the family to make it worthwhile carrying out precautionary measures, I would counsel that: (1) there shall be no smoking in the home, (2) both mother and baby be on a rigid diet, (3) breast feeding be carried out for at least four, preferably six months, and (4) environmental controls be implemented.

Michael Burr. The only thing I would add to that list is to arrange that the babies not be born in the autumn. However, in our South Wales study it was only important that the babies be breast-fed for a few weeks, since prolonged breast feeding did not produce any additional benefit.

John Warner. My main emphasis would be on (1) no smoking and (2) no pets. If you can breast feed then go ahead, but let's not be too dogmatic about it, because the present-day formula feeds are much less sensitizing than the older formulae. I would add that, like the South Wales study showed, we can find no advantage in substituting a soya-based formula for cows' milk. Presumably we must accept that there are different underlying abnormalities which pre-dispose to sensitization, for example, an IgA deficiency and opsonization or T-cell defects, and so different regimens will have to be used to protect babies in these different categories. It might very well be that low-dose exposure as may occur via breast milk could prove harmful in some situations but beneficial in others.

Robert Hamburger. We should certainly take the greatest care in avoiding the laying of a "guilt-trip" for those families who may be unable to comply with some of these regimens.

Alec Sehon. The message I am receiving is that allergy in the population increases from 16% in childhood to 30% among adults, due to environmental factors. If indeed this is so, have you any data to demonstrate some genetic predisposition among those individuals who will become more allergic as they grow older?

Robert Hamburger. I am sorry Alec, but that was not the message I was trying to get across. What I was trying to say was that regarding total serum IgE levels the environment is responsible for only 16% of the observed effect in infants, whilst this influence is increased to be responsible for 30% of the observed effect in adults.

Alec Sehon. There was also a statement about cord blood which I would like clarified. About 12 years ago Dr Miller stated that at 12 weeks of gestation he detected IgE in foetuses. Is that correct?

Sami Bahna. I think it was I who said that if the foetus is both predisposed and exposed to the right triggering antigen, then foetal IgE will be increased.

Alec Sehon. What would you say to the proposal that this increase could be due to an anti-idiotypic response?

John Warner. The same suggestion that anti-idiotypic antibody is transmitting information has been recently proposed in the "Lancet". However, in that case the transferral was via breast milk and not across the placenta.

Alec Sehon. But in the situations to which I refer, the baby has not yet been born.

Robert Hamburger. The foetal data to which you refer was published by Gitlin. What he showed was that the cells capable of producing IgE were present very early in embryonic development; he went no further than that.

Bruce Mitchell. If I may I should like to relate with the aid of just three slides some clinical studies concerning low dose exposure to antigens which we performed over five years with Jeremy Cogswell in England.

Sami Bahna. Please proceed Dr Mitchell.

Bruce Mitchell. The late Ellen Jarrett did some elegant experiments which showed that low dose exposure of rodents to egg albumen was likely to induce a specific IgE response. On that basis the small amount of egg-albumen that is passed in human breast milk should preferentially elicit an IgE response. The first slide shows that experimental model holds good because over the five years of this study the breast-fed babies are much more likely to give a positive skin test or IgE-RAST to egg albumen than children who had never been breast-fed. In contrast there was no difference in the response to house-dust mite nor to grass pollen in either of the groups. The second slide deals only with the amounts of house-dust mite allergen, *Der p I*, in bedroom dust to which our children were exposed during this study. Positive skin tests and IgE RAST to *Der p I* were much more common when the children were exposed to high concentrations of *Der p I*; the threshold value appeared to be 2 μg *Der p I* per gram of fine dust. Similar results were obtained for the assay of mite specific IgG antibodies, and so there is no doubt that exposure to inhaled allergens in early life alters your immune response. The third and final slide shows cumulative data for the 97 children. It is clear that eczema peaks at age 1-2 years and that this correlates with the detection of egg-specific IgE, so it would appear that breast-fed babies are more likely to suffer from eczema. The plot for wheezing episodes is somewhat different, as there is a gradual increase over the period of study and this correlates with *Der p I*-specific IgE levels. The wheeziness is also influenced by parental smoking habits, and at five years a synergistic effect between smoking and atopy is produced. On the basis of these results I would expect that the large study in South Wales reported by Mike Burr will find that exposure to mite antigen will correlate with symptoms when his children reach the ages of 4-7 years.

Terry Merrett. I think it could be dangerous to assume that only low doses of antigen will be present in breast milk. In our laboratory 30 specimens of human milk which had been sampled at random were assayed for their egg albumen and β-lactoglobulin content. The results formed two distinct groups, one with concentrations less than 200 μg/ml and the other with amounts ranging from 2-5 mg/ml. So far we have not found any milk which contains milligram quantities of both antigens. We would like to collect and assay specimens more systematically, but it does indicate that some babies will encounter substantial amounts of these antigens via their mothers' milk.

Alec Sehon. One final question which I have concerns the Hamburger pentapeptide. Could Bob tell us more about the results of clinical trials undertaken by Immunetech Pharmaceuticals? I understand they are quite remarkable.

Robert Hamburger. The *in vivo* data accumulated by Immunetech, a company in which I had an interest but am now dissociated from, is very impressive and it will be presented at the International Meeting in Montreux. The peptide has been given by injection, intranasally and as eye drops, and in some individuals it is remarkably effective in blocking major allergic disease. On the negative side I find it disturbing that *in vivo* data is not mirrored by changes in *in vitro* data. I have some of the data with me, but won't take the time to show it now.

Birgit Helm. I want to make the point that we have studied Dr Hamburger's peptide *in vitro* and *in vivo*, and in concentrations ranging from micromolar to millimolar. Unfortunately we were unable to show any inhibitory effects, whether the peptide contained an aspartic acid or asparagine residue; neither were active in our hands. We have also tested the material in competition and binding studies with the same low affinity receptors which Dr Hamburger originally reported, in 1980 or 81, to be blocked by the pentapeptide. Unfortunately, in our hands, it did not work either *in vitro* or *in vivo*.

Robert Hamburger. Nevertheless, I am still grateful that I have patents both on the preparation and use of the peptide.

Birgit Helm. I would like to comment on the effect of smoking and the possible association with atopic allergy. At the recent meeting in Anaheim a paper was given which suggested that the major glycoprotein from tobacco smoke is a very potent stimulus for IgE synthesis. This may explain why children of mothers who smoke have a high tendency to become atopic.

Sam Donta. Just a comment about smoking. It is thought that smoking might affect both the production of certain hormones such as ACTH and the release of enkephalins; on lymphocytes there are receptors for both of these substances. Together these substances might either augment T-helper, or reverse T-suppressor responses so as to increase the IgE synthesis to any antigens that are concurrently in the environment.

Larry Schwartz. If according to the Heisenberg Uncertainty Principle you don't know where you are, and you don't know how fast you are going - it seems like you are lost! However, as you know where you want to go you are not really lost, just unsure of the best way to get there. Along these lines I want to go back to Sami Bahna's data which indicated that oral food challenges and skin tests were not always concordant. In fact, probably the most disturbing thing was that in some cases the skin tests were negative yet a diagnosis of food allergy was made on the basis of a positive food challenge. Is this allergy, or is it some other form of adverse reaction, and how do you propose to get to the point where you can make that determination?

Sami Bahna. We would like to document immunologically all those adverse reactions to food that we call allergic reactions. Some are definitely allergic disease, but we cannot document an allergic basis for them. For example, we all see patients who have generalized urticaria and angioedema after food challenge; this cannot be just a food intolerance or an enzyme deficiency, it has to be an allergic reaction, yet skin tests and RAST results may all prove to be negative. Likewise there are patients who wheeze a few minutes after food challenge, yet their immunological tests will also be negative. These are examples of food allergy, although our immunological tools as yet do not detect them. This may not satisfy you completely Larry, but hopefully John Warner will.

John Warner. I would prefer the terminology "food intolerance" so as not to pre-suppose any mechanism as yet, although I agree that many of these children will subsequently be shown to be allergic. Tartraziné intolerance may be an example of an adverse reaction which results in the manifestation of allergic disease, yet is obviously not allergy. We know that large doses of tartrazine will non-specifically release histamine in patients with urticaria, and that very large doses will do the same in normal people. Since urticarial patients will develop a reaction to much lower levels of a histamine infusion than will normals, it seems that tartrazine will trigger their allergic symptoms non-specifically. This is obviously not allergy.

Sami Bahna. I wonder if David Marsh has any comments to make on the relative importance of heredity versus environment in the development of allergy, and whether heredity determines the allergic symptoms or the specific allergen as well?

David Marsh. My colleague Dr Myers tried to define the factors which might help to predict allergy in children, and she found that skin test positivity was a more reliable predictor in an elder sibling than in either parent. I think the specific question which you ask, heredity versus environment, is too tough a subject to answer at present, because there is just not enough data available.

Sami Bahna. Can a person inherit the same allergen as a parent?

David Marsh. The inheritance of a certain immune response gene is going to be an important factor in the specific immune response to a highly purified allergen. However, the inheritance of a particular HLA-D type from a parent is not a sufficient condition; for example, the presence of DR2, Dw2 in an individual doesn't always lead to a response against the *Amb a* V allergen from ragweed. To complicate the issue many ragweed allergic patients make IgE responses to allergens other than *Amb a* V. There are obviously other factors, perhaps its the T-cell repertoire or maybe the regulation of Ia expression, we don't really know. I think that studies of the T-cell receptor and the expression of Ia molecules together with the cytokines which control this regulation are going to be very fertile areas for study in the future.

John Warner. It is important to distinguish allergy from disease, because for allergy concordance in identical twins is high, whereas for disease it is quite low - as low as 20% for asthma. I have had the opportunity of studying just two cases of identical twins, both sets had almost identical skin test responses to a panel of allergens, but only one of each set was asthmatic as determined by monitoring bronchial reactivity after histamine challenge.

David Marsh. There are some very interesting studies by Mal Blumenthal's group which have not yet been published as a full paper. He is studying twins raised together and apart and finds an amazing concordance in allergic symptomatology and skin test positivity in monozygotic twins raised apart; in fact, the concordance is as high as twins raised together. I find that quite surprising, yet it goes along with other traits which the Minnesota people have observed, such as women being raised apart but wearing five rings on their fingers. If you can genetically control traits like that, which everybody used to think were environmental, I would imagine that the genetic aspects of the allergic response must be very high.

Session Six

ALLERGENS

Chairman: J. Pepys

Allergens and Allergy: Factors Involved in Allergy to Low Molecular Weight Chemicals

J. Pepys

34 Ferncroft Avenue, London NW3 7PE, UK

ABSTRACT

Common allergens are predominantly protein or glycoprotein. A role for polysaccharide allergens, especially in fungi, is suggested by the potent effects of the cell wall mannan of *Candida albicans*.

Low molecular weight inorganic and organic haptens are important as allergens. Induction of IgE antibody and elicitation of reactions are discussed in terms of hapten structure; carrier proteins and conjugates; simple leaving ligands such as chlorine ions, and the cross reactivity between IgE antibodies to new antigenic determinants on carrier proteins produced by one chemical hapten and the same or similar determinants produced by unrelated chemical haptens. Amongst the haptens illustrating these points are; the platinum, nickel and copper, salts; chlorhexidine and the striking influence on IgE antibody production of adding its conjugate with carrier protein to chlorinated water; quaternary ammonium salts; di-isocyanates and protein reactive anhydrides such as phthalic and trimelltic anhydrides.

KEYWORDS

IgE antibodies; polysaccharide allergens; low molecular weight chemicals; leaving chlorine ligands; new antigenic determinants.

INTRODUCTION

The terms used in allergy have precise meanings as well as less precise general usage.

Antigens by definition are agents capable of inducing the production of specific antibodies, humoral or cellular or both, capable of mediating the different types of allergic reaction, either singly or in association with one another. Sensitisation depends on the nature of the antigen on the one hand and the immunological response of the subject on the other. In the present context of common allergic disease the antibodies under consideration

Advances in the Biosciences Vol. 74

Printed in Great Britain

are those capable of mediating immediate and late reactions, exemplified by skin (Fig. 1) and bronchial reactions to tests (Fig. 2). The relevant antibodies are mainly, and not exclusively, in the mast cell sensitising IgE class. Heat-stable short-term sensitising IgG antibodies (IgG-STS) and precipitating antibodies may also be involved.

Allergens are antigens which elicit allergic reactions on combination with specific antibody. Allergenic properties are distinct from irritants in that only a proportion, usually minor, of exposed subjects react to the allergen and then to very small, non-irritant amounts.

Allergens will be discussed as follows: (1) common allergens; protein or glycoprotein are discussed elsewhere in detail, and polysaccharide type allergens; (2) low molecular weight, organic and inorganic, chemical haptens.

Atopy in immunological terms describes the production of IgE antibodies to common environmental allergens met in the ordinary way (Pepys, 1975). Atopy as defined originally by Coca (1923) consisted of a mixture of immediate type allergy and a number of clinical disorders which are often present in many subjects with no evidence of IgE mediated sensitivity. It is usually taken that about 10% of the population may have common allergic disorders. The prevalence, however, of atopy in immunological terms in asymptomatic subjects varies between 30 to 40% in our own and other investigators' reports. The presence of specific IgE antibodies can be shown by skin prick tests and/or serological tests.

Atopics are sensitised by allergens entering through mucosal pathways in relation to which lie the IgE producing plasma cells (Tada and Ishizaka, 1970). The induction of IgE antibody results from T and B cell cooperation mediated by C3 component of complement (Pepys *et al.*, 1977).

Common Allergens, Protein/Glycoprotein

The common allergens are mainly protein or glycoprotein and are described as globular proteins. They are usually between 10,000 to 40,000 daltons in molecular weight and most allergen preparations contain a diversity of allergenic moieties. They appear to have a high T-cell priming effect. There are, as yet, no known common structural features in allergens. The precise chemical nature of the reacting epitopes is known or being determined for only a few allergens. The availability of monomeric or univalent specific epitopes could be of clinical value by combining with IgE molecules without bridging them and thus blocking specific reactions to di- or multivalent allergens. This effect is more likely to be achieved by local than systemic use because of the presence of specific antibodies in the IgG or other classes which would combine preferentially as they have been found to do in the RAST. An approach of this sort to penicillin allergy by the use of a univalent penicilloyl hapten has been found effective (de Weck, 1972), but has been hampered by the acquisition of reactive properties by the treatment material, probably due to polymerisation.

Polysaccharide Type Allergens

A classical example of a polysaccharide type allergen is the cell wall mannan, a mannose polymer, of *Candida albicans*. This preparation has elicited immediate and late skin test and bronchial test reactions, giving positive skin test reactions in a high proportion of asthmatics. Further-

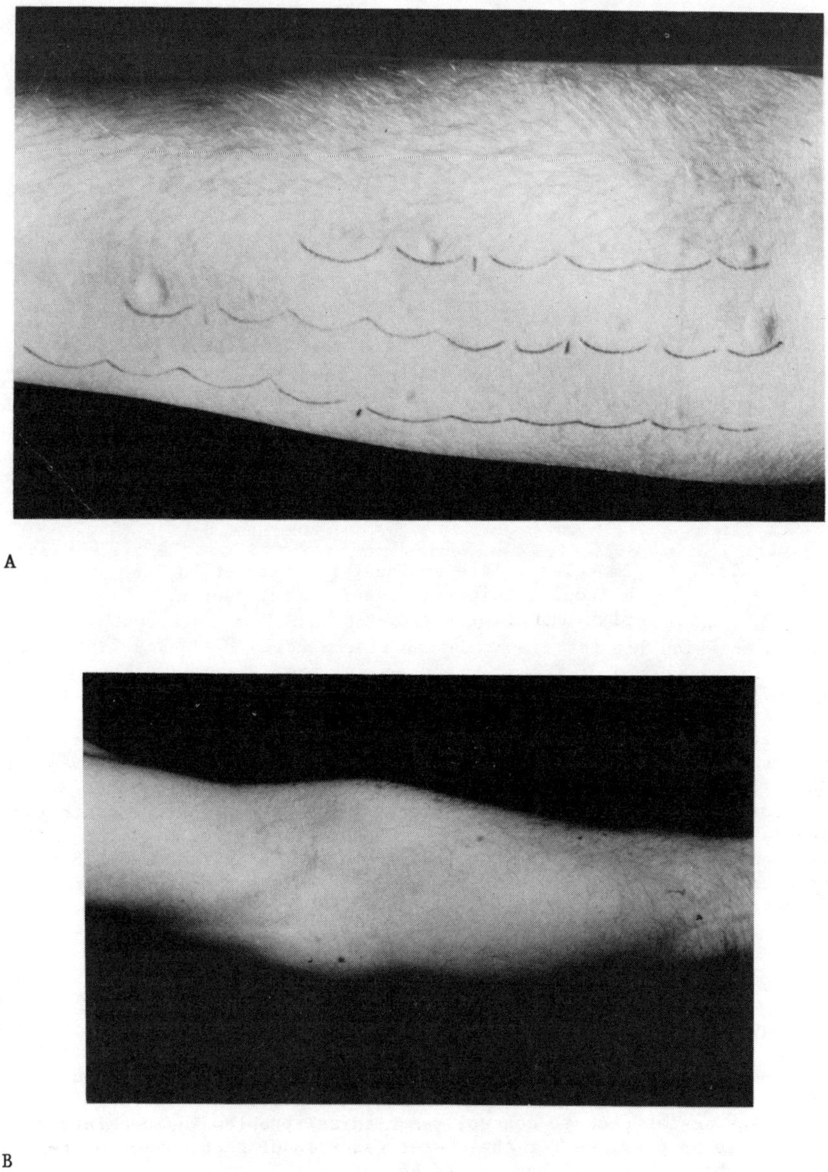

Fig. 1(a) Immediate skin prick test reactions to common allergens.

(b) Late - 3 to 6 hour skin test reaction to a common allergen.

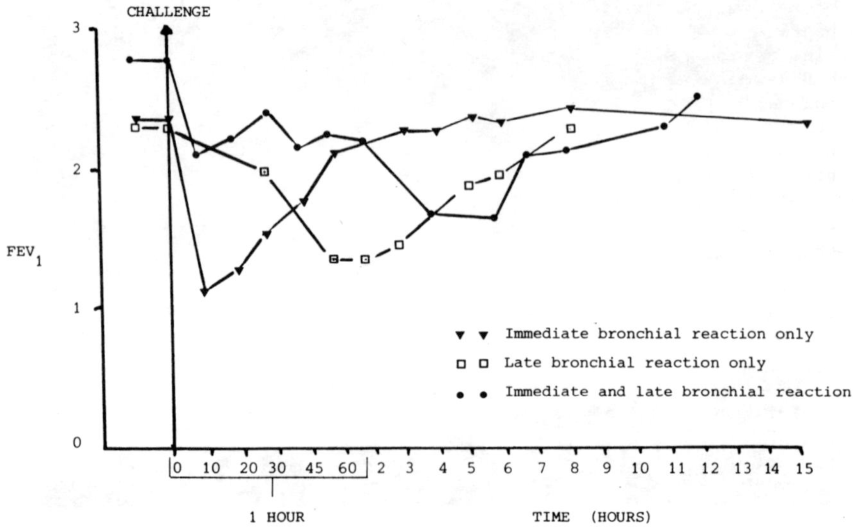

Fig. 2. Immediate, late and dual reactions to a low molecular weight inorganic metal hapten, platinum chloro-salt.

more, in control subjects giving negative reactions to intradermal tests, repeat tests over 1 to 3 weeks gave vigorous dual reactions, accompanied by increases in the amount of antibodies which are present at low level in all subjects (Pepys *et al.*, 1968). It is not known whether the potency of the mannan in enhancing sensitivity is a property of other polysaccharides, in fungi for example, but it is desirable that the polysaccharide content of allergen extracts be known and that it should be compared with the protein allergens present. Antibodies to the mannan in the IgE and IgG-STS class have been shown in skin and passive transfer tests and to these and other immunoglobulins in serological tests (Edge and Pepys, 1980).

Low Molecular Weight, Inorganic and Organic, Chemical Haptens

The production of common allergic diseases by low molecular weight chemicals is best shown by occupational exposures, though it is not limited to them. The presence of and a role for specific IgE antibody has been shown for a number of them. Their defined chemical nature provides basic information as to how they act and how to control the clinical problems. Such information is likely to be of value for the better understanding of common protein allergens whose chemistry has yet to be defined.

Immune Responses to Simple Chemicals

The many factors involved in immunological responses to chemical haptens, their carrier proteins and new antigenic determinants (NAD) and entering and leaving ligands in experimental animals and man are reviewed by Naor and Galili (1977). Their possible relationship in IgE terms to allergy is reviewed by Pepys (1986).

The haptens alone are tolerogenic and require combination with a carrier protein, mainly HSA it is assumed in man, so as to acquire immunogenicity via T and B cell cooperation for IgE antibody production. In addition to the heterogeneity of types of antibody production against the various possible combinations, antibodies may be produced with specificities for different regions on the same antigen. Heavy coupling by the hapten favours production of antibodies to it as against the carrier protein and vice versa. The nature of the coupling also has an effect. For example, DNP coupled to BSA via e-amino groups of lysine residues results in high immunogenicity for the hapten and poor for the carrier, whereas when coupled to tyrosine residues via azo-bonds there is weak immunogenicity for the hapten and strong for the carrier (Hanna *et al.*, 1972). Charge and structural configuration then play a part in antigen/antibody combination.

A further important factor for both immunogenicity and antigen/antibody combination, and to which increasing attention will need to be given in allergy, is the role of entering and leaving ligands, the classical example at present being that of chlorine ions. The influence of charge, structural configuration and ligands on antigen/antibody combination of simple chemicals is reviewed by Pressman (1971), who cites the influence of a simple ligand such as chlorine in the haptenic molecule. Substitution of chlorine for hydrogen in the paraposition of p-aminobenzoate results in more strongly bound combination with antibodies to it, whereas in the orthoposition it results in decreased combination, attributable to steric hindrance.

Chlorhexidine and Chlorine Ions

A remarkable effect of chlorine on immunogenicity and allergenicity in animals and man is provided by the results of simple addition of chlorinated water to chlorhexidine/carrier protein conjugates (Layton *et al.*, 1986). Chlorhexidine carrier conjugates elicited low level IgG antibody responses whereas N-chloro-chlorhexidine covalently bound to protein carriers induced both IgG and IgE responses. Immediate type anaphylactic reactions to chlorhexidine have been reported mainly in Japan (Ohtoshi *et al.*, 1986). Immediate skin prick test reactions, passive transfer reactions and high positive RASTs to chlorhexidine were found. In the light of the animal studies the possibility that chlorhexidine was diluted for clinical use with chlorine containing water has to be seriously considered.

Platinum Salts and Chlorine Ions

Platinum salts, low molecular weight inorganic metallic haptens, provide examples of the role of chlorine ion ligands on immunogenicity and allergenicity and also on the production of antibodies to new antigenic determinants in HSA. Highly chlorinated platinum salts enhance the secondary antibody response to proteins but only after prior priming with an adjuvant, such as *B. pertussis*, in the Norwegian brown rat. Specific IgE antibodies to the platinum determinants were induced by sensitisation with a heterologous carrier protein (Murdoch and Pepys, 1984a,b).

In man the chlorine salts of platinum are extremely potent sensitisers. Their allergenicity is directly related to the amounts of leaving chlorine ligands in the platinum salt, and complexes with strongly bound chlorine ligands with poor leaving abilities did not elicit reactions in sensitive subjects (Cleare *et al.*, 1976). The presence of platinum salt specific IgE antibodies is shown by carefully made prick tests and by passive transfer tests in the monkey and in man, the latter showing not only IgE but also IgG-STS antibodies mediating immediate reactions, and by the RAST (Pepys *et al.*, 1979; Cromwell *et al.*, 1979). Skin prick tests with platinum chloro-salt/HSA conjugates were negative in contrast to the high sensitivity to the platinum salt itself.

In platinum chloro-salt sensitive subjects IgE antibodies have been found against the platinum salt/HSA conjugate in the RAST in 4 out of 38 subjects and in 8 out of the 38 against DNP/HSA. Thus, an unrelated hapten has, in this case, been more effective in producing NAD's in HSA, capable of reacting with the IgE antibodies resulting from the platinum salt sensitisation.

Other Chemicals with Leaving Chlorine Ligands

Two chemicals, chloramine-T and amprolium dihydrochloride, both of which have leaving chlorine ligands, have been reported to cause respiratory allergy together with immediate skin test reactions to unconjugated material (Feinberg and Watrous, 1945; Greene and Freedman, 1976).

Nickel and Copper and IgE to NAD's in HSA

Nickel is another example of a low molecular weight metal hapten which has caused respiratory allergic disease by inhalation of soluble salts reported in 6 subjects. In one of these in whom nickel sulphate caused asthma together with immediate skin and bronchial test reactions, specific IgE and other antibodies were found (Malo *et al.*, 1982). RAST absorption tests were made with nickel, copper, cobalt, zinc and magnesium conjugated with HSA. The RAST values obtained paralleled in decreasing order the binding of these metals with HSA. The copper/HSA conjugate blocked the nickel/HSA RAST. Nickel and copper have a common transport site on HSA, thus showing specificity of NAD's produced by copper for the IgE antibodies against the NAD's induced in nickel/HSA. There was less blocking by cobalt/HSA and none for zinc/HSA and magnesium/HSA (Nieboer *et al.*, 1984).

Factors Influencing Haptenic Ammonium Ions

Anaphylactic reactions to quaternary ammonium myorelaxants such as succinylcholine, gallamine and pancuronium are associated with immediate skin test and RAST reactions for IgE antibody, mediated by the ammonium ions (Baldo and Fisher, 1983; Vervloet *et al.*, 1985). The latter worker has shown that the RAST is specific for diammonium salts in which there are 2 or more linking C ions, the latter probably related to allergen size necessary for bridging of IgE molecules on mast cell membranes. Females are more affected than males, possibly because of greater use of quaternary ammonium antibacterial cosmetic and deodorant agents such as cetrimide and benzalkonium which cross-react with the succinylcholine RAST. Whatever the case anaesthetists should make appropriate skin tests before using such myorelaxants and they need to have test material available.

Protein Reactive Anhydrides and IgE Antibodies to Hapten, Hapten-carrier Conjugates and NAD's

Allergic respiratory disease due to acid anhydrides such as phthalic anhydride and its modifications, and to trimellitic anhydride is associated with evidence of IgE antibodies in skin and RAST tests. Other reactions, mediated by IgG antibodies and possibly other mechanisms are reported. For both the phthalic and trimellitic anhydrides, animal models for specific IgE sensitisation are also reported.

IgE antibodies of heterogeneous specificities have been found against the haptens, the hapten carrier conjugates and NAD's for the phthalic anhydrides (Bernstein *et al.*, 1984).

In the case of trimellitic acid anhydride different patterns of allergic pulmonary reaction are reported associated with mainly either IgE or IgG antibodies. IgE antibodies to NAD's produced by TMA/HSA conjugates are present.

Di-isocyanates, IgE Antibodies to the Hapten, Hapten/Carrier Conjugate and NAD's

Respiratory allergy occurs to TDI and other di-isocyanates and is associated in some reports with immediate skin reactions (Cirla *et al.*, 1975) and immediate as well as late and dual bronchial reactions (O'Brien *et al.*, 1979). There is controversy about the RAST possibly due to the complex antibody and antigenic structures which may be involved (Bernstein, 1982).

IgE antibodies to NAD's in HSA as shown by RAST with TDI/HSA conjugate, also reacted in one subject with an acetic anhydride/HSA conjugate, that is with NAD's produced by an unrelated chemical hapten (Butcher *et al.*, 1982). Similarly, in experimental animals PCA antibodies were induced against di-isocyanate/homologous carrier protein conjugates. These antibodies could be absorbed by a DNP/homologous protein conjugate (Tse *et al.*, 1979), providing another example of the production of NAD's by an unrelated chemical hapten capable of combining with IgE induced by the sensitising hapten. In both of the above cases the haptens combined with lysine residues.

CONCLUSIONS

The ability to study the allergenic effects of well defined low molecular weight chemicals provides opportunities for basic, as well as applied analysis. For example, in the skin tests with platinum salts the availability of a range of these showed the importance of the amount and leaving ability of chlorine ligands in their allergenicity. Similar tests, in subjects sensitive to other agents, such as chloramine-T and amprolium dihydrochloride which also have leaving chlorine ions, and in those cases where other ligands are suspected, are likely to be informative.

The observations on IgE antibodies to NAD's created by one hapten yet combining well with NAD's created by an unrelated hapten await further studies in man and animals, to see whether these NAD's arising from related and unrelated sources, can elicit allergic reactions. An appropriate term for this possibility would be "extrinsically determined auto-allergy" (Pepys, 1986). Tests to establish this would need to pay attention to all the variables which could apply to the preparation of the test materials.

Possible relevance of the above findings to common allergens is suggested, for example, by the report of Bleumink (1970) of enhanced allergenicity of bovine lactoglobulin by incorporation of carbohydrate by N-glycosidic linkage, simply by incubation with lactose causing the "browning" or Maillard reaction, and of enhanced allergenicity resulting from mild heating of milk with lactose.

REFERENCES

Baldo, B.A. and Fisher, M. (1983). Substituted ammonium ions as allergenic determinants in drug allergy. Nature, 306, 262.

Bernstein, I.L. (1982). Isocyanate-induced pulmonary disease: a current perspective. J. Allerg. Clin. Immunol., 70, 24.

Bernstein, D.I., Gallagher, J.S., D'Souza, L. and Bernstein, I.L. (1984). Heterogeneity of specific IgE responses in workers sensitised to acid anhydride compounds. J. Allerg. Clin. Immunol., 74, 794.

Bleumink, E. (1970). Food allergy: the chemical nature of the substance eliciting symptoms. World Rev. Nutr. Diet., 12, 505.

Butcher, B.T., Mapp, C., Reed, M.A., O'Neil, C.E. and Salvaggio, J.E. (1982). Evidence for carrier specificity of IgE antibodies detected in sera of isocyanate exposed workers. J. Allerg. Clin. Immunol, 69, 123 (Abstr.).

Cirla, A.M., Inzoli, S., Aresini, G., Naur, C., Briatico, G. and Zedda, S. (1975). Evaluation of the diagnostic criteria in occupational asthma caused by isocyanates. Med. Lav., 66, 5.

Cleare, M.J., Hughes, E.G., Jacoby, B. and Pepys, J. (1976). Immediate (Type 1) allergic responses to platinum compounds. Clin. Allergy, 6, 183.

Coca, A.T. and Cooke, R.A. (1923). On the classification of the phenomena of hypersensitiveness. J. Immunol., 8, 763.

Cromwell, O., Pepys, J., Parish, W.E. and Hughes, E.G. (1979). Specific IgE antibodies to platinum salts in sensitised workers. Clin. Allergy, 9, 109.

Edge, G. and Pepys, J. (1980). Antibodies in different immunoglobulin classes to *Candida albicans* in allergic respiratory disease. Clin. Allergy, 10, 47.

Feinberg, S.M. and Watrous, R.M. (1945). Atopy to simple chemical compounds; sulphonechloramides. J. Allergy, 16, 209.

Greene, S.A. and Freedman, S. (1976). Asthma due to inhaled chemical agents - amprolium dihydrochloride. Clin. Allergy, 6, 105.

Hanna, N., Jarosch, E. and Leskowitz, S. (1972). Altered immunogenicity produced by change in mode of linkage of hapten to carrier. Proc. Soc. Exp. Biol. Med., 140, 89.

Layton, G.T., Stanworth, D.R. and Amos, H.E. (1986). Factors influencing the immunogenicity of the hapten drug chlorhexidine in mice. II. The role of the carrier and adjuvants in the induction of IgE and IgG anti-hapten responses. Immunol., 59, 459.

Malo, J-L., Cartier, A., Doepner, M., Nieboer, E., Evans, S.L. and Dolovich, J. (1982). Occupational asthma caused by nickel sulfate. J. Allerg. Clin. Immunol., 69, 55.

Murdoch, R.D. and Pepys, J. (1984a). Immunological responses to complex salts of platinum. I. Specific IgE antibody in the rat. Clin. Exp. Immunol., 57, 107.

Murdoch, R.D. and Pepys, J. (1984b). Immunological responses to complex salts of platinum. II. Enhanced IgE antibody responses to ovalbumin with concurrent administration of platinum salts in the rat. Clin. Exp. Immunol., 58, 478.

Naor, D. and Galili, N. (1977). Immune response to chemically modified antigens. Progr. Allergy, 22, 107.

Nieboer, E., Evans, S.L. and Dolovich, J. (1984). Occupational asthma from nickel sensitivity. II. Factors influencing the interaction of Ni^2HSA, and serum antibodies with nickel related specificity. Brit. J. Ind. Med.,

41, 56.

Ohtoshi, T., Yamauchi, N., Tadokoro, K., Migachi, S., Suzuki, S., Miyamoto, T. and Muranaka, M. (1986). IgE antibody-mediated shock reaction caused by topical application of chlorhexidine. Clin. Allergy, 16, 155.

O'Brien, I.M., Harries, M.G., Burge, P.S. and Pepys, J. (1979). Toluene di-isocyanate induced asthma. I. Reactions to TDI, MDI, HDI and histamine. Clin. Allergy, 9, 1.

Pepys, J., Faux, J.A., Longbottom, J.L., McCarthy, D.S. and Hargreave, F.E. (1968). *Candida albicans* precipitins in respiratory disease in man. J. Allergy, 41, 305.

Pepys, J. (1975). Atopy. In: Clinical Aspects of Immunology (Eds. P.G.H. Gell, R.R.A. Coombs and P.J. Lachmann), Blackwells, Oxford, pp.877-902.

Pepys, M.B., Brighton, W.D., Hewett, B.E., Bryant, D.E.W. and Pepys, J. (1977). Complement in the induction of IgE antibody formation. Clin. Exp. Immunol., 27, 397.

Pepys, J., Parish, W.E., Cromwell, O. and Hughes, E.G. (1979). Passive transfer in man and the monkey of Type 1 allergy due to heat labile and heat stable antibody to complex salts of platinum. Clin. Allergy, 9, 99.

Pepys, J. (1986). Autoantibodies induced by extrinsic low molecular weight chemical allergens. In: Proceedings of the XIIth International Congress of Allergology and Clinical Immunology (Eds. C.E. Reed *et al.*), C.V. Mosby, St. Louis, pp.204-208.

Pressman, D. (1971). The nature of antigen and antibody combining regions. In: Immunological Diseases, Vol.II, 2nd Edition (Ed. M. Samter), Little, Brown and Co., pp.11.

Tada, T. and Ishizaka, K. (1970). Distribution of γE forming cells in lymphoid tissues of the human and the monkey. J. Immunol., 104, 377.

Tse, C.S., Chen, S.E. and Bernstein, I.L. (1979). Induction of murine reaginic antibodies by toluene di-isocyanate. An animal model of immediate hypersensitivity reactions to isocyanates. Amer. Rev. Resp. Dis., 120, 829.

Vervloet, D., Amand, A., Didier, A., Furstoss, M. and Charpin, D. (1986). Allergy to quaternary ammonium cationic compounds. In: Proceedings of XIIth International Congress of Allergology and Clinical Immunology (Eds. C.E. Reed *et al.*), C.V. Mosby, St.Louis, pp.69.

de Weck, A.L. and Girard, J.P. (1972). Specific inhibition of allergic reactions to penicillin in man by a monovalent hapten. II. Clinical Studies. Int. Arch. Allerg. Appl. Immunol., 42, 798.

Monoclonal Antibodies as Structural Probes for Mite, Cat, and Cockroach Allergens

Martin D. Chapman
Division of Allergy and Clinical Immunology, Department of Internal Medicine, University of Virginia, Charlottesville, VA 22908, USA

ABSTRACT

The uses of monoclonal antibodies in the development of immunoassay procedures, allergen purification and epitope mapping are reviewed in detail with especial reference to house dust mite, cat dander and cockroach allergens.

KEYWORDS

Monoclonal antibodies; allergens; immunoassays; epitope mapping.

INTRODUCTION

The development of monoclonal antibodies (Mab) by Kohler and Milstein [1] in the mid-1970's had a tremendous impact in all areas of biomedical sciences and this technology is being increasingly used in the investigation of allergic disease. Over the past 5 years, panels of Mab have been raised against several common inhalant allergens, including dust mites, cat, grass and ragweed pollens, and insect allergens (cockroach and chironomids) and the principal applications of these Mab are outlined in Fig. 1. The production of allergen specific Mab has made it much easier to identify and define individual allergens, to compare allergens isolated in different laboratories, and to develop simple immunoassays to accurately measure allergen concentrations. In addition, Mab provide useful tools for allergen purification (by affinity chromatography) and for investigating the antigenic structure of allergen molecules. Clearly, the major advantage of Mab, as compared to polyclonal antisera, is that once their specificity has been defined, they provide a consistent marker for a given allergen and can be produced indefinitely.

Dust mites, cat and cockroach are among the most important allergens in house dust and IgE mediated reactions to these allergens are common in allergic diseases, such as perennial rhinitis, asthma and atopic dermatitis. Two groups of major mite allergens have been purified from *Dermatophagoides pteronyssinus* and *D. farinae* cultures; the 24 kD Group I allergens (*Der p I*

Printed in Great Britain

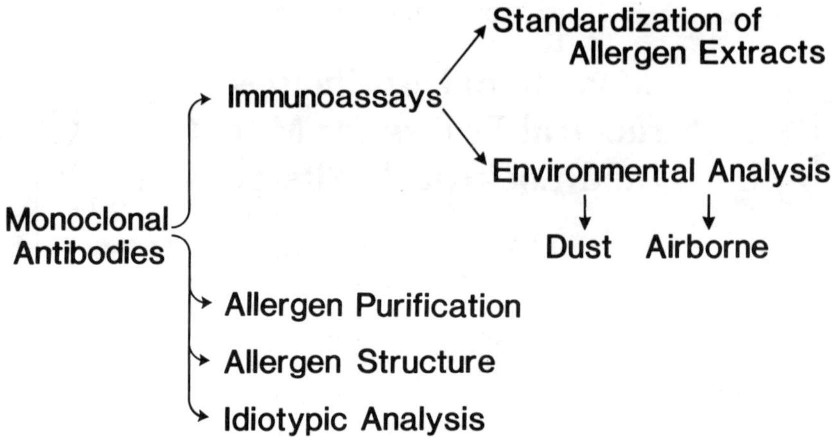

Fig. 1. Applications of allergen specific monoclonal antibodies in allergy research.

and *Der f I*) and the 15 kD Group II allergens (*Der p II* and *Der f II*) [2-11]; and a single major allergen, *Fel d I*, has been purified from cat dander extracts [12-15]. The cat and mite allergens have been used extensively in clinical and immunologic studies [see Refs. 10, 12, 16 for reviews]. While cockroach allergens are less well defined, the purification of allergens from German and American cockroaches (*Blattella germanica* and *Periplaneta americana*, respectively) has been reported and, clinically, a high prevalence of IgE mediated reactions to cockroaches has been reported among asthmatics in the USA [17-20]. This review focuses on three areas where Mab have important applications in studying mite, cat and cockroach allergens: i) immunoassay development; ii) allergen purification; and iii) antigenic analysis (epitope mapping).

IMMUNOASSAYS

Two-site Mab immunoassays have been developed for both the Group I and Group II mite allergens and for *Fel d I*. The principle of these assays is outlined in Fig. 2. A "capture" Mab is coated onto a solid phase, e.g. plastic microtiter well or CNBr activated paper disc and incubated with the test sample. Bound allergen is detected using a second Mab, directed against a non-overlapping epitope on the allergen, which is ^{125}I-labelled, enzyme labelled or biotinylated (Fig. 2). The main advantages of these assays are that they are technically straightforward (several hundred samples can be assayed within a few hours); they do not require purified allergens; and they are extremely sensitive (down to ~1 ng/ml). The assays are quantitated by interpolation from a control curve using a reference preparation of known allergen concentration. Several references are now available either as part of the WHO/IUIS Allergen Standardization Program (e.g. the *D. pteronyssinus* reference, NIBSC 82/518) [21,22] or from national standardization agencies (e.g. dust mite and cat standards from the Office of Biologics and Research Resources in the USA).

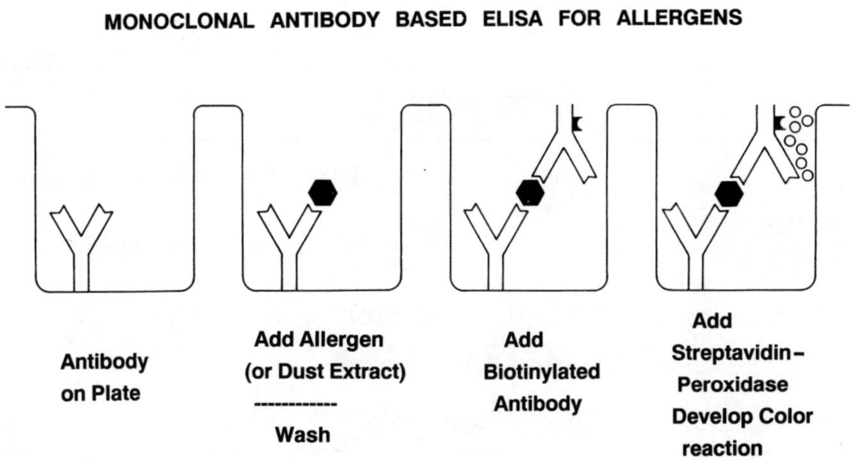

Fig. 2. Two-site monoclonal immunoassay for allergen detection and quantitation.

A list of the immunoassays for mite cat and cockroach allergens currently being used in our laboratory at the University of Virginia is shown in Table 1. These assays were originally developed as RIA's, but most have subsequently been modified as ELISA's using biotinylated second antibodies [25]. Most Mab raised against either *Der p I* or *Der f I* recognize species specific epitopes [26-28], therefore, two separate assays are required to measure the total Group I mite allergen content of an extract. In contrast, the Mab raised against Group II allergens bind to cross-reacting epitopes on *Der p II* and *Der f II* and either allergen can be measured in a single assay [11]. There is no cross-reactivity between assays for the Group I and Group II allergens, nor do these assays detect allergen in extracts of non-pyroglyphid mites (e.g. Tyrophagus sp., Acarus sp.). Similarly, the assay for *Fel d I* is cat specific and showed no cross-reactivity using non-feline derived animal dander extracts [14]. The assay for cockroach allergen is only part Mab based. Dr. Susan Pollart has produced a panel Mab to German cockroach extract, 5 of which bind components recognized by IgE antibodies in the majority of sera from cockroach allergic patients [29]. A "hybrid" radioimmunoassay for cockroach allergen has been developed using one of these Mab (clone 10A6) on the solid phase and serum from a cockroach allergic patient, followed by ^{125}I-labelled anti-IgE to detect bound allergen. This assay is currently being used to assess cockroach allergen levels in house dust samples, using an in house German cockroach reference preparation for quantitation. The 10A6 Mab recognizes a common allergen present in German, American and Oriental cockroach extracts, but not in other insect species, and further characterization of this allergen is in progress.

Monoclonal immunoassays for major allergens have two principal applications: i) standardization of commercial allergen extracts used for allergy diagnosis and treatment; and ii) assessment of environmental allergen exposure in house dust samples or airborne in patients' houses. Both of these areas have recently been reviewed in detail [10,24,30,31]. The key advantages of using

Table 1. Monoclonal immunoassays for common allergens in house dust*

Allergen	Solid Phase Mab	Second Antibody	Reference Standards	Assay
Der p I	5H8	4C1	NIBSC 82/518 UVA 87/03	RIA, ELISA [23-25]
Der f I	6A8	4C1	OBRR E1-Df UVA 87/02	RIA, ELISA [23-25]
Group II	7A1	6D6	*Der f II*	RIA [11]
Fel d I	6F9	3E4	OBRR Cat E3	RIA, ELISA [14]
Cockroach	10A6	Human IgE ab	UVA cockroach	RIA

*Assays and monoclonal antibodies currently being used in our laboratory. The published assay for *Fel d I* used Mab coupled to paper discs [14]. This assay now uses two different Mab (6F9 and 3E4) in either RIA or ELISA on plastic microplates. The present assay for cockroach allergen uses serum from cockroach allergic patients to bind Mab bound allergen and ^{125}I goat anti-IgE as detecting reagent. Once the 10A6 allergen has been purified it should be possible to produce more Mab and to develop a two-site assay. Similar Mab based assays for mite and cat allergens are in use in several other laboratories [27,32,46-48].

Mab immunoassays in allergen standardization are that they are fully quantitative and provide accurate and consistent measurements of specific allergens which can be reproduced over a period of many years. We expect that measurements of mite or cat allergens made with the present assays will be able to be reproduced within experimental error in 10 years time (some of the Mab have already been successfuly cryopreserved for 5 years). In using measurements of one or two major allergens to assess allergen potency, it is assumed that there is a good correlation between the total potency of an extract (e.g. as assessed by RAST inhibition) and major allergen content. Thus, measurements of *Fel d I* are generally accepted as giving a good estimate of the potency of cat dander extracts [12-16,32,33]. However, the situation may be more complicated if the source material for the extract varies and there is more than one major allergen, as is the case for dust mite extracts. While measurements of mite Group I allergens correlate with total potency in assays of whole mite culture extracts [21,22], they may underestimate the potency of extracts prepared from mite bodies, which contain less Group I allergen and a higher proportion of Group II allergens. Recent studies by Dr. Peter Heymann have shown that the Group I : Group II ratio in different mite extracts ranges from 35:1 - 1:1 [11]. The greatest differences were found in the two mite reference preparations, NIBSC 82/518 and OBRR E1 Dp. It is, therefore, hardly surprising that the use of these references has been somewhat controversial [34]! Until recently, sensitive assays for Group II allergens were not available and only a few laboratories had isolated these allergens. Now that a Mab immunoassay for Group II has been developed, levels of the major Groups of mite allergens can be compared and our recent data suggests that measurements of both Group I and Group II allergens provide a better assessment of the potency of mite extracts. Commercial allergen manufacturers will be able to use Mab assays to make better quantitative comparisons of extracts prepared from different source materials; to improve quality control; and to provide allergists with information about specific allergen levels, which may be useful in establishing effective doses for immunotherapy. Interestingly, we have found less variability in the major allergen content of commercial mite extracts than of cat extracts. Group I allergen levels usually vary over a 10-fold range (10-100 μg/ml), whereas the levels of *Fel d I* in 7 commercial cat dander skin test reagents varied by 300 fold (0.1-30 units/ml) [14].

The second important application of Mab immunoassays is to develop a system for measuring indoor allergen exposure which can be used to monitor patients' houses for specific allergens; to assess whether or not a given allergen level is likely to exacerbate allergic symptoms; and to be able to advise patients of appropriate avoidance procedures. This is analogous to using outdoor pollen counts to predict the severity of symptom days of patients with hay fever. Unlike the situation with pollens, many patients are often unaware that mites, cockroaches and fungi can cause asthma, and few patients take adequate steps to control allergen levels in their homes. Population surveys and case control studies have shown a higher prevalence of IgE antibodies to environmental allergens in asthmatics than in non-allergic controls. In addition, recent epidemiologic studies on seasonal allergen exposure and on the prevalence of IgE antibodies in patients admitted to hospital Emergency Rooms with asthma, emphasize the clinical significance of indoor allergens in the etiology of asthma [see Refs. 10 and 35].

Monoclonal assays have several features which are useful for environmental analysis: i) they are highly specific; ii) the ELISA assays can easily be carried out in routine clinical laboratories; iii) direct quantitative comparisons can be made of levels of different allergens [24]. The Mab immunoassays for *Der p I* and *Der f I* show a very good correlation with other methods of assessing mite allergen exposure, e.g. mite counts, RAST inhibition and inhibition RIA [23]. Because they are specific for either allergen, they also give a good guide to the proportions of *D. pteronyssinus* or *D. farinae* in a dust sample. These assays have been used to assess mite allergen levels in dust samples from several areas of the U.S.(Charlottesville, Atlanta, New Orleans, Cincinnati etc.). It is impossible to predict the mite allergen level of a house in a particular area, and results can range from <0.2 μg - >200 μg allergen/gram dust. However, based on epidemiologic studies of mite allergen exposure in various parts of the world, it has been possible to propose tentative levels which should be regarded as clinically significant. Recent recommendations are that Group I allergen levels of 2 μg/gram dust should be considered as risk for sensitization (i.e. development of IgE antibody responses) and that levels of 10 μg/gram should be considered a risk factor for acute asthma attacks [36]. The important point is not whether these particular values are correct, but that they provide a framework for future epidemiologic studies to test these values, and that Mab assays will play a key role in establishing mite allergen levels which should be considered as risk factors for allergic disease. It is not yet clear whether measurements of <u>both</u> Group I and Group II allergens are necessary to assess mite exposure. Production of the Group I allergens appears to be a constitutive part of mite growth and Group I assays have previously provided consistent data on mite levels. However, we need to know how the levels of Group I and II allergens in representative dust samples compare, and whether they are influenced to the same extent by environmental factors which affect mite growth, since it is possible that these factors could have different effects on the production of the two groups of allergens.

Clearly, assays of *Fel d I* are not a prerequisite for establishing exposure to the domestic cat! However, *Fel d I* assays are providing useful data about the quantities of cat allergen that accumulate in the home, the effect of cat removal on allergen levels, and the aerodynamic properties and distribution of cat allergen particles [14,37-40]. *Fel d I* levels in houses containing cats can be extremely high (up to 500 µg/gram dust) and do not show seasonal variation [14,37]. *Fel d I* can also often be detected in houses which do not contain cats, which testifies to the persistence and/or ubiquity of this allergen in the environment [38]. Wood and colleagues recently showed that *Fel d I* persisted for 4-6 months after cat removal and concluded that trials of cat allergen avoidance carried out for 2 months or less (a common clinical practice) would in most cases not be long enough to reduce *Fel d I* levels to those found in control houses without cats [38]. Dr. Christina Luczynska has used the Mab assay to study airborne *Fel d I* levels in houses and in cat vivarium rooms [40]. The sensitivity of the Mab assay is crucial for these studies, because even following disturbance, allergen levels in the air are low (usually <100 ng/m^3). Particle sizing experiments showed that most of the airborne *Fel d I* in vivarium rooms is associated with 5-20 u particles, however, in houses ~25% of the allergen was present on small (<5 u) particles and this may explain why cat allergic asthmatics often develop symptoms immediately on entering a house with a cat.

In future, it seems likely that immunological monitoring of exposure to indoor allergens will be an important part of the management of allergic diseases, particularly asthma. With the development of Mab assays for other common indoor allergens, it will be possible to make more objective assessment of allergen exposure in different geographic areas and to design clinical studies to determine the risk factors associated with exposure to a specific allergen. The effects of allergen reduction can then be compared and this should enable improved allergen avoidance regimes to be developed.

ALLERGEN PURIFICATION

The problems in using conventional biochemical techniques used to purify mite and cat allergens have been largely overcome by the use of Mab affinity chromatography. This single step procedure has particular applications where the source material is limited in supply (e.g. cat extract) or where the allergens themselves are heterogeneous and difficult to separate using standard protein purification procedures (e.g. mite allergens). Over the past 2 years, we have developed monoclonal antibody based affinity purification techniques for *Fel d I*, the Group I and Group II mite allergens and *Der f III*. *Fel d I* is a good example of an allergen which occurs in a relatively low concentration in the source material (<100 µg/ml in most cat dander extracts) and which has always been limited in supply. It can also be difficult to obtain sufficient quantities of cat dander extract from commercial suppliers for allergen purification purposes (especially extracts with a high *Fel d I* content). In order to purify this allergen, we were forced to turn to the most heterogeneous of allergen extracts - house dust, as a source of *Fel d I*! Many house dust extracts contain higher levels of *Fel d I* than commercial cat extracts and by using a selected dust extract with a high *Fel d I* content (~200 µg/ml), it was possible to purify mg quantities of *Fel d I* by Mab affinity chromatography [14] (Fig.3). This procedure, developed in collaboration with Dr. Rob Aalberse (Central Labora-

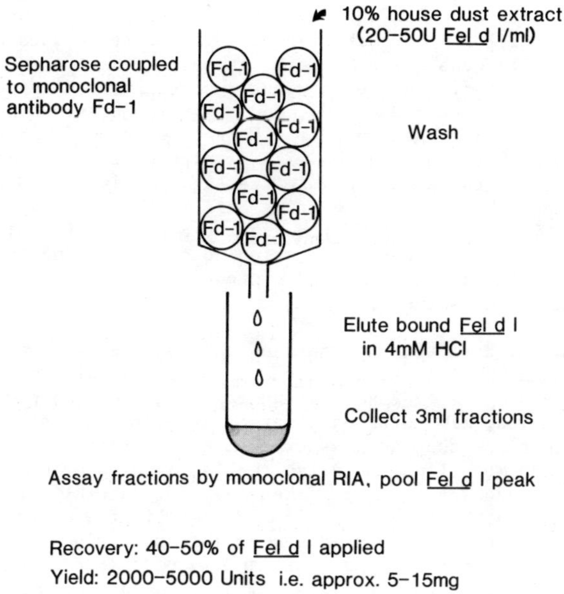

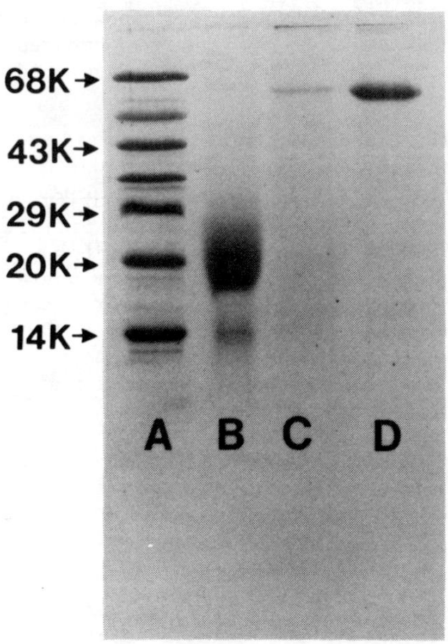

Fig. 3. Monoclonal affinity purification of *Fel d I* from house dust extract.

Top panel: outline of the purification protocol.

Left panel: SDS-PAGE analysis of a preparation of Mab purified *Fel d I*; lane A, molecular weight markers; lane B, *Fel d I*; lanes C and D, house dust extract pre- and post-absorption over the Fd-1 affinity column, respectively.

Modified and reproduced from the Journal of Immunology, 140, 812-818, (1988) with permission.

tory for Blood Transfusion, Amsterdam), involves simply eluting bound *Fel d I* from the monoclonal immunosorbent in 4 mM HCl. The allergen was ~90% pure as judged by SDS-PAGE or CIE and could be further purified by HPLC or a second cycle of affinity chromatography. Similar techniques have been used to purify mite allergens. The Group I mite allergens (*Der p I*, *Der f I* and *Der m I*) were each purified from aqueous, non-dialysed mite culture using the Group I cross-reactive Mab, 4C1, and a glycine/ethylene glycol eluting buffer at alkaline pH. The Group II allergens were purified using an anti-*Der p II* Mab (CLB-Dpx) and elution with glycine/NaCl. The main technical considerations with these methods are the antigen binding capacity of the immunosorbent and the choice of eluting buffer. We usually couple ~200 mg of partially purified Mab to 7 g CNBr activated Sepharose (total column volume ~20 ml) and determine the optimal elution conditions using buffers of different pH, ionic strength etc. to elute ^{125}I-labelled allergen from aliquots of the immunosorbent.

The use of Mab has dramatically improved the efficiency of allergen purification by reducing purification times and increasing allergen yields (typically, 10-20 mg of >90% pure Group I allergen can be purified within a week). In addition, Mab purified allergens appear to retain their native protein structure and show no loss of antigenic reactivity, as judged by binding to murine monoclonal or polyclonal antibodies or human IgG and IgE antibodies in immunoassays. This makes them very useful for biochemical and immunologic studies which have previously been limited by supplies of purified allergen [41], e.g. amino acid sequencing, isolation of immunoreactive peptides for epitope mapping, *in vitro* and *in vivo* studies in Man and in experimental animals. The N-terminal amino acid sequences have been determined for all the allergens described above and, in addition, ~90% of the protein sequence of *Der f I* has been determined using gas phase sequencing and tandem mass spectrometry (in collaboration with Drs. Pat Griffin and Don Hunt, Department of Chemistry, University of Virginia). The *Der f I* protein sequence shows approximately 70-80% homology with the deduced amino acid sequence of the *Der p I* cDNA clone isolated by Dr. Wayne Thomas and colleagues [42; see also this publication].

The Mab to mite and cat allergens were all raised using purified allergen for immunization. However, it is becoming increasingly clear that the production of Mab should be considered as a first step in allergen purification. We have adopted this approach for cockroach allergens, as an alternative to using biochemical purification procedures. Immunization with a crude allergen extract could potentially yield Mab directed against several different allergens which can then be affinity purified and tested for IgE binding activity using a panel of allergic sera. In the case of cockroach, we obtained 5 allergen specific Mab and purification of the allergens recognized by 10A6 and the other anti-cockroach Mab is in progress. This strategy is now being used by several laboratories, as well as our own, to isolate allergens from other complex allergen mixtures (such as molds and fungi), where the allergen may comprise only a small proportion of the extractable protein [43].

EPITOPE MAPPING

Monoclonal antibodies have been used extensively to determine the antigenic structure of globular proteins, e.g. cytochrome c, myoglobin, lysozyme [44, 45] and similar studies are being carried out using inhalant allergens. The aims of these studies are to determine the repertoire of B cell epitopes on allergen molecules (i.e. the number of epitopes, whether they are overlapping or non-overlapping determinants, whether there are "immunodominant" sites etc.) and to determine whether the Mab can be used to define epitopes recognized by human IgG or IgE antibodies. The goal is to determine the structural location and requirements of the antigen binding site(s) (particularly for IgE antibodies), however, this requires data on both primary structure and three dimensional structure, which is not yet available for most allergens. Nonetheless, considerable data has been obtained using panels of Mab to analyse epitopes on purified grass and ragweed pollen allergens (e.g. studies by Esch and Klapper, this publication) and also dust mite and cat allergens.

At least 4 laboratories have raised murine Mab against *Der p I*, with the remarkable finding that most of those Mab are *Der p I* specific and do not bind (or bind with 100-1000 fold lower affinity) to *Der f I* [26-28,46,47]. This "species specificity" is a feature of the murine immune response to *Der p I*: polyclonal murine anti-*Der p I* antisera also selectively bind *Der p I* and recent experiments have shown that different mouse strains (e.g. BALB/c, C57B16, C3H, CBA, SJL) produce similar antibody responses when immunized with *Der p I* in complete Freund's adjuvant (Li and Chapman, unpublished data). The majority of Mab raised in a single fusion against *Der f I* were *Der f I* specific, and only two Mab have been produced which bind equally to both allergens (both derived from the 4C1 hybrid in this *Der f I* fusion) [8,28]. Epitope mapping studies have been carried out using cross-inhibition radioimmunoassays to define the numbers of different sites on the Group I molecules. A number of different assays have been compared, using cold Mab either to inhibit binding of ^{125}I-labelled allergen to a solid phase Mab or to inhibit binding of ^{125}I-labelled Mab to allergen which is "presented" on a solid phase Mab (see Fig. 4). These assays defined 4 non-overlapping, species specific epitopes on *Der p I*, a single species specific epitope on *Der f I* and one common epitope on both allergens [10,28] (Fig. 5). The anti-*Der p I* Mab were derived from 5 fusions, and 70% had the same specificity, suggesting that they recognized an immunodominant epitope on *Der p I*. Interestingly, one of the *Der p I* specific Mab (clone 10B9) completely inhibited binding of the cross-reacting Mab, 4C1, to *Der p I*, suggesting that those two epitopes lie adjacent to one another on the molecule.

In contrast to murine antibody responses, human IgG and IgE antibodies to the Group I allergens are highly cross-reactive. There is a very good quantitative correlation between IgG and IgE antibody to Group I allergens in sera from most mite allergic patients and up to 95% of the IgE antibody to *Der f I* can be absorbed by *D. pteronyssinus* immunosorbent [8,10]. Because of these differences in the epitope specificity of murine and human antibody responses, it seemed unlikely that murine Mab would be useful probes for IgE binding epitopes. In cross-inhibition studies using high concentrations of Mab (1 mg/ml), mean inhibition by anti-*Der p I* Mab was generally <20%, whereas affinity purified rabbit antibodies to Group I cross-reactive epitopes consistently gave ~80% inhibition [28]. Only Mab which bound at or close to cross-reacting epitopes, i.e. 10B9 and 4C1, gave significant inhibition (mean 40%). These results suggest that there is considerable diversity of antigenic sites on the Group I allergens and that there may be multiple IgE binding sites. We have proposed that differences in mode of immunization may explain the dichotomy between the specificity of murine and human antibody responses, i.e. that mice immunized with high dose (10 μg) allergen in adjuvant, process antigen differently to humans sensitized by inhalation of ng allergen without adjuvant [28]. Preliminary studies have shown that mice immunized with 1-10 μg allergen in alum tend to produce more Group I cross-reactive IgG antibody and we are currently investigating whether this altered epitope specificity is linked to expression of murine IgE antibodies.

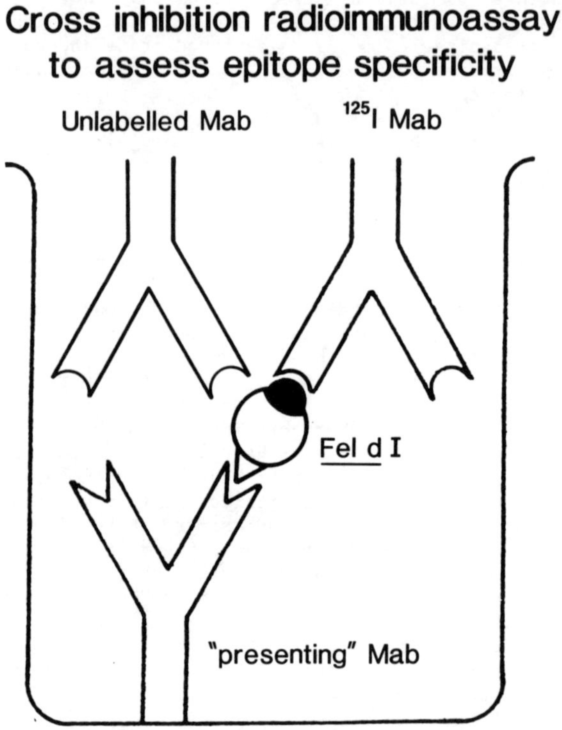

Fig. 4. Epitope mapping using monoclonal antibodies. Cold Mab is used to competitively inhibit binding of a ^{125}I-labelled Mab to allergen (e.g. *Der p I*, *Fel d I*) which is "presented" by a third Mab coupled non-covalently to a plastic microtiter well. The assay can be adapted for analysis of human IgE binding epitopes by using cold Mab to inhibit IgE antibody from binding to the allergen and using ^{125}I-labelled anti-IgE as the detecting reagent.

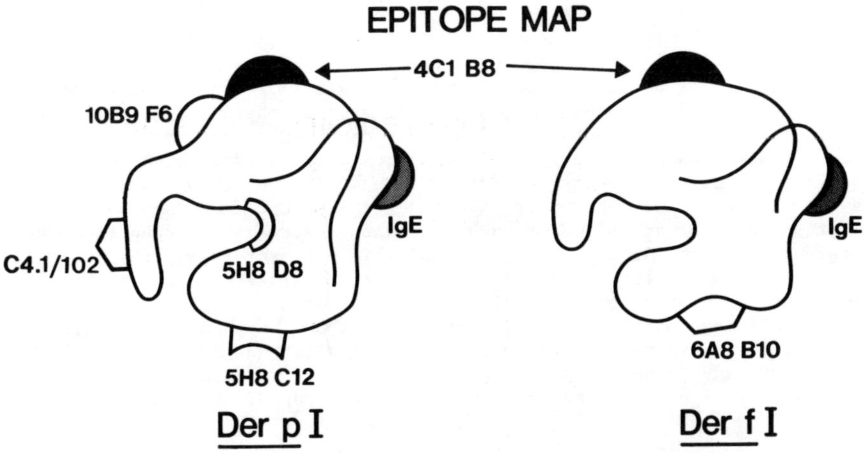

Fig. 5. Highly schematic view of antigenic sites on *Der p I* and *Der f I*, as seen by murine monoclonal antibodies! Reproduced with permission of the Journal of Allergy and Clinical Immunology, 80, (1987), 755-775. By comparing the structural homology between *Der p I* and thiol proteases of known X-ray crystallographic structure using molecular modelling techniques, it may be possible to predict the tertiary conformation of *Der p I* and to identify the Mab defined epitopes on the surface of the molecule.

Recent studies by Dr. Heymann have shown that both murine and human antibodies to *Der p II* and *Der f II* recognize cross-reacting epitopes [11]. Four anti-Group II Mab have been produced (1 against *Der p II*, by Dr. Rob Aalberse [46]; and 3 against *Der f II*, by Dr. Heymann) which define 3 non-overlapping epitopes. BALB/c mice failed to produce anti-Group II antibody responses and the Mab were derived from either BALB/b or A/J mice.

In similar studies, Mab have been raised against several non-overlapping epitopes on *Fel d I* [14,32,48]. The first Mab produced by Dr. Aalberse were directed against two non-overlapping sites, termed Fd1A and Fd1B [14,48]. We have recently produced 14 anti-*Fel d I* Mab which define two additional sites, Fd1C and Fd1D (Li and Chapman, unpublished data). Lombardero, Duffort and Carreira have also produced a series of Mab to non-overlapping sites on *Fel d I* [32]. The *Fel d I* specific Mab appear to be more useful for analysis of human IgE binding epitopes. Recent studies, using a small panel of cat allergic sera, have shown significant inhibition (40-75%) of IgE binding to *Fel d I* by some Mab. Another very interesting feature of these Mab is their specificity for *Fel d I* like molecules from other cat species. Experiments carried out by Drs. Hans de Groot and Rob Aalberse have shown that Mab directed against the Fd1B epitope bound strongly to hair and dander extracts from "big cats" (lion, tiger, panther, jaguar etc.), but Mab directed against Fd1A and Fd1C showed no significant binding. These results provide the first evidence for the existence of *Fel d I* related molecule(s) in other cat species and suggest that only one of the epitopes recognized by the Mab is conserved in these putative *Fel d I* homologues.

Monoclonal antibodies have been used to define overlapping and non-overlapping sites on several of the major inhalant allergens and it is clear that the next, and perhaps more difficult, step in these studies will be in locating the specific contact residues for Mab binding on the allergen molecules. The necessary amino acid sequence data is rapidly being obtained and this will allow more precise analysis of sequential or topographic epitopes, as has been done for other protein antigens. Current data shows a much greater degree of amino acid sequence homology between the Group II mite allergens than the Group I allergens and this may explain why the Group II allergens show greater antigenic cross-reactivity [11]. Recent thermal denaturation and reduction and alkylation experiments suggest that both murine IgG and human IgE binding epitopes on the Group I mite allergens are conformational, whereas those on the Group II allergens are sequential (Lombardero and Chapman, unpublished data). Thus, different epitope mapping strategies will probably be required for each group of allergens.

CONCLUSIONS

The production of monoclonal antibodies has made a significant contribution to basic and clinical research on mite, cat and cockroach allergens and will continue to play an important role in future studies. The term "structural probe" has been broadly interpreted to indicate that Mab provide researchers with a much better "handle" on house dust allergens and have multiple applications. Monoclonal immunoassays for mite allergens are now being used in 20-30 laboratories around the world and provide consistent allergen measurements for epidemiologic and clinical studies. The National Institutes of Health trial of mite desensitization in asthma is a good example of the clinical application of these assays. The mite extracts used in this study are being regularly assayed for major allergen content and at the end of the study, the allergen doses used for immunotherapy can be correlated with patients' symptomatic improvement. In addition, allergen levels in patients' homes are also being measured and this will not only provide comprehensive data on mite allergen distribution in the United States, but also allow comparison of allergen levels with patients symptoms, medication and response to treatment. Extension of these studies to other indoor allergens will require the production of more Mab to cockroach allergens and to other allergens, such as dog dander, molds and fungi.

Monoclonal antibodies have already provided considerable information about the antigenic structure of inhalant allergens and as more sequence data becomes available it will be possible to correlate Mab binding data with structural analyses. The isolation of immunoreactive peptides will allow easier comparison of epitopes defined by Mab with epitopes recognized by human IgG and IgE antibodies and also comparison of B cell epitopes with T cell epitopes. Better understanding of molecular structure of antigenic determinants on inhalant allergens and of factors which influence the allergen specific immune responses should increase our understanding of why these molecules commonly elicit IgE antibody responses and cause allergic disease.

ACKNOWLEDGEMENTS

I would like to thank my colleagues, Drs. Tom Platts-Mills, Peter Heymann,

Susan Pollart, Ying Li and Christina Luczynska for their collaboration and for allowing me to mention some recent, as yet unpublished data. I am also grateful to Dr. Rob Aalberse for his collaboration in our joint studies on *Fel d I*. This work was supported by National Institutes of Health Grants A120565, A124687 and A124261.

REFERENCES

1. Kohler, G. and Milstein, C. (1975). Continuous cultures of fused cells secreting antibody of undefined specificity. Nature, 256, 495-497.
2. Chapman M.D. and Platts-Mills, T.A.E. (1980). Purification and characterization of the major allergen from *Dermatophagoides pteronyssinus* - Antigen P_1. J. Immunol., 125, 587-592.
3. Stewart, G.A. and Turner, K.J. (1980). Physicochemical and immunochemical characterization of the allergens from the mite *Dermatophagoides pteronyssinus*. Aust. J. Exp. Biol. Med. Sci., 58, 259-274.
4. Dandeu, J.P., Le Mao, J., Lux, M., Rabillon, J. and David, B. (1982). Antigens and allergens in *Dermatophagoides farinae* mite. II. Purification of AG11, a major allergen from *D. farinae*. Immunology, 46, 679-687.
5. Holck, A., Dale, S. and Sletten, K. (1986). Purification and characterization of a major allergen from the house dust mite *Dermatophagoides farinae*. Allergy, 41, 408-414.
6. Lind, P. (1985). Purification and partial characterization of two major allergens from the house dust mite *Dermatophagoides pteronyssinus*. J. Allergy Clin. Immunol., 76, 753-761.
7. Lind, P. (1986). Demonstration of close physicochemical similarity and partial immunochemical identity between the major allergen, Dp42, of the house dust mite, *D. pteronyssinus* and corresponding antigens of *D. farinae* (Df6) and *D. microceras* (Dm6). Int. Archs. Allergy Appl. Immunol., 79, 60-65.
8. Heymann, P.W., Chapman, M.D. and Platts-Mills, T.A.E. (1986). Antigen *Der f I* from the dust mite *Dermatophagoides farinae*; structural comparison with *Der p I* from *D. pteronyssinus* and epitope specificity of murine IgG and human IgE antibodies. J. Immunol., 137, 2841-2847.
9. Yasueda, H., Mita, H., Yiu, Y. and Shida, T. (1986). Isolation and characterization of two allergens from *Dermatophagoides farinae*. Int. Archs. Allergy Appl. Immunol., 81, 214-223.
10. Platts-Mills, T.A.E. and Chapman, M.D. (1987). Dust mites: Immunology, allergic disease, and environmental control. J. Allergy Clin. Immunol., 80, 755-775.
11. Heymann, P.W., Chapman, M.D., Fox, J.W., Aalberse, R.C. and Platts-Mills, T.A.E. (1988). Antigenic and structural analysis of Group II allergens (*Der p II* and *Der f II*) from house dust mites (*Dermatophagoides* sp.). Submitted for publication.
12. Leiterman, K. and Ohman, J.L. (1984). Cat allergen 1: Biochemical, antigenic and allergenic properties. J. Allergy Clin. Immunol., 74, 147-153.
13. Lowenstein, H., Lind, P. and Weeke, B. (1986). Identification and clinical significance of allergenic molecules of cat origin. Allergy, 40, 430-438.
14. Chapman, M.D., Brown, M.B., Van Leeuwen, J., Aalberse, R.C. and Platts-Mills, T.A.E. (1988). Monoclonal antibodies to the major feline allergen *Fel d I*. II. Single step affinity purification of *Fel d I*, N-terminal sequence analysis and development of a sensitive two-site immunoassay to assess *Fel d I* exposure. J. Immunol., 140, 812-818.
15. Duffort, O., Carreira, J. and Lombardero, M. (1987). Characterization of the main IgE binding components of cat dander. Int. Archs. Allergy

Appl. Immunol., 84, 339-344.
16. Ohman, J.L., Marsh, D.G. and Goldman, M. (1982). Antibody responses following immunotherapy with cat pelt extract. J. Allergy Clin. Immunol., 63, 320-326.
17. Twarog, F.J., Picone, F.J., Strunk, R.S., So, J. and Colten, H.R. (1977). Immediate hypersensitivity to cockroach. Isolation and purification of the major antigens. J. Allergy Clin. Immunol., 59, 154-160.
18. Kang, B. (1976). Study on cockroach antigen as a probable causative agent in bronchial asthma. J. Allergy Clin. Immunol., 58, 357-365.
19. Kang, B. and Sulit, N. (1978). A comparative study of prevalence of skin hypersensitivity to cockroach and house dust antigens. Annals of Allergy, 41, 333-336.
20. Hulett, A.C. and Dockhorn, R.J. (1979). House dust mite (*D. farinae*) and cockroach allergy in a midwestern population. Annals of Allergy, 42, 160-165.
21. Ford, A.W., Rawle, F.C., Lind, P., Spieksma, F.Th.M., Lowenstein, H. and Platts-Mills, T.A.E. (1985). Standardization of *Dermatophagoides pteronyssinus*: Assessment of potency and allergen content in ten coded extracts. Int. Archs. Allergy Appl. Immunol., 76, 58-67.
22. Ford, A.W., Seagroatt, V., Platts-Mills, T.A.E. and Lowenstein, H. (1985) A collaborative study on the first international standard of *Dermatophagoides pteronyssinus* extract. J. Allergy Clin. Immunol., 75, 676-685.
23. Chapman, M.D., Heymann, P.W., Wilkins, S.R., Brown, M.B. and Platts-Mills, T.A.E. (1987). Monoclonal immunoassays for the major dust mite (*Dermatophagoides*) allergens, *Der p I* and *Der f I*, and quantitative analysis of the allergen content of mite and house dust extracts. J. Allergy Clin. Immunol., 80, 184-194.
24. Chapman, M.D. (1987). Allergen specific monoclonal antibodies: New tools for the management of allergic disease. Allergy, 43 (Suppl. 5), 7-14.
25. Luczynska, C.M., Arruda, K., Platts-Mills, T.A.E., Miller, J., Lopez, M. and Chapman, M.D. (1988). A two-site monoclonal antibody ELISA for the quantification of the major *Dermatophagoides* allergens, *Der p I* and *Der f I*. Submitted for publication.
26. Chapman, M.D., Sutherland, W.M. and Platts-Mills, T.A.E. (1984). Recognition of two *Dermatophagoides pteronyssinus* specific epitopes on antigen P_1 using monoclonal antibodies: Binding to each epitope can be inhibited by sera from dust mite allergen patients. J. Immunol., 133, 2488-2495.
27. Horn, N. and Lind, P. (1987). Selection and characterization of monoclonal antibodies against a major allergen in *D. pteronyssinus*. Species specific and common epitopes in three *Dermatophagoides* species. Int. Archs. Allergy Appl. Immunol., 83, in press.
28. Chapman, M.D., Heymann, P.W. and Platts-Mills, T.A.E. (1987). Epitope mapping of two major inhalant allergens, *Der p I* and *Der f I*, from mites of the genus *Dermatophagoides*. J. Immunol., 139, 1479-1484.
29. Pollart, S., Platts-Mills, T.A.E. and Chapman, M.D. (1988). Identification and quantitation of cockroach allergen using monoclonal antibodies. Abstract to be presented at the XIII International Congress of Allergology and Clinical Immunology, Montreux, Switzerland, October 1988.
30. Baldo, B.A. (1983). Standardization of allergens. Examination of existing procedures and the likely impact of new techniques on the quality control of extracts. Allergy, 38, 535-546.
31. Baldo, B.A. (1986). The use of monoclonal antibodies in allergen standardization. In: Proceedings of the XII International Congress of Allergology and Clinical Immunology, Washington, D.C. (ed. C.E. Reed), C.V. Mosby Co. St Louis, MO, pp.411-419.
32. Lombardero, M., Carreira, J. and Duffort, O. (1988). Monoclonal antibody

based radioimmunoassay for the quantitation of the main cat allergen (*Fel d I* or Cat-1). J. Immunol. Methods, 108, 71-76.

33. Anderson, M.C. and Baer, H. (1981). Allergenically active components of cat allergen extracts. J. Immunol., 127, 972-977.
34. Tovey, E.R. and Baldo, B.A. (1987). Comparison by electroblotting of IgE binding components in extracts of house dust mite bodies and spent mite culture. J. Allergy Clin. Immunol., 79, 93-102.
35. Chapman, M.D., Pollart, S.W., Luczynska, C. and Platts-Mills, T.A.E. (1988). Hidden allergic factors in the etiology of asthma. Chest, in press.
36. Platts-Mills, T.A.E. and De Weck, A. (1989). Dust mite allergens and asthma - A world wide problem. Report of an internation workshop. J. Allergy Clin. Immunol., in press.
37. Ohman, J.L., Lorusso, J.R. and Lewis, S. (1987). Cat allergen content of commercial house dust extracts: comparison of dust extracts from cat containing environment. J. Allergy Clin. Immunol., 79, 955-960.
38. Wood, R.A., Chapman, M.D., Adkinson, N.F. and Eggleston, P.A. (1988). The effect of cat removal on *Fel d I* content in house dust samples. J. Allergy Clin. Immunol., 81, 309 (Abstract).
39. Van Metre, T.E., Marsh, D.G., Adkinson, N.F., Fish, J.E., Kagey-Sobotka, A., Norman, P.S., Radden, E.B. and Rosenberg, G.L. (1986). Dose of cat (*Felis domesticus*) allergen I (*Fel d I*) that induces asthma. J. Allergy Clin. Immunol., 78, 62-71.
40. Luczynska, C.M., Chapman, M.D., Li, Y. and Platts-Mills, T.A.E. (1988). Aerodynamic size and airborne levels of cat allergen (*Fel d I*): relevance to asthma. J. Allergy Clin. Immunol., 81, 310 (Abstract).
41. Stewart, G.A. and Holt, P.G. (1987). Immunogenicity and tolerogenicity of a major house dist mite allergen, *Der p I* from *Dermatophagoides pteronyssinus*, in mice and rats. Int. Atchs. Allergy Appl. Immunol., 83, 44-51.
42. Chua, J., Stewart, G.A., Thomas, W.R., Simpson, R.J., Dilworth, R.J., Plozza, T.M. and Turner, K.J. (1988). Sequence analysis of cDNA coding for a major house dust mite allergen, *Der p I*. J. Exp. Med., 167, 175-182.
43. Sanchez, H. and Bush, R.K. (1988). Purification of an Alternaria (ALT) allergen by use of a monoclonal antibody. J. Allergy Clin. Immunol., 81, 184 (Abstract).
44. Benjamin, D.C., Berzofsky, J.A., East, I.J., Gurd, F.R.N., Hannum, C., Leach, S.J., Margoliash, E., Michael, J.G., Miller, A., Prager, E.M., Reichlin, M., Secarz, E.E., Smith-Gill, S.J., Todd, P.E. and Wilson, A.C. (1984). The antigenic structure of proteins: A reappraisal. Ann. Rev. Immunol., 2, 67-101.
45. Berzofsky, J.A. (1985). Intrinsic and extrinsic factors in protein antigenic structure. Science, 229, 932-940.
46. Van der Zee, J., van Swieten, P., Jansen, H.M. and Aalberse, R.C. (1988). Skin tests and histamine release with P1 depleted *Dermatophagoides pteronyssinus* body extracts and purified P1. J. Allergy Clin. Immunol., 81, 884-896.
47. Ley, V., Saiz, V. and Carreira, J. (1986). Isolation and characterization of the main allergen of *Dermatophagoides farinae* by monoclonal antibodies that recognize IgE related epitopes. Mol. Immunol., 23, 1311-1317.
48. DeGroot, H., van Swieten, P., van Leeuwen, J., Lind, P. and Aalberse, R.C. (1988). Monoclonal antibodies to the major feline allergen *Fel d I*. I. Biologic activity of affinity purified *Fel d I* and of *Fel d I* depleted extracts. J. Allergy Clin. Immunol., in press.

An Allergen and Antigenic Mapping Analysis of a Major Mite Allergen, *Der p I*

G. A. Stewart*, W. R. Thomas*, H. Y. Chua*, K. J. Turner* and H. M. Geysen†

*Clinical Immunology Research Unit, Princess Margaret Hospital, Thomas St, Subiaco, Perth;
†Commonwealth Serum Laboratories, Parkville, Victoria, Australia

ABSTRACT

The allergen *Der p I* has been cloned and its complete amino acid sequence determined. A complete set of solid phase overlapping octapeptides was synthesised and epitope mapping studies performed. The reactivity of IgG and IgE from mite allergic individuals, together with rabbit anti-*Der p I*, towards these peptides was assessed using anti-immunoglobulin-enzyme conjugates. Data from the limited number of human sera thus far examined showed that some epitopes were recognised by both isotypes, although isotype specific epitopes were also demonstrated. The spectra of epitopes recognised by both rabbit and human IgG were more extensive than those recognised by IgE. The relative positions of these epitopes have been assigned using, as models, the 3 dimensional structures of papain and actinidin, proteins which demonstrate significant amino acid homology with *Der p I*. The data thus far indicate that the epitopes are surface located. The epitopes determined experimentally were compared with those predicted using appropriate algorithms and some but not all predicted epitopes correlated. In addition, candidate T helper cell epitopes were predicted.

KEYWORDS

Mites; *Der p I*; homology; T helper cell; B cell; epitopes; predictive algorithms; papain; actinidin; allergy; cysteine protease.

INTRODUCTION

Pre-eminent amongst allergens involved in precipitating allergic disease are those derived from the house dust mite, particularly those of the genus *Dermatophagoides*. Because of this, they have been studied intensively and several allergens comprising *D. pteronyssinus* and *D. farinae* have been described and their physicochemical characteristics elucidated [8,10,11,32], although complete amino acid sequences of these allergens were, until recently, lacking.

To facilitate more detailed study of individual mite allergens from *D. pteronyssinus* in our laboratory, recombinant DNA technology has been utilized. These studies have resulted in the isolation of cDNA coding for two of the major allergens namely, *Der p I* and *Der p II* [9,12]. The cDNA clones have been sequenced and shown to code for mature proteins of 222 and 129 amino acid residues respectively.

With the availability of these sequences, it is now feasible to attempt to map those sequential or continuous (in contrast to topographical) epitopes that are recognised by the immune system, either by B cells or T cells. The number of clinically important allergens that have thus far been mapped in this way is limited although partial mapping studies have been performed [eg. 5,13]. As yet, only 1 allergen, namely ragweed Ra3, has been studied [1,22] with regard to both B and T cell epitopes, although several non-allergic proteins have been thus mapped.

The approach adopted in all these studies has been to synthesise a series of overlapping peptides usually of about 15 residues long, the first 5 of which correspond to the last 5 of the previous peptide, eventually giving rise to a set which covers the entire sequence of the protein under study. The resulting peptides were then tested with immune sera or primed T cells to locate the continuous B and T cell epitopes.

In this report, we describe preliminary data derived from continuous B cell epitope mapping studies on a major mite allergen *Der p I* using the overlapping peptide synthesis strategy described in references 14-16. In addition, we have analyzed its amino acid sequence using various predictive algorithms to define potential T and B cell epitopes and compare the theoretical data with those derived experimentally.

In this regard, the observations that *Der p I* is a cysteine protease demonstrating marked homology with proteases such as papain and actinidin [9], the 3 dimensional structures of which both have been determined to high resolution [3,20], will greatly facilitate the conceptualization of those epitopes determined experimentally.

The ultimate aims of this and similar studies currently being performed elsewhere are primarily designed to determine whether: (1) there are any intrinsic differences between antigenic and allergenic determinants, (2) T helper cell epitopes are different from B cell epitopes, (3) the route of exposure (i.e. inhalation versus parenteral) influences immunodominance and (4) T suppressor cell specific epitopes exist. The answers to these questions may help in understanding the mechanisms of allergic sensitization and, ultimately, may have practical application in the management of allergic disease.

MATERIALS AND METHODS

Overlapping Peptides from *Der p I*

Complete sets comprising 215 overlapping octapeptides representing the complete amino acid sequence of the mature *Der p I* molecule were synthesised using the Merrifield [27] solid phase method as described in detail elsewhere [14,15]. Briefly, a t-butyloxycarbonyl (BOC) protected amino group was coupled to polyacrylic acid grafted polyethylene rods by reacting them with a mixture of dicyclohexylcarbodiimide, hydroxybenzotriazole, mono-t-butyloxycarbonyl-1,6-diaminohexane hydrochloride and triethylene in dimethylformamide. The BOC group was deprotected with trifluoracetic acid and a BOC-β-alanine spacer was then coupled to the free amino groups. After deprotection, the appropriate C-terminal BOC amino acid of each desired octapeptide was coupled. This cycle was repeated until the full octapeptide sequences had been synthesised. The N-terminal amino group was then acetylated and side chain deprotection accomplished by treating the rods with boron tris (trifluoracetic acid). Schematic representations of the construction of the rod bound peptides and the overlapping peptide strategy are shown in Figs. 1 and 2 respectively.

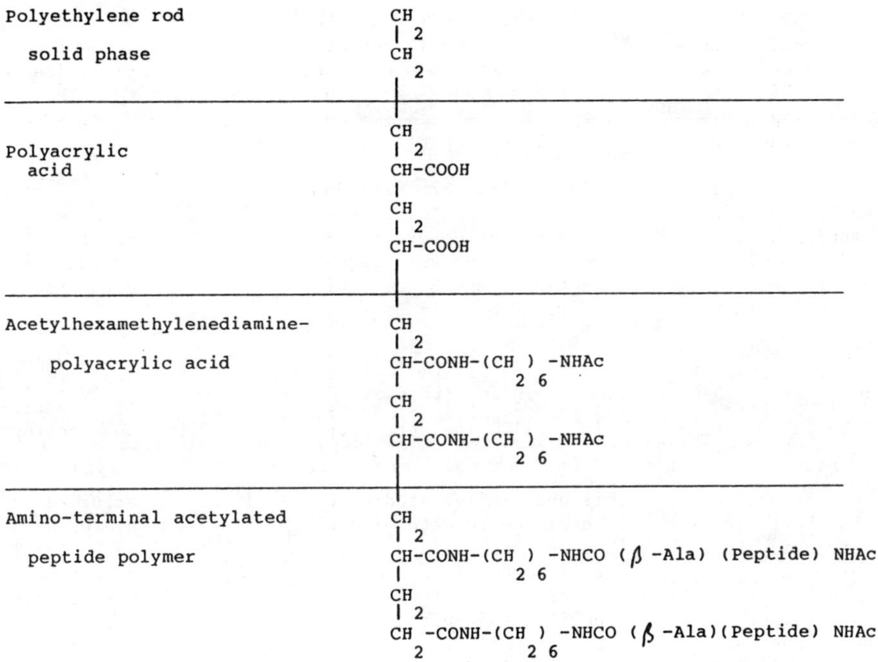

Fig. 1. Representation of the chemical composition of the rod coupled peptides used in epitope mapping studies. Adapted from reference 15.

```
1.........................13

1..............8
  2..............9
    3..............10
      4...............11
        5................12
          6.................13
```

Fig. 2. Schematic representation of the overlapping octapeptide synthesis strategy used in epitope mapping studies.

Enzyme Linked Immunosorbent Analysis (ELISA)

Prior to the ELISA, peptide coupled rods were washed in phosphate buffered saline and then immersed in an aqueous solution of 1% (w/v) sodium dodecyl sulphate, 0.1% (v/v) 2-mercaptoethanol and 0.1 M sodium phosphate buffer, pH 7.2 at 60°C. The rods were sonicated in this solution for 10 min, washed in water and then boiling methanol, and finally air dried.

The tips of the peptide coupled rods were then incubated with 200 µl of a solution containing 10% (w/v) bovine serum albumin, 1% (w/v) ovalbumin, 0.1% (v/v) Tween 20 and 0.85% (w/v) NaCl in 0.01 M phosphate buffer, pH 7.2 (diluting buffer) for 1 hr at 25°C. The rods were then transferred to trays containing 175 µl of an appropriate dilution of test serum and incubated at 4°C overnight. The rods were washed and then incubated with an appropriate anti-immunoglobulin-horse radish peroxidase conjugate in diluting buffer for 1 hr at 25°C. After washing, the rods were incubated with 2,2'-azinobis(3-ethylbenzthiazoline sulphonic acid) enzyme substrate and the reaction stopped by removing the rods and the absorbance determined. The rod-bound antibodies were removed by sonication as described above and further sera tested. The optical densities obtained by ELISA were plotted vertically against the number of the first amino acid comprising the octapeptide studied. The relative positions of peptides demonstrating reactivity with appropriate antisera were superimposed upon the basic three dimensional structure of papain and actinidin [3,20], taking into account the relative alignments.

Sera

The rabbit anti-*Der p I* antiserum used in this study has been described previously [32]. Human sera were obtained from mite allergic individuals (adult or children) which gave high radioallergosorbent scores and, in addition, were shown to react with *Der p I* in immunoblotting experiments (data not shown).

Predictive algorithms

The predictive algorithms used to analyze the sequence of *Der p I* are described in the Results section.

RESULTS AND DISCUSSION

Prediction of Antigenic Determinants

The delineation of amino acid sequences comprising B cell and T helper cell epitopes, of a small number of proteins has stimulated studies designed to determine whether they could be characterized by a set of common features, the determination of which might aid in predicting potentially antigenic regions of proteins in general. The reasons for this approach were essentially practical in that the precise delineation of all the continuous antigenic determinants on a protein was, until recently, a laborious and often expensive undertaking requiring either sophisticated immunochemical techniques or the synthesis of many short peptides (Table 1). The exponential increase in protein sequences generated by recombinant DNA technology further fuelled this search since the sequences of larger and more exotic proteins have been determined, inhibiting the adoption of such approaches on the basis of cost or availability of sufficient protein.

Table 1. Experimental Identification of B Cell and T Helper Cell Epitopes

Method	Requirements	References
Chemical or enzymatic cleavage	Protein	Elsayed *et al.* [13]
Monoclonal	Protein and range of antibodies	Chapman *et al.* [7]
Gene fragmentation	Gene and suitable expression vector	Mehra *et al.* [26]
Peptide synthesis	Amino acid sequence	Atassi and Atassi [1]

The references cited do not purport to assign priority to the use of these methods, rather they represent examples (where possible) of their use in the study of allergenic epitopes.

On the basis of these comparative studies, various authors have suggested that characteristics listed in Table 2 can be assigned to continuous B cell epitopes or T helper cell epitopes. For B cell epitopes these comprise surface accessibility, atomic mobility, hydrophilicity, hydropathicity, negative or positive electrostatic protential, and areas of high mutational changes. In contrast, it is currently assumed that T cells recognise structures characterised by properties different to those of B cell epitopes. Thus, the ability of a sequence to form an amphipathic helix or the possession of particular combinations (motifs) of polar and non polar amino acids have been promoted in this regard.

It can be seen from Table 2 that to predict continuous B cell epitopes, for example, using criteria such as surface accessibility or by comparing the test protein with a family of variants would require either X-ray crystallo-

Table 2. Putative Charactistics of Continuous B Cell and T Helper Cell Epitopes in Proteins

Parameter	Requirements	Reference
B Cell Epitopes		
Surface accessibility	3 Dimensional structure	Atassi and Lee [2]
Atomic mobility	3 Dimensional structure	Tainer *et al.* [33] Westoff *et al.* [34]
Sequence variability	Families of proteins with known amino acid sequences	Reichlin [29]
Charge distribution	Amino acid sequence	Hopp and Woods [18] Geysen *et al.* [16]
Hydrophilicity	Amino acid sequence	Hopp and Woods [18]
Hydropathicity	Amino acid sequence	Kyte and Doolittle [23]
T Helper Cell Epitopes		
Amphipathicity	Amino acid sequence	Delisi and Berzofsky[11]
Presence of C-terminal lysine	Amino acid sequence	Margalit *et al.* [25]
Motif	Amino acid sequence	Rothbard and Taylor [30]

graphy data or a series of amino acid sequences from similar proteins, information which would probably be unavailable. It is because of this that attention has been directed towards trying to extract information regarding epitopes from the primary amino acid sequences *per se*.

Thus, several groups have developed algorithms which have been reported to identify regions of proteins possessing characteristics such as hydrophilicity [17,18], hydropathicity [23], atomic mobility [21] and amphipathicity [11]. Although potentially useful, such algorithms have been shown to be only partially successful and it is clear that they will be modified as more information regarding antigenic epitopes emerges.

Structural Homology Between *Der p I*, Actinidin and Papain

Before describing the results obtained either using the predictive algorithms or experimentally using overlapping peptides, a brief description of the homology between *Der p I* and the cysteine proteases actinidin and papain is given since these data have been used to interpret the results described in subsequent sections.

The amino acid homology between *Der p I* and other cysteine proteases has been described previously [9]. However, the data presented in Tables 3 and 4 represent extensions of those given earlier in that the sequences of the 3 proteins have been aligned to take into account their likely 3 dimensional structural homologies. In addition, a more extensive breakdown of the percentage homologies in different sections of the molecule are given (Table 4).

Table 3. Alignment of Amino Acid Sequences from Papain, Actinidin and *Der p I*

```
                             10        20        30        40
PAP                     IPEYVDWRQKGAVTPVKNQGSCGSCWAFSAVVTIEGIIKIRTGNL

ACT 1                   LPSYVDWRSAGAVVDIKSQGECGGCWAFSAIATVEGINKITSGSL

Der p 1 1      TNACSINGNAPAEIDLRQMRTVTPIRMQGGCGSCWAFSGVAATESAYLAHRNQS
                             ___              - - - - -- - - - - -
                  ?                          63   C   CC  C

                 50        60        70        80        90
PAP            NQYSEQELLDCDR--RSYGCNGGYPWSALQLVAQY-GIHYRNTYPYEGVQRYCR

ACT 46         ISLSEQELIDCGRTQNTRGCDGGYITDGFQFIINDGGINTQENYPYTAQDGDCD

Der p 1 55     LDLAEQELVDCAS---QHGCHGDTIPRGIEYIQHNGVVQES-YYRYVAREQSCR
                       - -  -            - - - - - - -
                   C    C  95        22        CC CC     C           56

                100       110        120        130       140
PAP            SREKGPYAAKTDGVRQVQPYN-QG-ALLYSIANQPVSVVLQAAGKDFQLYRGGI

ACT 98         VALQDQKYVTIDTYDNVPYNN-EW-ALQTAVTYQPVSVALDAAGDAFKQYASGI

Der p 1 103    RPNAQRFGISNYCQIYPPNANKIREALAQPQRYCRHYWTIKDL-DAFRHYD-GR
                          ______      - - - - - - -    _____      - - -
                       C C ?               C   C   ? C C C       C

              150       160            170       180       190
PAP            FVGPCG---NKVDHAVAAVGYG----PGYILIKNSWGTGWGENGYIRIKRGTGN

ACT 148        FTGPCG---TAVDHAIVIVGYGTEGGVDYWIVKNSWDTTWGEEGYMRILRNV-G

Der p 1 154    TIIQRDNGYQPNYHAVNIVGYSNAQGVDYWIVRNSWDTNWGDNGYGYFAANIDL
               -           __________       _______       - _______
                  200

                200       210
PAP            SYGVCGLYTSSFYPVKN

ACT 197        GAGTCGIATMPSYPVKYNN

Der p 1 205    MMIEEYPYVVIL
                        _____
                     CC     C
                 153
```

The numbering above the sequences refers to papain (PAP). The corresponding numbers for actinidin (ACT) and *Der p I* are shown alongside their abbreviations. C represents residues in the hydrophobic cores of actinidin and papain. β pleated sheet and α helical structures are shown as —— and ---- respectively. The numbering beneath the sequences refers to the 1/2 cysteine residues involved in disulphide bond formation in papain and actinidin. Those numbers underlined indicate homologous 1/2 cysteines whereas those marked ? refer to the 1/2 cysteines in *Der p I* for which there are no homologous disulphide arrangements in either papain or actinidin. The amino acid sequences of the three proteins were obtained from references 28, 6 and 9.

Table 4. Homology Between Papain, Actinidin and *Der p I*

Sequence	Number of residues	Homologous Residues		
		Papain v Actinidin	Papain v *Der p I*	Actinidin v *Der p I*
1 - 81	81	44 (54.3)	28 (34.6)	25 (30.9)
82 - 116	35	6 (17.1)	3 (8.6)	4 (11.4)
117 - 212	96	50 (52.0)	22 (22.9)	31 (28.3)
1 - 212	212	98 (46)	53 (25)	60 (28.3)

Figures in parentheses represent percentage homologies in that section of sequence.

A schematic representation of the 3 dimensional structure of actinidin is shown in Fig. 3 where it can be seen that the molecule consists of two domains, each comprising characteristic secondary features [for reviews see references 4 and 19]. The L domain (adopting the convention originally described for papain) consists of 3 major (designated A, B and C) and 1 minor helical segments whilst the R domain consists mainly of β pleated sheet structure together with 2 small helical segments (designated D and E). The catalytic site (indicated by a star in Fig. 3) lies in the cleft separating the two domains. The L domain essentially comprises the N-terminal half of the molecule (residues 12-112 and 208-221), whereas the R domain comprises the C-terminal half (residues 1-11 and 113-207).

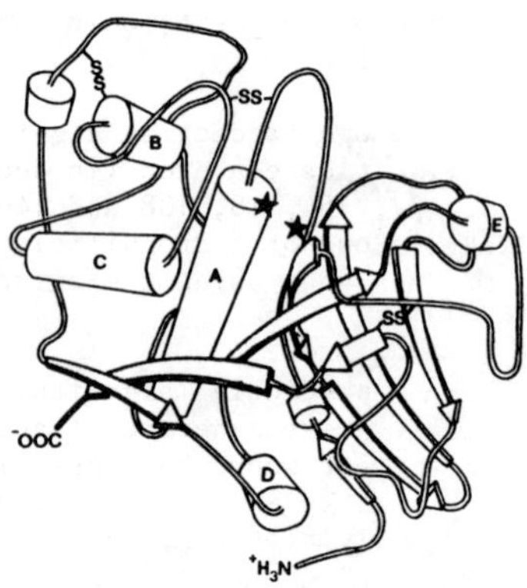

Fig. 3. Schematic representation of the general folding of actinidin together with the secondary structural features. Diagram taken from reference 4. The stars represent the catalytic sites.

In assessing the structural homology of *Der p I* with papain and actinidin, use has been made of similar analyses described previously [4,20]. In this regard, the major observations supporting the close structural similarities between the cysteine proteases have been the homology observed in (a) those residues comprising the core of the molecule, a region unlikely to be able to survive gross mutational events, (b) the disulphide bond arrangement and (c) the permitted variations in amino acid sequence in those regions distant from the catalytic site.

An examination of *Der p I* in this context shows that whilst its overall structure is likely to be similar to that shown in Fig. 3, there may be significant differences. For example, the majority of the core residues in *Der p I* are highly conserved, demonstrating either absolute homology, containing amino acids differing by only 1 codon base, or retaining hydrophobic characteristics. However, one region (there of course may be more) of the molecule namely, between residues 135-142 (actinidin numbering 128-135) is likely to be structurally different in that the hydrophobic nature of the putative core residues are not retained and, in addition, a non-homologous 1/2 cysteine residue is present. In support of this, hydrophilicity analyses revealed that this region is likely to be hydrophilic in *Der p I* but hydrophobic in actinidin and papain (data not shown). Other variations in sequence are also likely to influence local conformation in *Der p I*, for example, differences in the location of the 1/2 cysteine residues in *Der p I* are apparent. Thus, whilst 5 of the 1/2 cysteines in *Der p I* are homologous, 3 non-homologous residues are present. In addition, there are various deletions and insertions such as those located at the N and C-termini, and at positions 160-162. Despite these differences, it is probable that the structure of *Der p I* will be similar to that shown schematically in Fig. 3.

Prediction of B Cell and T Helper Cell Epitopes in *Der p I* Using Algorithms

Continuous B cell epitopes comprising *Der p I* were predicted using the hydrophilicity algorithm of Hopp and Woods as modified in reference 18 whereas T helper cell epitopes were predicted using the amphipathicity algorithm of Delisi and Berzofsky [11]. In this regard, the computer program described in reference 31 with the modifications suggested in reference 25 namely, an 11 residue block were used.

On the basis of the B cell epitope predicting algorithm, 7 candidate epitopes were detected with hydrophilicities centered (in decreasing order of magnitude) on residues 102, 59, 127, 154, 18, 108 and 146 (bottom panel of Fig. 4). The majority of these regions of hydrophilicity are to be found on exposed regions of the molecule when using the structure of actinidin shown in Fig. 3.

When the T helper cell epitope predicting algorithm was used to examine the sequence, 10 candidate epitopes were generated (Table 5), although some of these epitopes or sections thereof would be classified as doubtful by Margalit *et al.* [25] because of the inclusion of the potentially helix breaking proline residues, namely epitopes 1, 2, 6, 7 and 10. Interestingly, epitopes 2 (helix A in Fig. 3), 3 (helices A and B), 4 (helix B), 5 (helix C) and 7 (helices D and E) contain sequences likely to exist in helical conformation in the native molecule (see Fig. 3). Recent collaborative studies with Drs O'Hare and Lamb using cloned *Der p I* specific T cells have shown that one of the determinants predicted is active in this regard (unpublished observations).

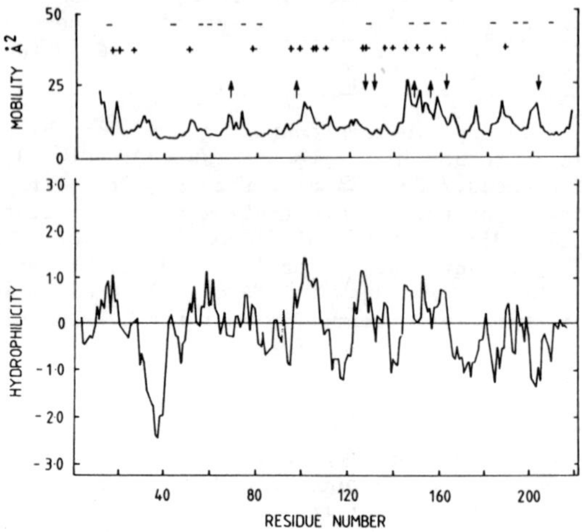

Fig. 4. Hydrophilicity plot (bottom panel) of *Der p I* obtained using the method of Hopp and Woods described in reference [17], together with negative and positive electrostatic charge distribution and atomic mobility profile (top panel) of actinidin adapted from ref. 4. The arrows indicate deletions (↑) and insertions (↓) in *Der p I* when compared with actinidin.

Prediction of B Cell Epitopes in *Der p I* Using Charge Distribution and Atomic Mobility Data

Recent studies have shown that continuous B cell epitopes are likely to be highly mobile structures, often on convex surfaces and possess negative electrostatic potential [16] but as mentioned previously, the exploitation of this information, particularly atomic mobility, will be dependent upon the possession of X-ray crystallography data. With regard to *Der p I*, such data exists and it is on the basis of the availability of atomic mobility data for actinidin [4] that the location of candidate B cell epitopes has been predicted.

The top panel of Fig. 4 shows that average mobility of the N, C_{α}, C and O atoms of actinidin plotted as a function of residue number and the arrows indicate deletions and additions to the *Der p I* sequence. The regions of low mobility correspond to the internal regions of actinidin and papain (data for the latter not shown) whereas the regions of high mobility correspond to the external loops. Approximately 10 regions can be classified as being of high mobility and, thus, potential epitopes. The negative and positive charge distribution is also shown in the upper panel of Fig. 4.

Experimentally Determined Continuous B Cell Epitopes

Sets of overlapping octapeptides representing the entire sequence of *Der p I* were synthesised and incubated with various allergic human sera and a rabbit anti-*Der p I* antiserum. A typical scan representing the continuous B cell epitopes of *Der p I* recognised by the IgG fraction of the rabbit anti-*Der p I*

Table 5. Predicted Amphipathic Segments in *Der p I*

Epitope No.	Residues	Sequence	Range of Angles
1.	14-30	IDLRQMRTVTPIRMQGG	121-134
2.	29-42	GGCGSCWAFSGVAA	117-122
3.	34-60	CWAFSGVAATESAYLAHRNQ SLDLAEQ	80-131
4.	56-72	DLAEQELVDCASQHGCH	81-91
5.	64-107	DCASQHGCHGDTIPRGIEYI QHNGVVQESYYRYVAREQSC RRPN	80-126
6.	100-112	EQSCRRPNAQRFG	80-132
7.	119-168	IYPPNANKIREALAQPQRYC RHYWTIKDLDAFRHYDGRTI IQRDNGYQPN	80-135
8.	180-201	AQGVDYWIVRNSWDTNWGDN GY	80-122
9.	193-208	DTNWGDNGYGYFAANI	80-127
10.	207-222	IDLMMIEEYPYVVILL	93-117

Eleven residue blocks as suggested in reference 25 were analysed using the computer program described in reference 31. Residues underlined are those which appear to be homologous with residues present in helices in the related proteins papain and actinidin. The hydropathicity scale of Kyte and Doolittle was used [23].

antiserum is shown in Fig. 5.

These studies showed that the most extensive reactivity was observed using the anti-*Der p I* specific rabbit IgG, followed by human IgG and then human IgE. Each of the two human isotypes appeared to recognise similar regions of the molecule but, in addition, recognised isotype specific regions. Thus, human mite allergic IgE recognised areas centered around residues 68, 121 and 151, whereas IgG recognised areas centered around residues 48, 161 and 190. In contrast, rabbit IgG reacted with all sites recognised by both human IgG and IgE. The relative location of these sites, together with those regions of high atomic mobility and hydrophilicity are shown in Fig. 6. This arrangement shows that, in general, the experimentally determined epitopes align with regions predicted using both of these parameters.

The relative location of some of the experimentally determined epitopes on the basic α carbon chain structure of the proteases is shown in Fig. 7. All the epitopes appear to be surface located.

SUMMARY

The availability of the amino acid sequence of *Der p I*, together with its observed homology with proteins of known 3 dimensional structure has enabled

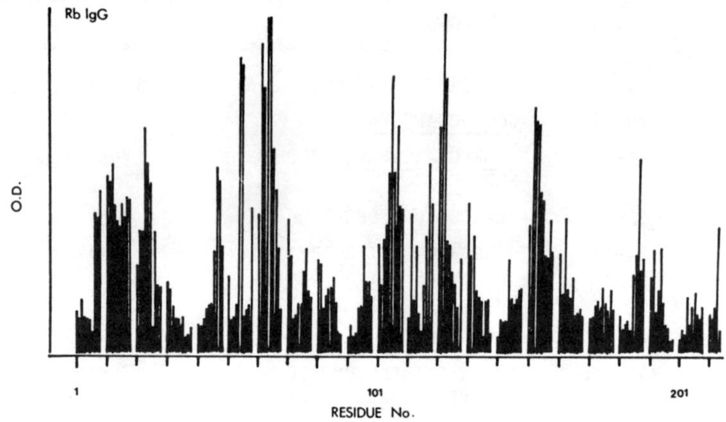

Fig. 5. Antigenic profile of *Der p I* recognised by a rabbit anti-*Der p I* antiserum. Each vertical bar represents the optical density (O.D.)obtained with each octapeptide.

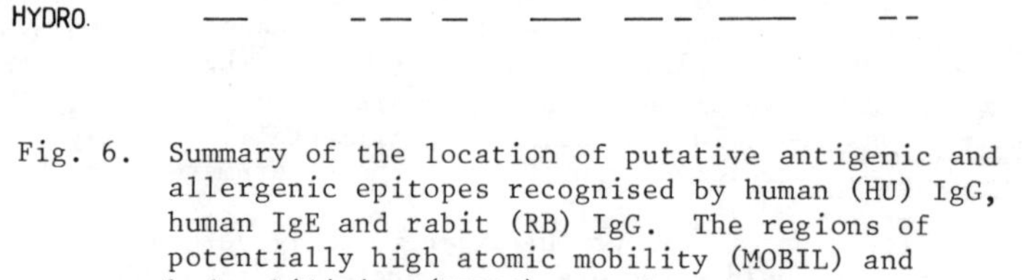

Fig. 6. Summary of the location of putative antigenic and allergenic epitopes recognised by human (HU) IgG, human IgE and rabit (RB) IgG. The regions of potentially high atomic mobility (MOBIL) and hydrophilicity (HYDRO) are also shown.

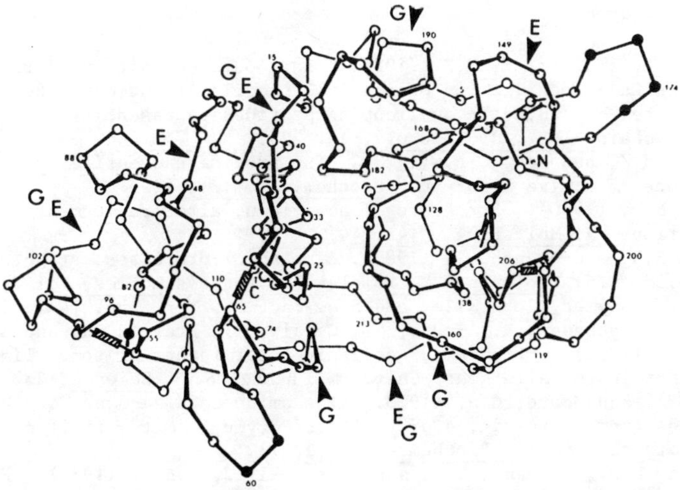

Fig. 7. Basic carbon chain of actinidin [3] showing possible locations of allergenic (E) and antigenic epitopes (G) of *Der p I*.

epitope mapping studies to commence. In this regard, the availability of a basic structure has proved to be of vital importance in studies designed to locate epitopes, including allergens such as midge haemoglobin [5].

We stress that the experimental data described herein are preliminary as very few sera have been examined. Much more work will be required to confirm or extend these very early observations. For example, such work must include inhibition experiments performed using native allergen, synthesis of peptides representing potential antigenic regions and analysis in solution phase and fine specificity determined experimentally using replacement nets [14,15]. Such experiments are presently underway. Despite these caveats, the approach adopted has great potential since it has been successfully applied to the determination of the epitopes comprising several other proteins.

The overlapping peptide strategy described herein will not provide information about all epitopes contained within *Der p I* or for that matter any protein, particularly those of a topographical nature. Thus, parallel or additional studies using, for example, gene fragmentation and expression, and monoclonal antibodies will be invaluable in piecing together the complete epitope map of this major mite allergen.

ACKNOWLEDGEMENTS

This study was supported by the Princess Margaret Children's Medical Research Foundation, the National Health and Medical Research Council, Commonwealth Serum Laboratories, and by a National Biotechnology Program Research grant from the Commonwealth Department of Industry, Technology, and Commerce. We thank Ms C. Jones and Ms C. Lange for typing the manuscript, Dr E.N. Baker for providing the original photographs used in Figs. 3 and 7, and O. Fragoulis and M. Bullas for performing the peptide-rod experiments. This is publication No. 290 of the Clinical Immunology Research Unit of the Princess Margaret Children's Medical Research Foundation.

REFERENCES

1. Atassi, H. and Atassi, M.Z. (1986). Antibody recognition of ragweed allergen Ra3: Localization of the full profile of the continuous antigenic sites by synthetic overlapping peptides representing the entire protein chain. Eur. J. Immunol., 16, 229.
2. Atassi, M.Z. and Lee, C-L. (1978). The precise and entire antigenic structure of native lyzozyme. Biochem. J., 171, 429.
3. Baker, E.N. (1980). Structure of actinidin, after refinement at 1.7 Å resolution. J. Mol. Biol., 141, 441.
4. Baker, E.N. and Drenth, J. (1987). The thiol proteases: structure and mechanism. In: Biological Macromolecules and Assemblies, Vol.3, Active sites of Enzymes, John Wiley and Sons Inc. New York, pp.314-368.
5. Baur, X., Aschauer, H., Mazur, G., Dewair, M., Prelicz, H. and Steigemann, W. (1986). Structure, antigenic determinants of some clinically important insect allergens: Chironomid hemaglobins. Science, 233, 351.
6. Carne, A. and Moore, C.H. (1978). The amino acid sequence of the tryptic peptides from actinidin, a proteolytic enzyme from the fruit of *Actinidia chinensis*. Biochem. J., 173, 73.
7. Chapman, M.D., Heymann, P.W. and Platts-Mills, T.A.E. (1987). Epitope mapping of two major inhalent allergens, *Der p I* and *Der f I* from mites of the genus *Dermatophagoides*. J. Immunol., 139, 1479.
8. Chapman, M.D. and Platts-Mills, T.A.E. (1980). Purification and characterization of the major allergen from *Dermatophagoides pteronyssinus* antigen P_1. J. Immunol., 125, 587.
9. Chua, K.Y., Stewart, G.A., Thomas, W.P., Simpson, R.J., Dilworth, R.J., Plozza, T.M. and Turner, K.J. (1988). Sequence analysis of a cDNA coding for a major house dust mite allergen, *Der p I*: Homology with cysteine proteases. J. Exp. Med., 167, 175.
10. Dandau, J.P., Le Mao, J., Lux, M., Rabillon, J. and David, B. (1982). Antigens and allergens in *Dermatophagoides farinae* mite. II. Purification of Ag11, a major allergen from *D. farinae*. Immunology, 40, 679.
11. Delisi, C. and Berzofsky, J.A. (1985). T-cell antigenic sites tend to be amphipathic structures. Proc. Natl. Acad. Sci. USA, 82, 7048.
12. Doyle, C.R., Chua, K.Y., Turner, K.J., Stewart, G.A. and Thomas, W.R. (1988). Isolation of cDNa clones coding for the major mite allergen *Der p II* using IgE specific immunoscreening. In preparation.
13. Elsayed, S., Aas, K., Sletten, K. and Johansson, S.G.O. (1972). Tryptic cleavage of a homologous cod fish allergen and isolation of two active polypeptide fragments. Immunochemistry, 9, 647.
14. Geysen, H.M., Meloen, R.H. and Bartelin, S.J. (1984). Use of peptide synthesis to probe viral antigens for epitopes to a resolution of a single amino acid. Proc. Natl. Acad. Sci. USA, 81, 3998.
15. Geysen, H.M., Rodda, S.J., Mason, T.J., Tribbick, G. and Schoofs, P.G. (1987). Strategies for epitope analysis using peptide synthesis. J. Immunol. Methods, 102, 259.
16. Geysen, H.M., Tainer, J.A., Rodda, S.J., Mason, T.J., Alexandra, H., Getzoff, E.D. and Lerner, R.A. (1987). Chemistry of antibody binding to a protein. Science, 235, 1184.
17. Hopp, T.P. (1986). Protein surface analysis. Methods for identifying antigenic determinants and other interaction sites. J. Immuno. Methods, 88, 1.
18. Hopp, T.P. and Woods, K.R. (1981). Prediction of protein antigenic determinants from amino acid sequences. Proc. Natl. Acad. Sci. USA, 78, 3824.
19. Kamphuis, I.G., Drenth, J. and Baker, E.N. (1985). Thiol proteases. Comparative studies based on the high-resolution structures of papain and actinidin, and on amino acid sequence information for cathepsins B

and H, and stem bromelain. J. Mol. Biol., 182, 317.

20. Kamphuis, I.G., Kalk, K.H., Swarte, M.B.A. and Drenth, J. (1985). Structure of papain at 1.65 Å resolution. J. Mol. Biol., 179, 233.
21. Karplus, P.A. and Schultz, G.E. (1985). Prediction of chain flexibility in proteins. A tool for the selection of peptide antigens. Naturwissenschaften, 72, 212.
22. Kurisaki, J., Atassi, H. and Atassi, M.Z. (1986). T cell recognition of ragweed allergen Ra3: Localization of the full T cell recognition profile by synthetic overlapping peptides representing the entire protein chain. Eur. J. Immunol., 16, 236.
23. Kyte, J. and Doolittle, R.F. (1982). A simple method for displaying the hydropathic character of a protein. J. Mol. Biol., 157, 105.
24. Lind, P. (1985). Purification and partial characterization of two major allergens from the house dust mite *Dermatophagoides pteronyssinus*. J. Allergy Clin. Immunol., 76, 753.
25. Margalit, H., Spouge, J.L., Cornette, J.L., Cease, K.B., Delisi, C. and Berzofsky, J.A. (1987). Prediction of immunodominant helper T cell antigenic sites from the primary sequences. J. Immunol., 138, 2213.
26. Mehra, V., Sweetner, D.S. and Young, R.A. (1986). Efficient mapping of protein antigenic determinants. Proc. Natl. Acad. Sci. USA, 83, 7013.
27. Merrifield, R.B. (1963). Solid phase peptide syntheses. 1. The synthesis of a tetrapeptide. J. Am. Chem. Soc., 85, 2149.
28. Mitchell, R.E., Chaiken, I.M. and Smith, E.L. (1970). The complete amino acid sequence of papain. Additions and corrections. J. Biol. Chem., 245, 3485.
29. Reichlin, M. (1975). Amino acid substitution and the antigenicity of globular proteins. Adv. Immunol., 20, 71.
30. Rothbard, J.B. and Taylor, W.R. (1988). A sequence pattern common to T cell epitopes. EMBO J., 7, 93.
31. Sette, A., Doria, G. and Adorini, L. (1986). A microcomputer program for hydrophilicity and amphipathicity analysis of protein antigens. Mol. Immunol., 23, 807.
32. Stewart, G.A. (1982). The isolation and characterization of the allergen Dpt 12 from *Dermatophagoides pteronyssinus* by chromatofocusing. Int. Archs. Allergy Appl. Immun., 69, 224.
33. Tainer, J.A., Getzoff, E.D., Alexander, H., Houghten, R.A., Olsen, A.J. *et al.* (1984). The reactivity of anti-peptide antibodies is a function of the atomic mobility of sites in a protein. Nature, 312, 127.
34. Westhof, E., Altschuh, D., Moras, D., Bloomer, A.C., Mondragon, A. *et al.* (1984). Correlation between segmental mobility and location of antigenic determinants in proteins. Nature, 311, 123.

IgE and Viruses

Sam T. Donta

Department of Medicine, University of Connecticut, USA

ABSTRACT

Most virus infections result in elevated total and virus-specific serum IgE levels. In some individuals, these elevated levels are associated with manifestations of allergic phenomena such as rashes or wheezing. The mechanisms by which virus infection induces these manifestations appears to involve a relative inhibition of T suppressor cells with resultant exaggerated increases in IgE levels and IgE-associated responses. Both interferon-dependent and interferon-independent enhancement of IgE-mediated release of histamine have been demonstrated. Measurements of IgE levels may be of diagnostic and prognostic value.

KEYWORDS

Virus; IgE; Respiratory Syncytial Virus; Parainfluenza Virus; Measles; Rubella; Hepatitis; EB Virus; Cytomegalovirus; Herpes Virus; AIDS.

Introduction

The relationship between IgE responses and various parasitic organisms is well-known. Less well-appreciated are the responses of IgE to other microbes, including bacteria and viruses. Reviewed here are the relationships between a number of viruses and IgE responses. And to give further focus to the review, the potential of the IgE response to serve as a diagnostic tool or prognosticator of the disease will be analyzed.

Respiratory Viruses

The association between respiratory infections and subsequent asthma, as well as its hereditary nature, was noted as early as Hippocrates. But it was not until the late 1970s and 1980s that the relationships between viral infections and IgE levels were delineated. Perelmutter *et al* noted increases in serum IgE levels during the acute phases of infection with a number of viruses, *Mycoplasma pneumoniae*, and *Chlamydia psittaci* (Perelmutter *et al.*, 1978). Stempel *et al* noted that children who wheezed during viral infections were more likely to have elevated serum IgE levels than infected children without wheezing (Stempel *et al.*, 1980). Eleven of twelve children with elevated IgE levels had wheezing, compared to two of twelve children who wheezed with normal IgE levels. The children with elevated IgE levels were no more likely to have recurring episodes of upper respiratory infection (URI), middle ear infections or infections due to respiratory syncytial virus (RSV), but appeared to have twice as many episodes of lower respiratory infections over a five year period of time compared to children with normal IgE levels. The authors' conclusions were that the elevated IgE levels may be predictive of bronchospastic responses to infection and to other allergenic stimuli.

In several studies, Welliver *et al.* demonstrated the kinetics of RSV-specific IgE responses in serum and on infected cells. RSV-specific IgE responses began as early as 2 days following the onset of infection and lasted long after RSV antigen could be detected (Welliver *et al.*, 1980; Welliver *et al.*, 1981; Welliver and Ogra, 1983). All patients with RSV-associated bronchiolitis or asthma had demonstrable virus-specific IgE compared to 33% of patients with pneumonia or URI. These results were more recently confirmed by Dang Bui, who showed that serum RSV-specific IgE and IgG_4 were both increased in infections associated with wheezing, but were not increased in patients with simple pneumonia or URI (Hong Dang Bui *et al.*, 1986).

The mechanisms by which viruses lead to elevated IgE levels and asthmatic episodes were studied by a number of investigators. It had been appreciated that most commonly, RSV and rhinovirus infection can lead to wheezing in susceptible individuals; less common are influenza, parainfluenza, and adenovirus infections (Chonmaitree *et al.*, 1988). And as interferon can enhance IgE-mediated release of histamine from most cells, it was assumed that interferon was the link between virus infection and wheezing. RSV, however, is a poor inducer of interferon, and Chonmaitree *et al.* recently demonstrated that certain strains of RSV and rhinovirus which did not induce interferon production were nonetheless effective enhancers of IgE-mediated histamine release (Chonmaitree *et al.*, 1988). The factors responsible for this interferon-independent mechanism remain to be identified, but may be the virus itself or a virus-induced cellular product.

Treatment of patients who have RSV infections with the antiviral agent ribavirin has been thought to reduce the likelihood of complications such as bronchiolitis and wheezing. And Rosner *et al.* demonstrated that ribavirin treatment of RSV infections led to lower RSV-specific IgE and IgA responses compared to placebo-treated or untreated control patients (Rosner *et al.*, 1987). The effect of treatment and correlation of IgE responses with outcome was not, however, evaluated.

In parallel studies of parainfluenza virus (PV) infections, Welliver *et al.* again demonstrated an association between infected children who had croup and increased PV-specific IgE responses (approx. 3-fold) compared to infected children without croup (Welliver *et al.*, 1985). Further studies demonstrated increased responses of croup patients to PV-associated lymphocyte transformation assays, leading the authors to hypothesize a deficiency of T suppressor cell populations in these patients induced by the virus (Welliver *et al.*, 1985). Supporting this hypothesis are the results of successful desensitization studies in atopic individuals which are associated with increased numbers of T suppressor cells. Long-term studies of PV-infected children who develop bronchiolitis demonstrate an increased predisposition to wheezing secondary to other unrelated stimuli, such as smoking and RSV infection, and protection associated with breast feeding in infancy (Welliver *et al.*, 1986).

During the acute phase of measles, there are modest elevations in serum IgE levels in 50-60% of patients (Shalit *et al.*, 1984; Griffin *et al.*, 1985). Children with measles complicated by encephalitis have higher levels than do children with or without pneumonia. Measles virus can induce interferon, but the virus is a strong inducer of immunosuppression, especially of T helper and B cell function. Allergy to tuberculin and other skin-reactive antigens is suppressed during measles. Of particular relevance is the case report of a five year old child with hyperimmunoglobulin E syndrome whose rash went into remission during measles infection (Lin, 1986). Concurrently, serum IgE levels were reduced from 10,000 U/ml to 840 U/ml along with marked decreases in T cells (T_4 decreases greater than T_8) and B cells. With recovery from measles, IgE levels, T cell and B cell levels, and the rash returned to their pre-infection states.

Neurotropic Viruses

As with measles encephalitis, in one study of patients with Japanese encephalitis virus infection, serum IgE levels were increased during the acute phase of infection (Pavri *et al.*, 1980).

With herpes simplex virus (HSV) infections, relationships between atopic eczema and symptomatic virus infection have been noted (David and Longson, 1985). Serum IgE levels were increased in such patients, although virus-specific IgE levels were not studied. HSV infection of mice was shown to lead to HSV-specific IgE which appeared 8-14 days following infection (Ida *et al.*, 1983). Mice infected with HSV or Friend leukemia virus had suppressed spleen cell responses to Ascaris antigen (Yoo and Bennett, 1981). Whereas infection with the Friend retrovirus had to occur prior to Ascaris testing in order to be suppressive, infection with HSV suppressed the Ascaris response even after the initiation of the Ascaris testing.

In patients with recurrent HSV (type 1 or 2) infections, serum IgE appears to be elevated, when compared to patients with few recurrences, over a year's period of time (Legace-Simard *et al.*, 1986). No differences between the groups were observed in their IgG or IgM levels. These results suggest that IgE levels may be of value in predicting those patients more prone to recurrences. The possibility that HSV-specific IgE levels may be a better diagnostic prognosticator of recurrent infection has not, however, been evaluated.

Lymphotropic Viruses

In studies of other herpes viruses, especially the lymphocytotropic Epstein-Barr (EB) virus and cytomegalovirus (CMV), increases in total and virus-specific serum IgE levels have been noted during the acute illness (Bahna *et al.*, 1984; van Loon *et al.*, 1987). With EB virus infection, the serum IgE responses during the first week of infection appear to be greater than those of other immunoglobulins. IgE levels appear to be reduced 30-400 days following the onset of illness (Bahna *et al.*, 1984). Patients with chronic EBV infections appear to have increased responses to other allergens, and have increased IgE levels (Olson *et al.*, 1986), suggesting that such levels may be of help in supporting the diagnosis of chronic EBV infection. *In vitro*, EBV-transformed B lymphocytes, regardless of source (ie. blood, tonsil, bone marrow), demonstrate IgE productions accounting for 0.1 to 1.0% of EBV-responsive B lymphocytes (Stein *et al.*, 1986). While the mechanisms involved in the activation of IgE synthesis secondary to EBV infection and in the production of various allergic-type symptoms (eg. ampicillin-associated rash) remain to be delineated, EBV infection does lead to defective immunoregulation against environmental antigens.

In apparent contrast to EBV infection, recurrent CMV infections are not typically associated with elevated CMV-specific IgE responses. Compared to IgM and IgG responses, CMV-specific IgE levels are elevated in 0 to 8% of patients with recurrent CMV infection (van Loon *et al.*, 1987). As all immunoglobulin classes are elevated in primary CMV infection, it would appear that CMV-specific IgE levels would be a better diagnostic indicator of recent (primary) infection than are IgM, IgA or IgG antiviral titers.

In patients with symptomatic human immunodeficiency virus (HIV) infections (ie. AIDS), four of fourteen patients studied had increased serum IgE levels compared to two of eleven patients with asymptomatic infection (Pederson *et al.*, 1987). The relationships between elevated IgE levels and the various accelerated allergic reactions (eg. sulfonamide allergy) seen in patients with AIDS have not been studied.

Miscellaneous Viruses

In rubella, virus-specific IgE levels rise in the first few days, peak at 6-9 days, then stay constant for 56-63 days before some decline (Salonen *et al.*, 1985). Rubella specific IgE or IgD levels could be used to diagnose acute rubella, but there would appear to be no advantages over IgM, and the longer IgE response could limit its value.

Acute infection with either hepatitis A or hepatitis B viruses results in elevated serum IgE levels, and the levels may be greater in more severe disease (Levo and Shalit, 1981; Rioche, 1986). Patients with chronic active hepatitis and chronic asymptomatic carriers have little if any increases (Levo and Shalit, 1981). However, virus-specific IgE levels have not been determined.

Summary and Conclusions

In a variety of virus infections, serum IgE levels, especially virus-specific IgE levels, are generally increased during the acute phase of infection. In patients with a predisposition to atopy, and perhaps even in some of those without such a predisposition, the elevated IgE levels are often associated with allergic phenomena. In the respiratory tract, this is manifested by bronchiolitis and wheezing; in generalized infections, with skin manifestations. The mechanisms by which virus-induced, IgE-mediated histamine release occurs may be interferon dependent or independent. While not proven, the available evidence suggests that infection with certain viruses may lead to decreases in T suppressor cell populations, resulting in relatively heightened T helper cell function and exaggerated IgE responses.

The results of several studies suggest that use of IgE as a diagnostic tool for certain virus infections may be valuable, perhaps especially if virus-specific IgE levels are measured. With certain virus infections, IgE measurements may be of prognostic value as well. More research is needed to further delineate these potentials.

REFERENCES

Bahna, S.L., Heiner, D.C. *et al*. (1984). Sequential changes of the five immunoglobulin classes and other responses in infectious mononucleosis. Int. Archs Allergy Appl. Immunol., 74, 1.

Chonmaitree, T., Lett-Brown, M.A. *et al*. (1988). Role of Interferon in leucocyte histamine release caused by common respiratory viruses. J. Infect. Dis., 157, 127.

David, T.J. and Longson, M. (1985). Herpes simplex infections in atopic eczema. Arch. Dis. Childhood, 60, 338.

Griffin, D.E., Cooper, S.J. *et al*. (1985). Changes in plasma IgE levels during complicated and uncomplicated measles virus infections. J. Allergy Clin. Immunol., 76, 206.

Hong Dang Bui, R., Molinaro, G.A. *et al*. (1986). Virus-specific IgE and IgG4 antibodies in serum of children infected with respiratory syncytial virus. J. Pediat., 110, 87.

Ida, S., Siraganian, R.P. *et al*. (1983). Cell-bound and circulating IgE antibody to Herpes simplex virus. J. Gen. Virol., 64, 533.

Lagace-Simard, J., Portnoy, J.D. *et al*. (1986). High levels of IgE in patients suffering from frequent recurrent herpes simplex lesions. J. Aller. and Clin. Immunol., 77, 582.

Levo and Shalit (1981). Serum IgE in liver disease. Ann. Allergy, 47, 457.

Lin, C.Y. (1986). Transient remission after intercurrent measles infection in a patient with hyperimmunoglobulin E syndrome. Pediat. Res., 20, 685.

van Loon, A.M., Heessen, F.W.A. *et al*. (1987). Antibody isotope response after human cytomegalovirus infection. J. Virol. Methods, 15, 101.

Olson, G.B., Kanaan, M.N. *et al*. (1986). Correlation between allergy and persistent Epstein-Barr virus infections in chronic-active Epstein-Barr virus-infected patients. J. Aller. Clin. Immunol., 78, 308.

Pavri, K.M., Ramamoorthy, C.L. *et al*. (1980). Immunoglobulin E in patients with Japanese Encephalitis. Infect. Immun., 28, 290.

Pederson, M., Permin, H. *et al*. (1987). Histamine release from basophil leucocytes induced by microbial antigen preparations in patients with AIDS. Allergy, 42, 291.

Perelmutter, L., Phipps, P. and Potvin, L. (1978). Viral infections and IgE levels. Ann. Allergy, 41, 158.

Rioche, M. (1986). Relation entre la concentration serique des IgE totales et L'evolution des hepatites virales. Gastroenterol. Clin. Biol., 10, 521.

Rosner, I.K., Welliver, R.C. *et al*. (1987). Effect of ribavirin therapy on respiratory syncytial virus-specific IgE and IgA responses after infection. J. Infect. Dis., 155, 1043.

Salonen, E.M., Hovi, T. *et al*. (1985). Kinetics of specific IgA, IgD, IgE, IgG and IgM antibody responses in rubella. J. Med. Virol., 16, 1.

Shalit, M., Ackerman, Z. *et al*. (1984). Immunoglobulin E response during measles. Int. Archs. Allergy Appl. Immun., 75, 84.

Stein, L.D., Chan, M.A. *et al*. (1986). Epstein-Barr virus-induced IgE production in limiting dilution cultures of normal human B cells. Eur. J. Immunol., 16, 1167.

Stempel, D., Clyde, W.A. *et al*. (1980). Serum IgE levels and the clinical expression of respiratory disease. J. Pediat., 97, 185.

Welliver, R., Kaul, T.N. *et al*. (1980). The appearance of cell-bound IgE in respiratory-tract epithelium after respiratory-syncytial-virus infection. N.E.J.M., 303, 1198.

Welliver, R.C. and Ogra, P.L. (1983). The role of IgE in pathogenesis of mucosal viral infections. Ann. NY Acad. Sci., 321.
Welliver, R.C., Sun, M. *et al.* (1985). Defective regulation of immune responses in croup due to parainfluenza virus. Pediat. Res., 19, 716.
Welliver, R.C., Wong, D.T. *et al.* (1981). The development of respiratory syncytial virus-specific IgE and the release of histamine in nasopharyngeal secretions after infection. N.E.J.M., 305, 841.
Welliver, R.C., Wong, D.T. *et al.* (1986). Parainfluenza virus bronchiolitis: Epidemiology and pathogenesis. A.J.D.C., 140, 34.
Yoo, T.J. and Bennett, M. (1981). Effect of herpes simplex virus Type 2 and Friend erythroleukemia virus infection of IgE antibody responses to ascaris antigen in mice. Int. Arch. Allergy Appl. Immunol., 65, 235.

Discussion: Session Six

Chairman: Jack Pepys

1. Questions to Jack Pepys

David Marsh. Since a T-cell response must precede the B-cell response, I wonder if you would like to comment on this initial processing step which now has to recognize the HSA as a foreign substance?

Jack Pepys. The foreigness of the HSA is the result of the combination of the hapten with it. Then the body sees it as foreign. I have not looked at whether one can see T-cell priming to that new antigenic determinant.

David Marsh. You were emphasizing that it was not so much the DNP or the platinum, it's more a change in the structure of the self antigen. I'm just wondering how you visualize that in terms of a changed T-cell determinant which is recognized at the processing stage.

Jack Pepys. It must be recognized because they produce IgE against it, but that is all I can say. The HSA-hapten conjugate is presented to T-cell receptors which now recognize it as a foreign substance; then a complex series of events (involving cytokines) eventually triggers a population of B-cells to secrete humoral antibody. What else could one postulate?

David Marsh. Well the point is, if your contention is correct, namely that whether you put on DNP or platinum or whatever, isn't really the critical issue; your saying that it's really the modification of the structural relatedness of the residues of the native molecule. Now we know that the native molecule is being processed by peptide degradation, that of course is dramatically changing the configuration of the peptides and your saying that this must be reflected in T-cell epitopes. I agree with you. I just wondered if you might speculate on how the T-cell epitopes are presented with the aid of MHC proteins?

Jack Pepys. I emphasized the new antigenic determinant on the HSA because that is the thing that the IgE is recognizing, but *in vivo* presumably the conjugated platinum (and the DNP) may still be playing a role in priming T-cells. I'm not assuming that the new antigenic determinant is produced and then separated from the hapten in the body, because I don't think one

Advances in the Biosciences Vol. 74

Printed in Great Britain

can answer that.

Alec Sehon. Only the Almighty can demonstrate a mechanism is correct. We can only postulate a plausible explanation and expect that it will have to be modified as our understanding grows. What we are really considering here is the breakdown of tolerance, apparently brought about by perturbing the immune balance so as to favour suppression or up-regulation. My explanation is that here we have a modified protein which the immune system sees as foreign and so is triggered to produce antibodies. Now, although an antibody is a rigid molecule, HSA is rather flexible and so can adapt itself to cause the phenomenon which you observe. I admit it could take many years to determine whether or not this explanation is correct.

Sola Alaba. Your last slide focussed attention on the carbohydrate content of allergenic extracts. At DPC we recognize that some extracts are mostly composed of carbohydrate, and so I would like to assure you that in the AlaSTAT Allergy tests even these moieties are adequately labelled with ligand.

Jack Pepys. What I was trying to say was, not only should it be there, but it should be there separately from the protein, because they may be very different in what they do.

Sola Alaba. We do have allergens which are composed exclusively of carbohydrate, and through another pathway in our synthesis these are accommodated in the test kits.

2. Questions to Martin Chapman

Jack Pepys. I was at a meeting in Acapulco talking about various allergens. One of the doctors in the audience, a young woman, had a patient whose job it was to raise the lion cubs in the zoo and after a year or two, if she came near them when they were urinating, she got asthma. This once again provides the evidence for the presence of allergen in the urine of these animals. Have you examined the urines yet?

Martin Chapman. Well a lot of the data suggests that there is very little *Fel d I* in cat urine, but Rob Aalberse hasn't collected big cat urine yet. I must get on to him about that! There are several anecdotal stories about big cat allergy, and it will be interesting to see how it turns out.

Sola Alaba. Since *Fel d I* is so ubiquitous to the cat, I wonder if anybody has ever tested it for enzymatic activity?

Martin Chapman. Not to my knowledge.

Sola Alaba. What do you think *Fel d I* is doing in cat saliva?

Martin Chapman. Immunofluorescence studies indicate that it is produced in the sebaceous glands of the skin, as well as saliva. Currently we are doing some airborne studies on *Fel d I*, because nobody knows exactly how this allergen gets from the cat into the air. Christina Luzinska in our lab is working in patients' houses trying to see at what level they actually become provoked and what particle sizes are involved.

Sola Alaba. I think the reason is that they lick their fur, shed the dander and that is the source of airborne allergen.

Martin Chapman. Yes, that's right, although I think it's a bit more complicated. However, the reason for looking at cat allergy is that it is one very good example of where patients can enter a house and immediately get allergic symptoms. This is quite different from the mite situation where most patients do not report a direct correlation between asthma development and exposure; they don't know about mites so they don't associate them with symptoms. Similarly, most people don't realise they are allergic to cockroaches.

3. Questions to Geoff Stewart

Alec Sehon. First of all I would like to congratulate you on an excellent piece of work which was most elegantly presented. The conclusion which you are tentatively suggesting at the moment is that the epitopes which are interacting with IgE antibodies in the same patient may be different from those which interact with IgG. In principle, I don't see any rationale for this, although in terms of the molecular biology of antibody formation we know that exons are being shuffled in the expression of a given immune class.

Geoff Stewart. The data indicates that IgG might be recognizing more epitopes than IgE, but that the epitopes recognized by E are also recognized by G. I think what we are seeing, and this is purely speculative, is that the selection of epitopes may be determined by the route of exposure. Perhaps it's not surprising that we see more epitopes recognized by IgG than IgE, simply because there are more IgG B-cells and so more chance of picking up the epitopes.

Robert Hamburger. On the other hand, that does fit quite nicely with independent regulation of IgE doesn't it?

Geoff Stewart. Yes.

Alec Sehon. Did you look at IgM as well?

Geoff Stewart. No, we didn't, but that is one of the things which we shall be doing.

Birgit Helm. Did you look at which IgG subtypes were elicited?

Geoff Stewart. No, that is also in the pipeline. However, let me hasten to add that the solid phase technique we use is very robust, up to fifty determinations can be done with the same set of pegs and peptides because antibody can be sonicated off. So although the method can have a few problems, I anticipate that subclass experiments will be shortly completed - we have already synthesized four of the epitopes picked up by our method.

Birgit Helm. I am rather intrigued by the possibility that certain epitopes will cause IgE antibodies to be formed. For example, like the tobacco leaf protein, I think parasitic surface antigens will specifically induce the IgE isotype. Do you have any opinion on this?

Geoff Stewart. Our data does not indicate any gross differences between IgE and IgG responses. I think the IgG response will be more extensive than the IgE response, but I don't think that you will get epitope selection.

David Marsh. A question in regard to the apparently greater number of epitopes being recognized by IgG than IgE. Firstly, have any of your patients been on therapy with mite antigens? Secondly, do you have the levels of IgG and IgE in the same patient, and if so were they measured by radioimmunoassay?

Geoff Stewart. It is possible that some of our adult patients had been hyposensitized. This is one of the things we will control as we investigate more patients. The patients were originally selected because of their reactivity to dot blot analysis, but we still have the sera and intend to estimate IgE and IgG levels.

Jim Braun. Do you have any indication about the effects of some of these peptides in blocking biological responses like skin tests?

Geoff Stewart. We have synthesized four of them, but we have not yet had time to use them.

4. Questions to Sam Donta

Sola Alaba. This was all very fascinating. We often see sera with high total serum IgE levels, but no apparent specificity for commonly inhaled or ingested allergens. So the very high total serum IgE associated with viral infection could be very important in explaining this dilemma. Can I ask you if there have been any reports demonstrating a cross-reactivity between a viral antigen and other more familiar allergens?

Sam Donta. I think you make some valid points. From my review it would certainly appear that viral infections can be associated with elevated total serum IgE. With regards to cross-reactivity: among viruses it is usually rather restricted as for example among *Herpes simplex* I and II, but not among other Herpes viruses including the new *Herpes simplex* V. Nor is there any cross-reactivity among the *Varicella*, the *EB* or the *Cytomegalovirus*. I have no knowledge of cross-reactivity between viruses and other antigens.

Sola Alaba. Could you tell us also what is the time lapse between the acute phase where the IgE peaks and when it comes down to normal?

Sam Donta. The lapse time is different according to the virus. The IgE behaviour is either akin to an IgG response, or intermediate between the IgG and IgA response. With some viruses, for example RSV, the IgE response was far greater than 50-60 days and the decline was very slow, totally unlike the IgM response.

Birgit Helm. I was intrigued by your observation that EBV transformation is associated with high IgE levels. EBV transformation of course also leads to the induction of the low affinity receptor of IgE on B-cells; furthermore it is found that whereas normal T-cells don't express the low affinity IgE receptor HIV transformed T-cells do so. Preliminary evidence from Gordon's laboratory suggests that when the low affinity IgE receptor is shed as a proteolytic fragment it has B-cell stimulating activity, which could lead to clonal expansion of further IgE producing cells. Are you aware of any other viruses which give rise to a number of cells bearing low affinity IgE receptors?

Sam Donta. No, I'm not aware of other viruses, but thank you for your observations.

Aftab Ansari. There have been several reports showing that interferon suppresses IgE production in general and I was wondering if prolonged infection, which would induce interferon production, would have any generalized effect on the level of IgE, in terms of lowering IgE levels?

Sam Donta. No, I didn't find any such reports. I'm trying to think of an animal model in which you could do that, usually interferon production is not sustained over long periods of time and so I don't know how you would end up making those correlations. Even in chronic or prolonged EBV infections I don't think there have been any correlations made between IgE and interferon levels.

Wayne Thomas. I wonder if you could tell us of your findings with bacteria? There are still diagnostic problems with, say, pertussis and some lower respiratory tract infections.

Sam Donta. I am sorry, but I have not reviewed that area and cannot answer you directly. It seems that some microbes, for example *B. pertussis* and *H. influenzae* not only trigger responses in susceptible individuals, but also somehow enhance their survival in the lung by avoiding a bronchodilatory response. Our own work points to an involvement by a 650 dalton inhibitor of isoproterenol induced activation of adenylate cyclase, beta adrenergic and dopaminergic binding.

Louis Perelmutter. Your paper reminds me that 8 or 9 years ago we showed that increases in total IgE was influenced by a number of viral agents. What we did observe, and what you indicated, is that the IgE precedes other immunoglobulin changes. We would never have expected that if one considers the ontogeny of immunoglobulin production. I think therefore it is important from a diagnostic point of view to study these early changes in relation to various infectious agents. We also showed that T-suppressor cells decreased during the acute phase when the total IgE level was raised, and increased again during convalescence as the IgE level dropped.

Sheldon Spector. My question has to do with some of the work Jim Jones in Denver has been doing with chronic EB virus. He has found a correlation between an increase in allergy in these patients which goes along with the increase in IgE. One of the postulates was that this was a way of perpetuating the chronic status, in other words some of the allergic triggers actually had something to do with keeping this infection going. Has anybody looked at this in any of the other models that you mentioned such as hepatitis?

Terry Merrett. I may be able to help answer Sheldon's question. A year or two ago now, Ross Barnetson and I reported in the Lancet the case of a patient with infectious mononucleosis. Fortunately it was a patient that Ross had been following in his clinic for a number of years, so we had sequential blood samples to look at. During the infection this particular lady became exquisitely clinically sensitive to cows' milk, and it was only during the infection that she became IgE-RAST and skin test positive to cows' milk. I for one thank you Sam for opening this "can of worms", because it reminds us that allergy may cohabit with infectious diseases.

Herbert Mansmann. Speaking of worms, I had occasion recently to review the literature on the occurrence of IgE stimulation via parasitic infection, and was greatly impressed with the number of occasions in which specific IgE was only a small percentage of the total IgE. Was there much data in your reviews which indicated that there was only a small response to the specific with a high response to the total IgE? This is based on another observation and that is that in experimental animals if you immunize them with ovalbumin for example, and then subsequently subject them to a parasitic infection, then indeed they do turn on some very significant amounts of anti-ovalbumin IgE.

Sam Donta. No data that I recall, but it seems that somebody should try to do these experiments.

Session Seven

MAST CELLS AND ALLERGY

Chairman: J. Pepys

Induction of Immunological Tolerance by Monomethoxypolyethylene Glycol (mPEG) Conjugates of Xenogeneic Immunoglobulins

Alec H. Sehon

MRC Group for Allergy Research, Department of Immunology, The University of Manitoba, 730 Williams Avenue, Winnipeg, Manitoba, Canada R3E OW3

ABSTRACT

To overcome complications due to the immunogenicity of xenogeneic monoclonal antibodies (xIg), tolerogenic derivatives of xIg were synthesized by coupling xIg to an optimal number of mPEG chains of 6400 Da. The results of this study demonstrate that, in spite of multiple injections of xIg, a long-lasting suppression (>95%) of the anti-xIg antibody response was induced for periods in excess of 300 days by administration of tolerogenic $xIg(mPEG)_n$ conjugates. It is proposed that patients receiving immunogenic immunotoxins (xIg-Tx) should be pretreated with tolerogenic $(xIg\text{-}Tx)(mPEG)_n$ conjugates so as to eliminate or minimize the formation of anti-xIg antibodies by the patients.

KEYWORDS

Xenogeneic immunoglobulins; immunotoxins; tolerogenic conjugates; immunological tolerance; monomethoxypolyethylene glycol (mPEG) conjugates.

INTRODUCTION

It was shown in this and other laboratories (Lee and Sehon, 1977; Sehon, 1988; Savoca *et al.*, 1984; Kawamura *et al.*, 1985) that conjugates of a variety of protein antigens and allergens with an appropriate number of mPEG chains were nonimmunogenic and, most importantly, tolerogenic. Thus, these conjugates lacked most of the antigenic and allergenic activity of the unmodified proteins and were capable of inducing immunologically specific suppression of the corresponding IgE and IgG responses in mice and rats.

In spite of the elegant successes achieved in recent years by the application of recombinant DNA methodologies to the synthesis of tailor-made chimeric or "humanized" antibodies (Riechmann *et al.*, 1988), as also exemplified in the papers presented at this Symposium by Drs. M. Brüggemann and B. Helm, it is likely that — even if these antibodies were nonimmunogenic in man — their derivatives with toxins would be immunogenic. Consequently,

Advances in the Biosciences Vol. 74

Printed in Great Britain

the effectiveness of these immunotoxins, as well as that of xenogeneic monoclonal antibodies (mAb) or of their toxin conjugates (mAb-Tx), which are increasingly used for diverse therapeutic purposes in man, will be undermined by their inherent immunogenicity (Dillman *et al.*, 1982). In effect, it has been demonstrated that, on repeated injections, mAb or mAb-Tx were rapidly eliminated from the patients' circulation in the form of immune complexes consisting of the injected xenogeneic "magic bullets" and the patients' antibodies directed to them.

The results presented in this paper demonstrate that a long-lasting state of tolerance to xenogeneic immunoglobulins (xIg) was established in mice by administration of the corresponding $xIg(mPEG)_n$ conjugates, in spite of repeated injections of the immunogenic unmodified xIg; the subscript n represents the degree of conjugation which corresponds to the number of mPEG chains coupled per xIg molecule. For these studies, the xIg consisted of human monoclonal immunoglobulins (HIgG) isolated from the plasma of a patient suffering from multiple myeloma.

MATERIALS AND METHODS

Mice. Female, 6-8 weeks old B6D2F1 mice, bred and maintained in the central animal care facility of the University of Manitoba, were used in all experiments.

Synthesis of $HIgG(mPEG)_n$ Conjugates

The HIgG used in these experiments, which originated from a myeloma patient (PK), was precipitated with ammonium sulphate as described earlier (Wilkinson *et al.*, 1987) and was purified by TRISACRYL DEAE chromatography and gel filtration through AcA-34 (LKB, Bromma, Sweden). For the synthesis of the chemically activated mPEG intermediate the method described previously was used (Jackson *et al.*, 1987). The mPEG intermediate was then reacted with HIgG in 0.025 M borate buffer, pH 9.6, for 1 h at 4°C. For the separation of $HIgG(mPEG)_n$ from the hydrolysed mPEG intermediate, the reaction mixture was applied to a 2.5 x 90.0 cm column of AcA-34 equilibrated in PBS, pH 7.4. The protein content of each conjugate was determined by the biuret method and the average number of mPEG chains was estimated by nuclear magnetic resonance.

Preparation of Immunogens

To increase the immunogenicity of the HIgG without the use of adjuvants, it was aggregated by heating the DEAE purified fraction at 6 mg/ml for 25 min at 63°C followed by placing it on ice for 1 h. This procedure was repeated and the partially aggregated preparation was stored at -20°C until required.

Measurement of Serum Antibodies to HIgG and OA

Anti-HIgG antibody titres were assessed by ELISA in 96-well microtitre plates (Costar, Cambridge, MA, USA) precoated with 100 μg/ml of the DEAE purified monoclonal HIgG. The antibody was quantitated with the aid of alkaline phosphatase conjugated to rabbit antibody preparations specific for IgG2a, or IgG2b (Zymed, San Francisco, CA, USA), using as substrate p-nitrophenyl phosphate. The antibody titre was expressed as the reciprocal of the serum dilution that gave an optical density reading of 0.3 at 405 nm using a

Dynatech MR 600 ELISA reader (Alexandria, VA, USA).

RESULTS

Since in earlier experiments (Wilkinson *et al.*, 1987) the suppression induced by $HIgG(mPEG)_n$ had been shown to be dose dependent and the highest degree of tolerance had been achieved with a dose of 100 µg, all the new conjugates were screened for their tolerogenic strength at this dose. The protocol involved the i.v. injection of each conjugate into a group of 3 mice, followed 7 days later by an i.p. immunization with 20 µg of haHIgG. The control groups, as usual, received PBS in lieu of the mPEG conjugate and were treated otherwise in an identical manner as the test groups. All conjugates used in the experiments to be described were shown by this procedure to induce a marked suppression, of the order of 85-96%, of the IgG1 anti-HIgG antibody response. A diversity of protocols, differing in the number of doses of a given conjugate and of haHIgG injected over differing lengths of time, as well as in the intervals between injections, were used for testing the capacity of the conjugates to induce long-term suppression. Below are given only some representative regimens and, gratifyingly, all of these led to a high degree of tolerance. Thus, the data in Table 1 illustrate that a single i.v. injection of 100 µg of $HIgG(mPEG)_{31}$ administered on day 0 was capable of suppressing the IgG1 anti-HIgG response up to 133 days in animals which received three i.p. immunizing doses of 20 µg of haHIgG each on days 7, 49 and 119 (this experiment was not continued beyond day 133). The control groups received PBS in lieu of $HIgG(mPEG)_{31}$.

To test the kinetics of induction of suppression by $HIgG(mPEG)_{31}$, the tolerogen was injected into groups of mice 10 or 30, or 60 days prior to immunization with a single dose of 20 µg of haHIgG. With respect to the response of the control groups, it is obvious from the results plotted in Fig. 1 that the IgG1 and anti-HIgG antibodies of the test groups were reduced profoundly (>92%) even by one injection of the tolerogen given 60 days prior to immunization.

Table 1. Suppression of primary, secondary and tertiary responses by a single i.v. injection of 100 µg of $HIgG(mPEG)_{31}$ on day 0

1° Immunization on day 7; bleedings on days 23 and 37.

2° Immunization on day 49; bleedings on days 63 and 70.

3° Immunization on day 119; bleedings on days 126 and 133.

Treatment	Primary IgG1 ELISA Titres		Secondary IgG1 ELISA Titres		Tertiary IgG1 ELISA Titres	
	d23	d37	d63	d70	d126	d133
PBS	22,000	12,000	188,000	200,000	60,000	80,500
$HIgG(mPEG)_{31}$	800	600	800	800	900	1,290

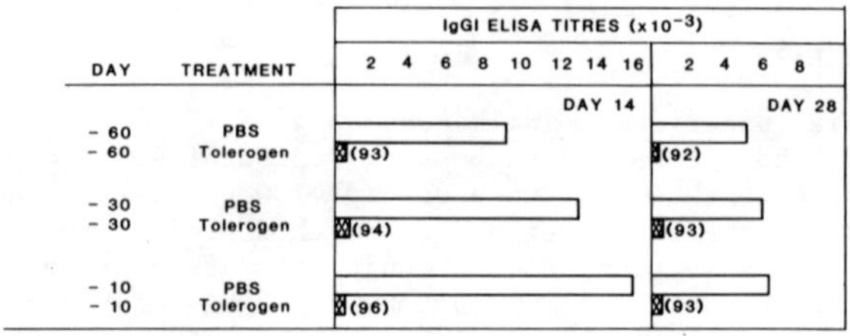

Fig. 1. Kinetics of suppression induced by HIgG(mPEG)$_{31}$ conjugate. Mice received an i.v. injection of 100 μg of HIgG(mPEG)$_{31}$ or PBS on day -60, or -30, or -10. All mice were given on day 0 an i.p. injection of 20 μg of haHIgG. The numbers in brackets represent the o/o suppression of the IgG1 antibody response.

The results of the experiment illustrated in Fig. 2 demonstrate that injections of the tolerogen on days -7, 42 and 55 resulted in suppression of the anti-HIgG response in excess of 95% in animals which received 5 doses of the antigen over a period of 270 days. Moreover, it is to be pointed out that these tolerogenic conjugates suppressed also the IgG2a and IgG2b anti-HIgG responses.

In other experiments (data not shown here), it was demonstrated that injection of an unrelated antigen, i.e. ovalbumin (OA), on day 146 into mice immunosuppressed with regard to HIgG, induced an anti-OA IgG1 response which was essentially identical to that induced in mice which had not been treated with HIgG(mPEG)$_{22}$.

DISCUSSION

The results presented in this paper demonstrate that profound antigen-specific suppression (>90%), which lasted for at least 300 days, was achieved by HIgG(mPEG)$_n$ conjugates in spite of multiple injections of the immunogenic form of HIgG. Moreover, the suppression encompassed multiple IgG subclasses, i.e. IgG1, IgG2a and IgG2b anti-HIgG antibodies.

It is also to be mentioned that, as in earlier experiments (Wilkinson *et al.*, 1987), specific suppression of the HIgG response could be also transferred by spleen cells of animals which had received the mPEG conjugate, i.e. by suppressor cells of animals which had received these conjugate (data not shown).

Although our results hold the promise that mPEG conjugates of xIg may prove useful for clinical applications, it is obvious that it remains to be demonstrated that these conjugates also suppress the immune response to the idiotypic determinants of cell-bound xIg. Moreover, in support of the likelihood that a host tolerized to an xIg of a given isotype by the corresponding xIg(mPEG)$_n$ conjugate would be also tolerant to other xIgs of the same isotype but differing in their idiotypes, one may quote the earlier findings in

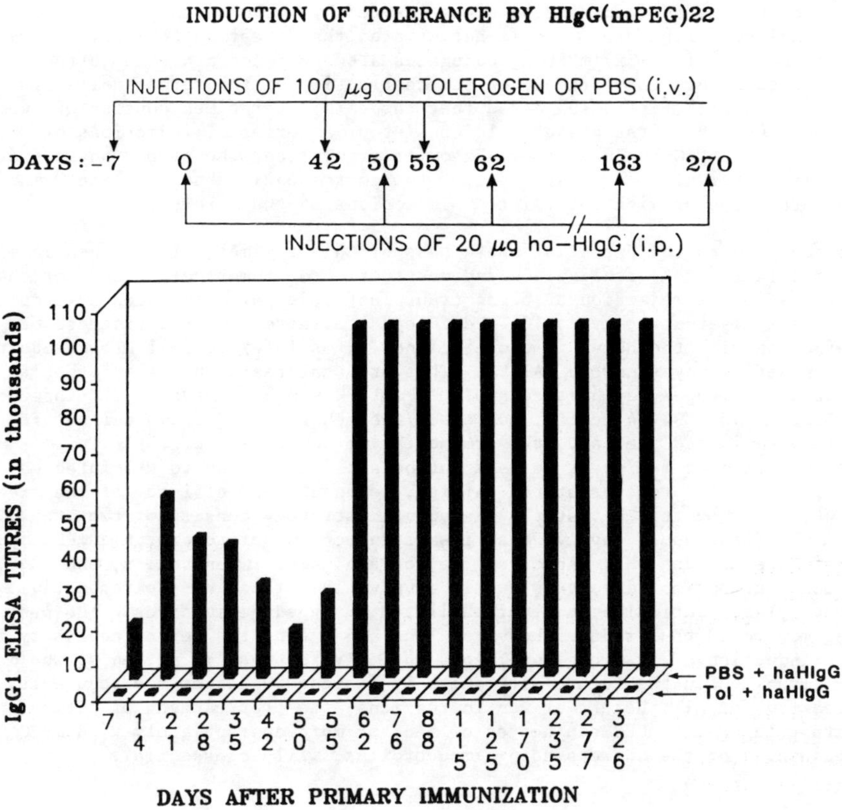

Fig. 2. Induction of tolerance to HIgG by the treatment of mice with HIgG(mPEG)$_{22}$. The control group received PBS and the test group three injections of 100 μg each of HIgG(mPEG)$_{22}$ on days -7, 42 and 55. Both groups were immunized five times with 20 μg of haHIgG on days 0, 50, 62, 163 and 270. (The days are to be read vertically.)

this laboratory (Lee and Sehon, 1978). Thus, it was demonstrated that tolerization of mice by OA(mPEG)$_n$ resulted in the abrogation of not only the anti-OA response, but also of an anti-hapten response, as long as the hapten in question was presented to the suppressed animals in the form of a covalent conjugate with OA. This finding may be considered as an indication that the suppressor cells induced by tolerogenic mPEG conjugates exerted their down-regulating effect not only on the lymphocytes recognizing the protein incorporated in the mPEG conjugate, but also on lymphocytes recognizing additional epitopes when these were coupled to the same protein.

As stated in the Introduction, coupling of multiple chains of mPEG to an antigen, or allergen resulted, respectively, in a substantial decrease in antigenicity or allergenicity — probably due to conformational changes and/or masking of the antigen-binding sites exerted by the mPEG chains coupled to the antibody. Therefore, it is anticipated that treatment of patients with $xIg(mPEG)_n$ or $(xIg-Tx)(mPEG)_n$ conjugates alone may not prove too effective. Hence, to ensure the efficacy of the destruction of the target cells by "magic bullets", it is suggested that the clinical regimen consist of two phases, i.e. the first phase would consist of a series of injections of tolerogenic mPEG conjugates of xIg or of xIg-Tx, and the second phase — at optimal intervals — of injections of the corresponding "nonpegylated" magic bullets alone or with intermittent injections of the tolerogens.

In addition to the potential usefulness of this regimen with xenogeneic mAb or mAb-Tx for the localization and destruction of tumor cells, and for the suppression of rejection of organ transplants, it has been recently proposed that a similar strategy could be developed for arresting the progress of infection with the human immunodeficiency virus (HIV) to full blown acquired immunodeficiency syndrome (AIDS). This proposal rests on the hypothesis that HIV epitopes may be accessible to anti-HIV mAb *in vivo* on HIV-infected cells, i.e. on CD4(T4) cells, monocytes/macrophages, and brain cells. If this were indeed the case, the armamentarium developed for the *in vivo* annihilation of malignant cells might be applicable also to HIV-infected cells. For the reasons stated earlier, to ensure the efficacy of the treatment, it is suggested that the therapeutic strategy consist of two steps, i.e. the first would involve a series of injections of tolerogenic mPEG conjugates of anti-HIV mAb or of the corresponding immunotoxins, and the second (at optimal intervals) would involve injections of the respective nonpegylated anti-HIV mAb or of their toxin-tagged derivatives. The proposed regimen would thus reduce the viral load and extend the period necessary for the induction of a strong autologous anti-HIV response in patients, who may be immunized in the future with suitably designed vaccines and adjuvants. Clearly, for all these therapeutic regimens, the pretreatment of patients with tolerogenic mPEG conjugates of mAb, or more appropriately with mPEG conjugates of the corresponding immunotoxins, will be essential.

REFERENCES

Dillman, R.O., Shawler, D.L., Sobol, R.E., Collins, H.A., Beauregard, J.C., Wormsley, S.B. and Royston, I. (1982). Murine monoclonal antibody therapy in two patients with chronic lymphocytic leukemia. Blood, 59, 1036-1045.

Jackson, C-J.C., Charlton, J.L., Kuzminski, K., Lang, G.M. and Sehon, A.H. (1987). Synthesis, isolation and characterization of conjugates of ovalbumin with monomethoxypolyethylene glycol using cyanuric chloride as the coupling agent. Mol. Biochem., 165, 114-127.

Kawamura, K.T., Igarashi, T., Fujjii, T., Kamisaki, T., Wada, H. and Kishimoto, S. (1985). Immune responses to polyethylene glycol modified L-asparaginase in mice. Int. Arch. Allergy Appl. Immunol., 76, 324-330.

Lee, W.Y. and Sehon, A.H. (1977). Abrogation of reaginic antibodies with modified allergens. Nature, 267, 618-619.

Lee, W.Y. and Sehon, A.H. (1978). Suppression of reaginic antibodies with modified allergens. II. Abrogation of reaginic antibodies with allergens conjugated to polyethylene glycol. Int. Arch. Allergy Appl. Immunol., 56, 193-206.

Riechmann, L., Clark, M., Waldmann, H. and Winter, G. (1988). Reshaping human antibodies for therapy. Nature, 332, 323-327.

Savoca, K.V., Davis, F.F. and Palczuk, N.C. (1984). Induction of tolerance in mice by uricase and monomethoxypolyethylene glycol. Int. Arch. Allergy Appl. Immunol., 75, 58-67.

Sehon, A.H. (1982). Suppression of IgE antibody responses with tolerogenic conjugates of allergens and haptens. Prog. Allergy, 32, 161-202.

Wilkinson, I., Jackson, C-J.C., Lang, G.M., Holford-Strevens, V. and Sehon, A.H. (1987). Tolerance induction in mice by conjugates of monoclonal immunoglobulins and monomethoxypolyethylene glycol. J. Immunol., 139, 326-331.

Tryptase, an Allergy Attack Enzyme from Human Mast Cells

Stephen C. Alter and
Lawrence B. Schwartz

Department of Internal Medicine, Medical College of Virginia, Box 263, MCV Station, Richmond, VA 23298–0263, USA

ABSTRACT

Tryptase, a serine endoprotease with trypsin-like specificity for small synthetic substrates, is the major protein component of human mast cell secretory granules. Tryptase is distinct from trypsin, however, in that it is not inhibited by several known trypsin inhibitors and does not cleave several biological trypsin substrates. Active tryptase, a tetramer of 134,000 Mr, rapidly dissociates to inactive monomers in the absence of heparin or certain other polyanions. Its binding to mast cell heparin proteoglycan, or to chondroitin sulfates and heparan sulfate in the extracellular matrix, limits the diffusion of tryptase away from the site of inflammation and regulates the duration of tryptase activity. Tryptase has been shown to generate C3a from C3, destroy high molecular weight kininogen and fibrinogen, and to activate latent collagenase *in vitro*, but it has no effect on a number of other proenzymes or cofactors involved in coagulation or kinin production. Tryptase is found in both types of human mast cells, and is selectively located in mast cell granules. Tryptase levels in normal human plasma are at or below 5 ng/ml, while levels of 10 ng/ml or greater have been found in essentially all subjects within one to four hours after the onset of systemic anaphylaxis. Tryptase levels in bronchoalveolar lavage, nasal lavage and skin chamber fluids have also been measured in controlled clinical studies on the *in vivo* response to allergen challenge. Tryptase, a specific marker for mast cell activation, is useful as a clinical marker for mast cell mediated events.

KEYWORDS

Tryptase; mast cells; mast cell subsets; proteoglycan; allergy.

INTRODUCTION

Human mast cells originally were demonstrated to be rich in trypsin-like activity by histochemical techniques. Kinetic analyses of the cleavage of N-α-benzoyl-arginine-β-napthylamide *in situ* suggested that a single enzyme accounted for the activity under these conditions [1,2]. Later a tryptic

Advances in the Biosciences Vol. 74

Printed in Great Britain

neutral protease was purified from dispersed lung mast cells and named tryptase [3]; this enzyme was shown to be the major protein component of mast cells in skin as well as lung [4]. Tryptase is found in both of the two known types of human mast cells [5], but is not present in the other cell types in lung, bowel, skin and peripheral blood, with the exception of basophils in which only trace amounts (500-fold less than in mast cells) have been detected [6]. The abundance and selective presence of tryptase in mast cells suggests that the expression of its activity *in vivo* is an important component of mast cell function. Two additional neutral proteases, chymase [7] and carboxypeptidase [8] have been definitively localized to mast cells, but are found in only one of the two types of human mast cells. This review will summarize what is known of the physico-chemistry, function and regulation of tryptase and the utilization of tryptase as an indicator for the participation of mast cells in human health and disease.

PHYSICO-CHEMICAL CHARACTERIZATION

Tryptase has been purified to homogeneity from human lung [3,9], skin [10] and pituitary [11] tissue and characterized (Table 1). The purified enzyme, like that extracted from unstimulated mast cells, is fully active; no proenzyme form or activation requirement has been observed. The enzyme is a serine endoprotease with trypsin-like specificity in that synthetic peptide and ester substrates are cleaved on the carboxy side of lysine or arginine residues. However, the activity of tryptase on biologic substrates and its lack of susceptibility to inhibition by biologic protease inhibitors clearly distinguishes tryptase from trypsin. Tryptase accounts for greater than 90% of the tosyl-L-arginine methyl ester (TAMe) hydrolytic activity extracted from dispersed human lung and skin mast cells. By immunoassay, 35 pg tryptase per skin mast cell and 12 pg per lung mast cell have been found [4].

Table 1. Biochemical properties of tryptase

Molecular weight	134,000
Subunit structure	tetramer 33,000 to 37,000 Mr 1 active site per subunit common antigenic determinants
Cellular levels	
T mast cells (adult lung)	10 pg/cell
TC mast cells (adult foreskin)	35 pg/cell
Basophils	0.04 pg/cell
Response to protease inhibitors	
diisopropylfluorophosphate tosyl-L-lysine chloromethylketone	sensitive
plasma and tissue protease inhibitors	resistant
soybean, lima bean and ovomucoid trypsin inhibitors	resistant

The active enzyme has a Mr of 134,000 by gel filtration. Either under denaturing conditions (with or without alkylation or reduction) or under physiologic conditions in the absence of heparin the enzyme dissociates into monomers of 33,000 to 37,000 Mr, indicating that the holoenzyme is a tetramer of noncovalently bound monomers [3]. The subunits migrate over a range of electrophoretic mobilities in SDS polyacrylamide gels, usually with two major forms of 33,000 to 35,000 Mr and 34,000 to 37,000 Mr [3,9,11,12]. Under non-denaturing, non-dissociating conditions the enzyme does not enter polyacrylamide gels, presumably because of aggregation at low ionic strength. In the presence of urea at acid pH, however, the enzyme migrates as a single band of protein [3]. Several murine monoclonal anti-tryptase antibodies, each recognizing distinct epitopes, interact with subunits over the entire range of their electrophoretic mobilities, indicating shared antigenic determinants [12]. Comparison of tryptase extracted from skin and lung mast cells by Western blot analysis and calculation of respective specific activities indicated that the same enzyme is present in both tissues [4]. Each subunit of tryptase also contains an active site as determined by titration with radioactive diisopropyl fluorophosphate [3]. A single unambiguous NH_2-terminal amino acid sequence has been found for tryptase isolated from both lung (Schwartz *et al*, unpublished data) and pituitary [11] showing that heterogeneity does not arise in the NH_2-terminal amino acid sequence. Tryptase monomers probably are products of a single gene with subunit heterogeneity developing from post-translational modifications.

REGULATION

Tryptase is stored fully active in the secretory granules of human mast cells, where presumably tryptase, like chymase [13], is bound to the sulfated proteoglycans, heparin [14,15] and chondroitin sulfate E [16,17,18], which also are stored there. The acidic environment within the mast cell granule [19,20] is suboptimal for the expression of tryptase activity, whereas after degranulation the near neutral pH of the extracellular milieu is optimal. The complex of tryptase and proteoglycan most likely persists after release; experiments performed *in vitro* show that tryptase binds tightly to heparin, requiring 0.8 M NaCl to break the association. This is analogous to the situation in rodents, where rat mast cell proteases also have been shown to remain associated with proteoglycan [21] after degranulation. Active tetrameric tryptase dissociates to inactive monomers within minutes in the absence of heparin, whereas tryptase activity as well as the tetrameric form is stabilized for several hours by the presence of heparin under physiologic conditions (Fig. 1) [22]. Physiologic protease inhibitors such as aprotinin [9,10], α_1-trypsin inhibitor [9,10,22], α_2-macroglobulin [9,10], protease nexin (Johnson and Schwartz, unpublished results), antithrombin III and anti-leukoprotease (Alter, unpublished results) as well as trypsin inhibitors present in human urine [23] fail to inhibit tryptase activity. Therefore, the interaction of tryptase with heparin appears to be the primary mechanism available for regulation of tryptase activity after release.

The mechanism by which tryptase activity is lost in the absence of heparin was shown to involved dissociation of subunits rather than autodegradation [22,24]. DFP-inhibited tryptase is tetrameric in the presence of heparin, but converts to the monomeric form in the absence of heparin. Inactive monomeric tryptase, obtained by incubation of the active enzyme without heparin, showed neither a decrease in molecular weight by gel filtration nor the formation of low molecular weight degradation products by SDS PAGE. Active monomeric tryptase has not been detected. Tryptase stability in the absence of heparin is increased by increasing the concentration of tryptase,

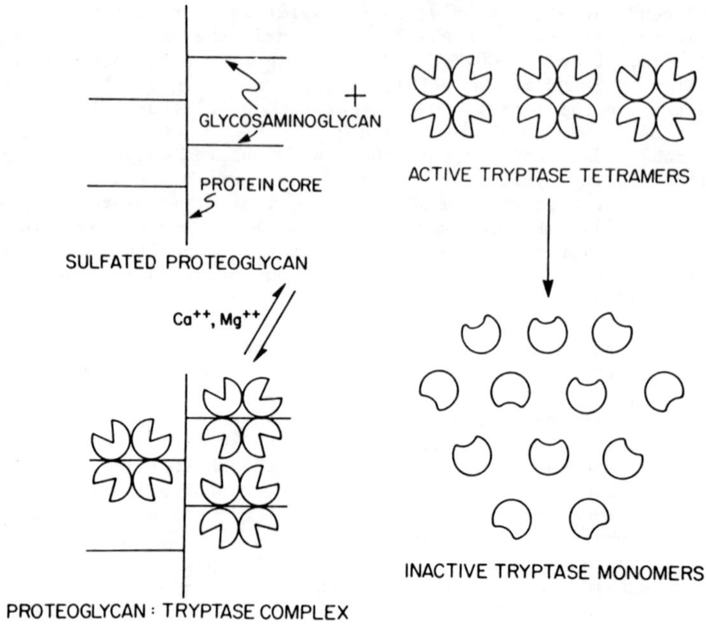

Fig. 1. Regulation of tryptase by sulfated proteoglycans.

which should shift the equilibrium toward preservation of the tetrameric form. If tryptase inactivation were due to autodegradation, an increased rate of inactivation would have been expected. Thus, stabilization of tryptase involves preventing conversion of the active tetramer to inactive monomers.

The interaction of tryptase with heparin depends primarily upon the negative charge density of the polyanion [24] rather than on a specific carbohydrate sequence. Other polysaccharides, such as dextran sulfate, chondroitin sulfates and heparan sulfate, bind to tryptase and stabilize its activity to varying extents. Even polyglutamic acid is able to protect tryptase activity, showing that the interaction does not require carbohydrate. Hyaluronic acid and keratan sulfate, with lower negative charge densities than any of the other polyanions listed, do not provide any appreciable stabilization of tryptase activity under physiologic conditions. Thus it appears that the persistence of tryptase activity in tissues can be regulated to a great extent by the interaction with highly sulfated proteoglycans.

The size of the tryptase/proteoglycan complex also would tend to limit diffusion of tryptase away from the site of inflammation. Tryptase not bound to heparin might bind to chondroitin sulfates and heparan sulfate in the extracellular matrix, also limiting diffusion. Whether or not tryptase enters the circulation in an active form is not known. However, active enzyme has been detected after allergen challenge in nasal lavage fluid [25] and in skin chamber fluid overlying sites of exposed dermis [26]. Thus, in the absence of a mechanism for separating tryptase from heparin and other proteoglycans *in vivo*, tryptase activity may be expressed for prolonged periods of time in the area of inflammation. Physiologic concentrations of calcium and magnesium have been shown to decrease the stability of tryptase in the presence of heparin (S. Alter, unpublished), and may serve a regulatory function *in vivo*.

BIOLOGICAL FUNCTION

The ultimate physiological role of tryptase, as for the mast cell, is unknown. To examine this problem the interactions of tryptase with human factors involved in complement, kinin production, coagulation and tissue metabolism have been observed *in vitro* (Table 2). For example, tryptase cleaves C3 (but not C5) to produce active C3a anaphylatoxin [27], a potent mediator that may activate mast cells, increase vascular permeability and cause bronchospasm. C3a is generated *in vivo* during allergen challenge [28], however, its activation by tryptase occurs only at a slow rate *in vitro*. Tryptase does not appear to enhance kinin production because it does not cleave plasma [29] or tissue [23] prekallikreins or low molecular weight kininogen [30] under physiologic conditions. High molecular weight kininogen is cleaved by tryptase; both the procoagulant activity and the ability to produced bradykinin after treatment with kallikrein are destroyed [31]. Cleavage/consumption of high molecular weight kininogen has been observed as a result of anaphylaxis induced by antigen [32] or radiocontrast dye [33], and may be a consequence of released tryptase. Tryptase also has been shown to destroy the thrombin-catalyzed clotting ability of fibrinogen [34] which further supports an anticoagulant function. Other proteins involved in the regulation of the clotting mechanism, including plasminogen [34], Hageman Factor (Factor XII) [35] and antithrombin III (S. Alter, unpublished results),

Table 2. Functional properties of human mast cell tryptase

Pathway Component	Effect
Kinin Pathway	
Bradykinin	no effect
High molecular weight kininogen	inactivation
Low molecular weight kininogen	no effect
Plasma prekallikrein	no effect
Tissue prekallikrein	no effect
Complement Pathway	
C3	C3a generation
C5	no effect
Coagulation Pathway	
Fibrinogen	fibrinogenolysis
High molecular weight kininogen	inactivation
Plasminogen	no effect
Factor XII	no effect
Connective tissue metabolism	
Procollagenase	activation

do not serve as substrates for tryptase *in vitro*. The anticoagulant activity of tryptase, plus the participation of mast cell heparin [36] and chondroitin sulfate E [37] in thrombin inhibition and the ability of prostaglandin D_2 to block platelet aggregation [38], suggest that the inhibition of coagulation by mast cell products is an important aspect of mast cell-mediated events. This might explain, in part, the lack of fibrin deposition observed in areas of immediate hypersensitivity reactions (like urticaria), as opposed to Arthus or delayed type reactions.

Another potential role for mast cell proteases concerns the metabolism of connective tissue. Such an activity in moderation might be a normal part of a tissue repair mechanism, but when excessive may play a role in inflammatory disease states such as rheumatoid arthritis, where mast cell numbers are markedly elevated in rheumatoid synovium. A recent study has demonstrated the capacity of tryptase to activate latent collagenase derived from rheumatoid synovial tissue [39]. This exciting observation may provide a link between the increased concentration of mast cells and the connective tissue damage commonly associated with this disease. New therapeutic options directed at the accumulation and activation of mast cells and the activity of tryptase and other mast cell proteases may prove useful in dealing with this often crippling disease.

LOCALIZATION AND IDENTIFICATION OF MAST CELLS AND MAST CELL SUBSETS

Tryptase was originally shown to be contained in mast cell secretory granules by virtue of its IgE-mediated release in parallel with histamine [40]. Murine monoclonal anti-tryptase antibodies [12] were developed and used to demonstrate by immunohistochemical techniques that all mast cells (identified by positive staining with Alcian blue) from lung, skin and bowel [12,41] contain tryptase. Other than mast cells, only human basophils contain a detectable amount of tryptase [6]. However, the small amount of tryptase per basophil (300 to 700-fold less than in lung and skin mast cells) results in a lack of staining of basophils by standard techniques, and implies that basophils are a relatively insignificant source of the enzyme. Thus, tryptase can be used as a sensitive and specific antigenic marker for mast cells. Electron microscopy using immunogold labeling techniques [42] has verified that tryptase is found exclusively in the secretory granules of mast cells and is distributed throughout all granules of each mast cell, preferentially over electron dense regions.

Immunohistochemical staining for tryptase and chymase (another neutral protease found in mast cell granules) has allowed the identification of two types of human mast cells [5]. Their distributions in normal human tissues are summarized in Table 3. T mast cells contain tryptase and not chymase and predominate in lung and mucosa of normal small bowel. The presence of T mast cells appears to be dependent upon a functional T lymphocyte system because patients with acquired (AIDS) or congenital (SCID) immunodeficiency diseases show a marked lack of T mast cells in the bowel [43]. TC mast cells, the other mast cell type, contain tryptase together with chymase (4.5 per TC mast cell derived from adult foreskin) in each secretory granule. TC mast cells are found primarily in the skin and bowel mucosa of normal subjects and occur at normal concentrations in immunodeficient patients. Functional [44,45,46] and ultrastructural [42] features also distinguish T from TC types of human mast cells. The T and TC subsets of human mast cells appear to be somewhat analogous to the mucosal and connective tissue types of rodent mast cells, respectively. The functional justification for mast cell diversity is unknown, but the numerous characteristics distinguishing

Table 3. Distribution of human T and TC types of mast cells in normal tissues

Property	Subset T	Subset TC
Neutral protease content	Tryptase	Tryptase Chymase Carboxypeptidase
Tissue localization		
Small intestine		
Mucosa		
epithelium	–	–
subepithelium	++	+
Submucosa	+	++
Skin		
epidermis	–	–
dermis	+/–	++
Lung		
Bronchi	++	++
Alveoli	++	–
Nasal mucosa		
epithelium	+	–
subepithelium	++	++
Conjunctiva		
epithelium	+	–
subepithelium	++	++

Legend: ++ = strongly present
+ = weakly present
– = absent

these mast cell types implies to us a purposeful function for each one.

ASSESSMENT OF MAST CELL ACTIVATION *IN VIVO*

Tryptase levels in complex biologic fluids serve as an indication of mast cell activation (Table 4). A specific immunoassay [47] has been used to detect tryptase levels as low as 2.5 ng/ml in neat fluids. Normal levels in serum or plasma are less than 5 ng/ml, usually less than 2.5 ng/ml, while levels greater than 10 ng/ml have been found in essentially all patients within one to four hours after the onset of systemic anaphylxis with hypotension [48,49] and may reach as high as 800 ng/ml. Elevated tryptase levels are first detected within 15 to 30 minutes post challenge, and decline with a half time of about two hours (unpublished observations, Schwartz *et al.*). By contrast, histamine levels peak within five to ten minutes after challenge and return to baseline by about 30 minutes [32,50]. The delayed appearance of tryptase probably reflects a slower rate of diffusion through tissues due to the difference in molecular size of the tryptase/proteoglycan complex over histamine, as well as the binding of tryptase to proteoglycans in the

Table 4. Biologic fluids and conditions in which tryptase levels indicate mast cell activation

Serum/plasma
Systemic anaphylaxis
Allergen/IgG-mediated
Aspirin/nonsteroidal antiinflammatory drugs
Systemic mastocytosis
Bronchoalveolar lavage fluid
Atopic asthma
Nasal lavage fluid
Allergic rhinitis
Tears
Atopic conjunctivitis
Vernal conjunctivitis
Giant papillary conjunctivitis
Skin chamber fluid
IgE-mediated/Allergen
Codeine

extracellular matrix. Basophils also may release large amounts of histamine, making if difficult to determine the cellular origin of this biogenic amine when detected in plasma. Release of histamine by basophils can occur during blood coagulation, prohibiting the use of histamine levels in serum as an indicator of mast cell/basophil activation. Because the amount of tryptase in basophils is negligible, the level of tryptase in either serum or plasma is useful as an indication of mast cell involvement. Thus, tryptase levels provide a useful diagnostic correlate in the evaluation of suspected anaphylactic events.

Plasma tryptase levels also were evaluated in 11 of 17 patients with systemic mastocytosis, even though they were not acutely symptomatic [48]. Histamine levels in these patients were linearly related to tryptase ($p < 0.01$), suggesting that a steady state equilibrium had been reached between the release and removal of both mediators. High levels of tryptase indicated ongoing mast cell activation in patients with nonacute mastocytosis.

In controlled clinical studies, elevated levels of plasma tryptase and histamine after food challenge correlated with the appearance of clinical symptoms [51], indicating mast cell involvement in food hypersensitivity reactions. However, whereas tryptase levels rose in seven of eleven subjects, histamine levels rose in eleven of eleven subjects, indicating that a more sensitive assay for tryptase will be needed to detect less severe reactions. The role of mast cells in non-immunologic reactions also can be studied using tryptase as a marker. For example, systemic reactions induced by cyclooxygenase inhibitors in the three subjects studied thus far, resulted in elevated serum tryptase levels in each case (Schwartz *et al.*, unpublished).

Tryptase levels also have been utilized in studies of cutaneous, pulmonary and nasal models of allergic diseases. For example, tryptase and histamine appear together in skin chamber fluid overlying exposed dermis after allergen challenge [26]. Maximal histamine levels occur earlier (about 30 minutes after challenge) than do maximal tryptase levels (30 to 60 minutes). If allergen is then replaced with buffer, levels of both mediators steadily decline and do not rise again over a six to eight hour interval. However, a second challenge with the same allergen several hours after the initial challenge, induces release of histamine alone. Basophils have been observed on the dermal surface just prior to this second allergen challenge. During continuous allergen challenge over a six hour time course histamine alone is released during the later stages of the reaction [52]. Challenge with codeine (which activates TC mast cells but not basophils) after an immediate allergic reaction or after continuous allergen challenge results in the appearance of substantial amounts of tryptase and histamine, suggesting the global depletion of tryptase had not occurred after the allergen challenge. These results argue that the basophil is the cell source of histamine in late cutaneous allergic reactions, and for the mast cell as the major cell source of mediators in the immediate reaction.

Measurement of tryptase in bronchoalveolar lavage fluid (BAL) before and after endobronchial allergen challenge has provided new insight into the pathogenesis and interrelationship of atopy and asthma [53]. Nonatopic subjects with or without asthma showed the same low level of tryptase in BAL fluid before and after allergen challenge. Atopic nonasthmatics, like those in the nonatopic groups, had low levels of tryptase before challenge. In contrast, atopic asthmatics had high baseline levels, suggesting ongoing mast cell activation. A corrolary to the increased mast cell activation found at baseline in atopic asthmatics is the observation that adenosine, which augments allergen-mediated release of preformed granule mediators (but not newly generated lipids) from mast cells *in vitro*, when inhaled causes bronchospasm in atopic asthmatics, but not in nonasthmatic subjects [54]. This bronchospasm is inhibitable by antihistamines, consistent with the response being mediated primarily by histamine release from pulmonary mast cells. The atopic asthmatic and nonasthmatic groups each responded to allergen challenge by releasing increased levels of tryptase in BAL fluid, these levels being higher in the atopic asthmatics. These results suggest that ongoing mast cell activation may play a role in the persistence of bronchial hyperreactivity in atopic asthmatics, and may also explain the hyperreactivity of atopic asthmatics to adenosine.

Levels of tryptase in nasal lavage fluid are consistently elevated above baseline after a clinical response to nasal allergen challenge. The allergen dose that elicits the maximal clinical response typically elicits the maximal level of tryptase [55]. Histamine levels also tend to rise after allergen challenge, but histamine baseline levels are quite variable, and the dose response relationship to allergen does not always follow the clinical response. Therefore, mast cell activation, as indicated by the level of tryptase, appears to be the primary determinant of an immediate allergic reaction.

CONCLUDING COMMENTS

Of the three neutral proteases thus far localized to human mast cells tryptase appears to be the most abundant. Each one is stored in secretory granules, fully active, and together they are released during degranulation events. Although the *in vitro* observations concerning the capacity of tryptase to generate C3a from C3, destroy fibrinogen and high molecular weight kininogen and promote the activation of latent collagenase are of great potential biologic interest, the evidence for tryptase performing these processes *in vivo* is speculative. The function of released tryptase *in vivo*, certainly when derived from the TC mast cell type, will reflect the concerted actions of chymase, carboxypeptidase and tryptase as part of the released proteoglycan-protease complex. Because these enzymes are selectively concentrated in mast cells, even without knowledge of their exact biologic activities they serve as indicators of the occurrence of mast cell activation when detected in the complex milieu of biologic fluids. Tryptase is released from each mast cell type and already has proven to be a clinically useful marker of allergic activity. The objectivity gained by measurement of allergy attack enzymes will be critical to the diagnosis and monitoring of allergic reactions.

REFERENCES

1. Glenner, C.G. and Cohen, L.A. (1960). Histochemical demonstration of a species-specific trypsin-like enzyme in mast cells. Nature (London), 185, 846-847.
2. Hopsu-Havu, V.K. and Glenner, C.G. (1963). A histochemical enzyme kinetic system applied to the trypsin-like amidase and esterase activity in human mast cells. J. Cell. Biol., 17, 503-520.
3. Schwartz, L.B., Lewis, R.A. and Austen, K.F. (1981). Tryptase from human pulmonary mast cells: purification and characterization. J. Biol. Chem., 256, 11939-11943.
4. Schwartz, L.B., Irani, A.M., Roller, K., Castells, M.C. and Schechter, N.M. (1987). Quantitation of histamine, tryptase, and chymase in dispersed human T and TC mast cells. J. Immunol., 138, 2611-2615.
5. Irani, A.M., Schechter, N.M., Craig, S.S., DeBlois, G. and Schwartz, L.B. (1986). Two types of human mast cells that have distinct neutral protease compositions. Proc. Natl. Acad. Sci. USA, 83, 4464-4468.
6. Castells, M.C., Irani, A.M. and Schwartz, L.B. (1987). Evaluation of human peripheral blood leukocytes for mast cell tryptase. J. Immunol., 138, 2184-2189.
7. Schechter, N.M., Choi, J.K., Slavin, D.A., Dersienski, D.T., Sayama, S., Dong, G., Lavker, R.M., Proud, D. and Lazarus, G. (1986). Identification of a chymotrypsin-like proteinase in human mast cells. J. Immunol., 137, 962-970.
8. Goldstein, S.M., Kaempfer, C.E., Proud, D., Schwartz, L.B., Irani, A.M. and Wintroub, B.U. (1987). Detection and partial characterization of a human mast cell carboxypeptidase. J. Immunol., 139, 2724-2729.
9. Smith, T.J., Hougland, M.W. and Johnson, D.A. (1984). Human lung tryptase, purification and characterization. J. Biol. Chem., 259, 11046-11049.
10. Fraki, J.E., Schechter, N.M., Harvima, I.T. and Lazarus, G.S. (1985). Human skin tryptase. In: The Biological Role of Proteinases and their Inhibitors in Skin (eds. H. Ogawa, G.S.Lazarus, V.K. Hopsu-Havu), University of Tokyo Press, Tokyo, Japan, p.127.
11. Cromlish, J.A., Seidah, N.G., Marcinkiewicz, M., Hamelin, J., Johnson, D.A. and Chretein, M. (1987). Human pituitary tryptase: molecular forms, NH_2-terminal sequence, immunocytochemical localization, and specificity with prohormone and fluorogenic substrates. J. Biol. Chem., 262, 1363-1373.
12. Schwartz, L.B. (1985). Monoclonal antibodies against human mast cell tryptase demonstrate shared antigenic sites on subunits of tryptase and selective localization of the enzyme to mast cell. J. Immunol., 134,

526-531.
13. Sayama, S., Iozzo, R.V., Lazarus, G.S. and Schechter, N.M. (1987). Human skin chymotrypsin-like chymase, subcellular localization to mast cell granules and interaction with heparin and other glycosaminoglycans. J. Biol. Chem., 262, 6808-6815.
14. Metcalfe, D.D., Lewis, R.A., Silbert, J.E., Rosenberg, R.D., Wasserman, S.I. and Austen, K.F. (1979). Isolation and characterization of heparin from human lung. J. Clin. Invest., 64, 1537-1543.
15. Metcalfe, D.D., Soter, N.A., Wasserman, S.I. and Austen, K.F. (1980). Identification of sulfated mucopolysaccharides including heparin in the lesional skin of a patient with mastocytosis. J. Invest. Dermatol., 74, 210-215.
16. Eliakim, R., Gilead, L., Ligumski, M., Okon, E., Rachmilewitz, D. and Razin, E. (1986). Histamine and chondroitin sulfate E proteoglycan released by cultured human colonic mucosa: indication for a possible presence of E mast cells. Proc. Natl. Acad. Sci. USA, 83, 461-464.
17. Stevens, R.L., Fox, C., Lichtenstein, L.M. and Austen, K.F. (1988). Identification of chondroitin sulfate E proteoglycans and heparin proteoglycans in the secretory granules of human lung mast cells. J. Allergy Clin. Immunol., 81, 519 (Abstract).
18. Thompson, H.L., Schulman, E.S. and Metcalfe, D. (1988). Identification of chondroitin sulfate E in human lung mast cells. J. Immunol., 140, in press.
19. Lagunoff, D. and Rickard, A. (1983). Evidence for control of mast cell granule protease *in situ* by low pH. Exp. Cell. Res., 144, 353-360.
20. DeYoung, M.B., Nemeth, E.F. and Scarpa, A. (1987). Measurement of the internal pH of mast cell granules using microvolumetric fluorescence and isotopic techniques. Archs. Biochem. Biophys., 254, 222-233.
21. Schwartz, L.B., Riedel, C., Caulfield, J.P., Wasserman, S.I. and Austen, K.F. (1981). Cell association of complexes of chymase, heparin proteoglycan, and protein after degranulation. J. Immunol., 126, 2071-2078.
22. Schwartz, L.B. and Bradford, T.R. (1986). Regulation of tryptase from human lung mast cells by heparin, stabilization of the active tetramer. J. Biol. Chem., 261, 7372-7379.
23. Alter, S.C., Yates, P., Margolius, H.S. and Schwartz, L.B. (1987). Tryptase and kinin generation, tryptase from human mast cells does not activate human urinary prokallikrein. Int. Archs. Allergy Appl. Immunol. 83, 321-324.
24. Alter, S.C., Metcalfe, D.D., Bradford, T.R. and Schwartz, L.B. (1987). Regulation of human mast cell tryptase: effects of enzyme concentration, ionic strength and the structure and negative charge density of polysaccharides. Biochem. J., 248, 821-827.
25. Baumgarten, C.R., Nichols, R.C., Naclerio, R.M., Lichtenstein, L.M., Norman, P.S. and Proud, D. (1986). Plasma kallikrein during experimentally induced allergic rhinitis: role in kinin formation and contribution to TAME-esterase activity in nasal secretions. J. Immunol., 137, 977-982.
26. Schwartz, L.B., Atkins, P.C., Bradford, T.R., Fleekop, P., Shalit, M. and Zweiman, B. (1987). Release of tryptase together with histamine during the immediate cutaneous response to allergen. J. Allergy Clin. Immunol., 80, 850-855.
27. Schwartz, L.B., Kawahara, M.S., Hugli, T.E., Vik, D., Fearon, D.T. and Austen, K.F. (1983). Generation of C3a anaphylatoxin from human C3 by human mast cell tryptase. J. Immunol., 130, 1891-1895.
28. Nagata, S. and Glovsky, M.M. (1987). Activation of human serum complement with allergens. I. Generation of C3a, C4a, C5a and induction of human neutrophil aggregation. J. Allergy Clin. Immunol., 80, 24-32.
29. Schwartz, L.B., Maier, M. and Spragg, J. (1986). Interaction of human

low molecular weight kininogen with human mast cell tryptase. Adv. Exp. Med. Biol., 198A, 105-111.
31. Maier, M., Spragg, J. and Schwartz, L.B. (1983). Inactivation of human high molecular weight kininogen with human mast cell tryptase. J. Immunol., 130, 2352-2356.
32. Smith, P.L., Kagey-Sabotka, A., Bleeker, E.R., Traystman, R., Kaplan, A.P., Gralnick, H., Valentine, M.D., Permutt, S. and Lichtenstein, L.M. (1980). Physiologic manifestations of human anaphylaxis. J. Clin. Invest., 66, 1072-1080.
33. Zuraw, B.L., Christiansen, S.C., Sugimoto, S.L. and Lang, D.L. (1988). Radiographic contrast media reactions associated with hypotension are accompanied by cleavage of high molecular weight kininogen in plasma. J. Allergy Clin. Immunol., 81, 218 (Abstract).
34. Schwartz, L.B., Bradford, T.R., Littman, B.L. and Wintroub, B.U. (1985). The fibrinolytic activity of purified tryptase from human lung mast cells J. Immunol., 135, 2762-2767.
35. Alter, S.C., Lammle, B., Griffin, J.B. and Schwartz, L.B. (1988). Tryptase from human mast cells does not activate purified human Hageman Factor. Clin. Allergy, 18, 151-156.
36. Rosenberg, R.D. and Lam, L. (1979). Correlation between structure and function of heparin. Proc. Natl. Acad. Sci. USA, 76, 1218-1222.
37. Scully, M.F., Ellis, V., Seno, N. and Kakkar, V.V. (1986). The anti-coagulant properties of mast cell product, chondroitin sulfate E. Biochem. Biophys. Res. Commun., 137, 15-22.
38. Mills, D.C. and MacFarlane, D.E. (1974). Stimulation of human platelet adenylate cyclase by prostaglandin D_2. Thromb. Res., 5, 401-412.
39. Gruber, B.L., Schwartz, L.B., Ramamurthy, N.S., Irani, A.M. and Marchese, M.J. (1988). Activation of rheumatoid synovial collagenase by mast cell tryptase. J. Immunol., in press.
40. Schwartz, L.B., Lewis, R.A., Seldin, D. and Austen, K.F. (1981). Acid hydrolases and tryptase from secretory granules of dispersed human lung mast cells. J. Immunol., 126, 1290-1294.
41. Craig, S.S., DeBlois, G. and Schwartz, L.B. (1986). Mast cells in human keloid, small intestine and lung by an immunoperoxidase technique using a murine monoclonal antibody against tryptase. Amer. J. Pathol., 124, 427-435.
42. Craig, S.S., Schechter, N.M. and Schwartz, L.B. (1988). Ultrastructural analysis of human T and TC mast cells identified by immunoelectron microscopy. Laboratory Invest., in press.
43. Irani, A.M., Craig, S.S., DeBlois, G., Elson, C.O., Schechter, N.M. and Schwartz, L.B. (1987). Deficiency of the tryptase-positive, chymase negative mast cell type in gastrointestinal mucosa of patients with defective T lymphocyte function. J. Immunol., 138, 4381-4386.
44. Tharp, M.D., Kagey-Sobotka, A., Fox, C.C., Marone, G., Lichtenstein, L.M. and Sullivan, T.J. (1987). Functional heterogeneity of human mast cells from different anatomic sites: *in vitro* responses to morphine sulfate. J. Allergy Clin. Immunol., 79, 646-653.
45. Lawrence, I.D., Warner, J.A., Cohan, V.L., Hubbard, W.C., Kagey-Sobotka, A. and Lichtenstein, L.M. (1987). Purification and characterization of human skin mast cells. Evidence for human mast cell heterogeneity. J. Immunol., 139, 3062-3069.
46. Lowman, M.A., Rees, P.H., Benyon, R.C. and Church, M.K. (1988). Human mast cell heterogeneity: histamine release from mast cells dispersed from skin, lung, adenoids, tonsils, and colon in response to IgE-dependent and non-immunologic stimuli. J. Allergy Clin. Immunol., 81, 590-597.
47. Wenzel, S., Irani, A.M., Sanders, J.M., Bradford, T.R. and Ashwartz, L.B. (1986). Immunoassay of tryptase from human mast cells. J. Immunol.

Methods, 86, 139-142.

48. Schwartz, L.B., Metcalfe, D.D., Miller, J.S., Earl, H. and Sullivan, T. (1987). Tryptase levels as an indicator of mast-cell activation in systemic anaphylaxis and mastocytosis. New England J. Med., 316, 1622-1626.
49. Miller, J.S. and Schwartz, L.B. (1988). Tryptase levels as an indication of mast cell activation in a patient with hymenoptera anaphylaxis and a patient with Salmonella enteritis. J. Allergy Clin. Immunol., 81, 512 (Abstract).
50. Sheffer, A.L., Soter, N.A., McFadden, E.R. Jr. and Austen, K.F. (1983). Exercise-induced anaphylaxis: a distinct form of physical allergy. J. Allergy Clin. Immunol., 71, 311-316.
51. Ebbeling, W.L., Schwartz, L.B. and Buckley, R.H. (1988). Comparison of plasma histamine and tryptase levels following double-blind and placebo-controlled food challenges in patients with asthma and/or atopic dermatitis. J. Allergy Clin. Immunol., 81, 69 (Abstract).
52. Shalit, M., Schwartz, L.B., von Allmen, C., Valenzano, M., Atkins, P., Kaliner, M. and Zweiman, B. (1988). *In vivo* histamine and tryptase release during continuous and interrupted antigen challenge. J. Allergy Clin. Immunol., 81, 516 (Abstract).
53. Wenzel, S.E., Fowler, A.A. and Schwartz, L.B. (1988). Activation of pulmonary mast cells by bronchoalveolar allergen challenge: *in vivo* release of histamine and tryptase in atopic subjects with and without asthma. Am. Rev. Respir. Dis., in press.
54. Cushley, M.J., Tattersfield, A.E. and Holgate, S.T. (1984). Adenosine-induced bronchoconstriction in asthma. Antagonism by theophylline. Am. Rev. Respir. Dis., 129, 380-384.
55. Castells, M.C. and Schwartz, L.B. (1988). Tryptase levels in nasal lavage fluid as an indicator of the immediate allergic response. J. Allergy Clin. Immunol., in press.

Interleukin-4 Induced IgE Formation

Hans L. Spiegelberg
Department of Immunology, Research Institute of Scripps Clinic, La Jolla, CA 92037, USA

ABSTRACT

It has recently been shown and confirmed in this study that interleukin-4 (IL-4) stimulates and interferon-γ (IFN-γ) inhibits lipopolysaccharide (LPS) activated murine B cells to secrete IgE and IgG1 *in vitro*. In man, a polyclonal B cell activator analogous to LPS that induces IL-4 dependent IgE secretion has not yet been found. To learn more about the role of the polyclonal activator and IL-4 in IgE formation, we stimulated murine B cells with IL-4 and either LPS or polyribosyl-ribitol-phosphate (PRP). B cells activated with PRP and IL-4 formed the same amount of IgG1 as LPS activated cells. In contrast, PRP activated cells secreted <10% of the IgE formed by LPS and IL-4 activated cells. These data suggest that factors in addition to IL-4 may be necessary for IgE secretion and that the mechanism of IL-4 induced IgE secretion may be different from IL-4 induced IgG1 formation. To determine whether IgE and IgG1 formation is linked *in vivo*, we infected mice with four different helminth parasites (*N. brasiliensis*, *M. corti*, *T. crassiceps* and *T. spiralis*). All parasites induced a concomitant increase of the IgE and IgG1 serum levels and usually a decrease of the IgG2a level, suggesting that the polyclonal IgE and IgG1 responses are linked *in vivo* and most likely involve T helper cells that secrete IL-4. Although most of the specific antibody was of IgG1 isotype, significant IgG2a anti-parasite antibodies were also detected to the *T. crassiceps* parasite despite a decline of the IgG2a serum level. This suggests that the polyclonal and antigen specific responses to the parasites may be regulated differently. Furthermore, the effect of the parasitic infection on the IgE, IgG1 and IgG2a serum levels was dependent on the anatomical site of infection. Mice infected subcutaneously with *T. crassiceps* formed 10 fold less IgE and more IgG2a than IgG1 compared to intraperitoneally infected mice. These data suggest that T helper cells may differentiate under the influence of accessory cells varying in cytokine production at different anatomic locations into either IL-4 or IFN-γ secreting cells and suggest a role of accessory cells in the isotype restricted responses to helminth infections and allergens.

KEYWORDS

Cytokines; T-helper cells; helminths; IgG, IgE responses.

INTRODUCTION

The mechanism that regulates the switch from IgM to IgE secretion in B cells is of particular interest because over-production of IgE antibodies is the major cause of allergic diseases. T cells have long been known to play a role in this process [1]; however, the lymphokines produced by the T cells that determine whether a B cell switches from IgM to IgE or IgG formation have only recently been identified. In 1982, Isakson *et al.* [2] first showed that interleukin-4 (IL-4, previously called B cell differentiation factor-γ or B cell stimulatory factor-1) causes lipopolysaccharide (LPS) activated murine B cells to secrete IgG1. Subsequently, Coffman *et al.* [3] demonstrated that IL-4 does not only induce LPS activated B cells to secrete IgG1 but also IgE and Snapper and Paul showed that interferon-γ (IFN-γ) induces IgG2a secretion [4]. Furthermore, IFN-γ acts antagonistically to IL-4 in that it inhibits the LPS and IL-4 induced IgE/IgG1 secretion [5]. Because antigen specific cloned T helper cells can be divided into two types, one called TH_1, secreting IFN-γ, and the other TH_2, producing IL-4 [6], it was postulated that the magnitude of an IgE response may be the result of the interaction of B cells with T helper cells that produce either IL-4 or IFN-γ. Evidence that IL-4 and IFN-γ also function antagonistically *in vivo* has also been obtained. Mice infected with helminth parasites show a concomitant increase of the IgE and IgG1 and a decrease of the IgG2a serum levels [7,8] and the antibody response to viral infections is restricted almost entirely to IgG2a antibodies [9]. Finkelman and colleagues demonstrated that injections of large amounts of monoclonal antibodies to IL-4 inhibit polyclonal IgE responses [10] and antibodies to IFN-γ enhance the IgE and IgG1 secretion [11]. Although these findings strongly suggest that IL-4 and IFN-γ are important factors in the regulation of IgE formation, many questions remain regarding the role of IL-4 and IFN-γ in the regulation of IgE antibody formation. First, it has not yet been shown whether IL-4 is directly involved in eliciting the μ to ε/γ1 chain rearrangement on the DNA level or whether it induces expansion of IgE committed cells. However, since IL-4 acts on $sIgM^+/sIgD^+$ cells, it is likely that it is somehow involved in the μ to ε/γ1 switch mechanism. Second, it is unknown whether IL-4 is the only lymphokine that is needed to induce IgE/IgG1 secretion. Third, it is not clear whether the regulation of IgE formation is always linked *in vitro* and *in vivo* to the IgG1 subclass in mice and whether human IgE antibody formation is also linked to an IgG subclass. Fourth, it is controversial whether two classes of T helper cells exist [5] or whether T helper cells are multipotential and differentiate after interaction with antigen into either IL-4 or IFN-γ secreting cells [12]. Finally, it is not understood how certain antigens such as allergens selectively interact with T cells secreting IL-4.

The experiments described in this review were undertaken to gain more insight into the role of the polyclonal activator in the IL-4 induced IgE and IgG1 secretion *in vitro* and to determine whether IgE and IgG1 formation is coregulated *in vivo*. For this purpose, mouse B cells were activated with two different polyclonal activators and the changes in the serum levels of the different isotypes were quantitated in mice infected with four different helminth parasites. Furthermore, to obtain clues regarding the presumably preferential interaction of IgE inducing allergens with T helper cells secreting IL-4, we also investigated whether infection with the same parasite at a different anatomical location has the same or a different effect on IgE, IgG1 and IgG2a responses.

RESULTS

The Effect of Different Polyclonal Activators on IL-4 Induced IgE and IgG1 Secretion

After the report by Coffman *et al.* [3] demonstrating that IL-4 induces mouse B cells to secrete IgE after LPS and IL-4 stimulation, we attempted to induce IgE secretion by human B cells activated with *S. aureus* Cowan I, LPS or Epstein Barr Virus in the presence of commercially available recombinant IL-4 or polyvalent supernatants from mix lymphocyte cultures. Although a monoclonal anti-IFN-γ antibody was added to neutralize any inhibitory effect of IFN-γ present in the culture systems, neither of the activators induced IgE secretion by human B cells in the presence of recombinant IL-4 [13]. Because LPS is not a potent activator for human B cells and LPS may be crucial for the murine IL-4 induced IgE secretion, we tested B cell activators other than LPS in the murine system. In preliminary experiments, anti-IgM and IL-4 induced only a small IgG1 and no IgE secretion. In contrast, the polyclonal B cell activator polyribosyl-ribitol-phosphate (PRP), the capsular carbohydrate of *H. influenzae* b [14] induced a large IgG1 secretion in the presence of IL-4. For these experiments, IL-4 was purified from supernatants of the IL-4 secreting T cell line D-10 [15] and the B cells were cultured by the method of Coffman *et al.* [3]. As shown in Table 1, LPS and PRP alone induced predominantly IgM secretion. The quantities of IgG subclasses secreted after LPS or PRP stimulation followed in descending order the gene order $\gamma 3 > \gamma 1 > \gamma 2b > \gamma 2a$. Normal B cells did not secrete IgE under these conditions. Addition of IL-4 24 hours after LPS stimulation induced IgE and IgG1 in a dose dependent manner. It also caused a slight decrease of the IgM secretion and a large suppression of IgG3, IgG2a and IgG2b formation. Addition of IL-4 to PRP activated cells caused a slightly higher IgG1 secretion than that of LPS activated cells. However, the IgE secretion was <10% of that of LPS and IL-4 activated cells. IL-4 also caused a decrease of the IgG3, IgG2a and IgG2b formation induced by PRP alone; however, it had little effect on IgM secretion. The lack of IgE production was not the effect of PRP induced IFN-γ secretion by small numbers of contaminating T or NK cells in the B cell preparation because addition of a monoclonal antibody to IFN-γ did not enhance IgE secretion. Removal of NK cells and macrophages also did not result in higher IgE secretion by PRP and IL-4 activated B cells and addition of PRP to LPS and IL-4 stimulated cells did not inhibit the LPS and IL-4 produced IgE formation. The reason why PRP was much less effective than LPS in inducing IgE secretion is unknown. Although the IL-4 dose response curves for IgG1 secretion were similar for LPS and PRP activated cells, it is possible that more IL-4 is necessary to induce IgE secretion by PRP than LPS activated cells. However, a 10 fold higher IL-4 concentration than that used in experiments of Table 1 still did not induce significantly more IgE secretion. Another possibility that could explain the difference between LPS and PRP activated cells could be the need for an as yet unknown B cell derived lymphokine in the IgE but not IgG1 switch mechanism that is induced by LPS but not PRP. Addition of IL-1 to PRP and IL-4 activated cells had no effect on IgE secretion; however, additional lymphokines or supernatants from LPS activated B cells will have to be analyzed to test this possibility. Activation of different B cell subpopulations by LPS and PRP could also explain the low IgE production by PRP and IL-4 activated cells. Therefore, we fractionated B cells into high and low density (HD, LD) cells by Percoll gradient centrifugation. The HD and LD cells represented about 50% and 22%, respectively, of the total B cell population. Both fractions were >90% $sIgM^+/sIgD^+$ and the LD cells were ~15% larger than the HD cells as determined by FACS analyses. HD and LD B cells from normal mice formed similar amounts of IgE. In contrast, LD cells formed about 10 fold more IgG1

Table 1. Ig secretion by normal balb/c spleen B cells stimulated with LPS, PRP and IL-4

Activator	μg/ml					ng/ml
	IgM	IgG3	IgG1	IgG2b	IgG2a	IgE
LPS (10 μg/ml)	30	3.8	1.3	0.4	0.2	<1
LPS + IL-4 (250 pM)	22	0.3	11.8	<0.1	<0.1	380
LPS + IL-4 (80 pM)	19	0.2	7.2	<0.1	<0.1	130
PRP (30 μg/ml)	39	2.3	0.7	0.4	0.3	<1
PRP + IL-4 (250 pM)	35	0.2	15.8	<0.1	<0.1	11
PRP + IL-4 (80 pM)	36	0.2	11.2	<0.1	<0.1	7

Effect of IFN-γ	μg IgG1/ml	ng IgE/ml
LPS + IL-4 (125 pM)	12.1	280
LPS + IL-4 + IFN-γ (200 U/ml)	0.3	5
LPS + IL-4 + anti-IFN-γ (1 μg/ml)	13.3	220

whether activated with LPS or PRP and IL-4. The LD cells also formed more IgM and IgG3 than the HD cells. These data showed that LPS and PRP with or without IL-4 act on both HD and LD cells. They did not explain why PRP is ineffective in inducing IgE secretion because PRP and IL-4 neither induced IgE secretion by HD nor LD cells and the IL-4 dose response curves for the two B cell populations were similar. Although these experiments did not exclude the possibility that LPS and PRP acted on different B cell subpopulations, they demonstrated that not all polyclonal activators prepare B cells to respond to IL-4 with IgE secretion and may explain why the LPS unreactive human B cells cannot easily be induced to secrete IgE with IL-4. LPS may provide a membrane activation signal essential for IgE but not IgG1 secretion that is not provided by PRP. Whatever the explanation for the lack of PRP and IL-4 induced IgE secretion is, the data suggest that IL-4 induced IgE formation *in vitro* is regulated differently from IgG1 formation. This conclusion has recently also been reached by Snapper *et al.* [16] based on differences in the IL-4 dose response and time required during culture of IL-4 in the LPS and IL-4 induced IgE and IgG1 secretion *in vitro*. In summary, IL-4 is essential for the induction of IgE secretion *in vitro* by murine B cells. However, factors in addition to IL-4 are required that are induced with LPS but not necessarily other polyclonal B cell activators. The reason why human B cells could as yet not be induced with IL-4 to secrete IgE may be because a potent human B cell activator such as LPS is missing.

The Effect of Helminth Infections in the Polyclonal and Antigen Specific IgE and IgG Subclass Response in Mice

Since IL-4 induces both IgE and IgG1 secretion *in vitro*, we investigated whether the regulation of these two isotypes is also coupled *in vivo*. In the past, it was reported that infections of rats and mice with the parasite *N. brasiliensis* (NBr) induces specifically IgE [17,18]. Conversely, the parasite *Mesocestoides corti* was shown to induce a polyclonal increase of the IgG1 serum level [19] and predominantly IgG1 anti-parasite antibodies [20,21]. Because the increases in polyclonal and parasite specific IgE and IgG1 responses have not been studied simultaneously in these experiments, we compared the two parameters in mice infected with four different helminth parasites. As shown in Fig. 1, infection with NBr induced a concomitant increase of both the IgE and IgG1 and a decrease of the IgG2a serum levels. This was observed in both high and low IgE responder strains of mice [7].

Athymic nude mice did not show the increases in the IgG1 and IgE serum levels, indicating the T cell dependency of the IgG1 and IgE formation to the parasite infection. SJL mice were unique in that they also showed a very large increase of the IgG2b but not IgG2a serum level. In all other strains tested, the IgG2b levels decreased like the IgG2a levels or remained unchanged. The NBr infection had relatively small effect on the IgG3 serum level. Essentially the same observations were made in mice infected with *M. corti*, *Taenia crassiceps* and *Trichinella spiralis* (Fig. 2). The increases in the IgE serum levels were measurable as soon as three days after infecttion, whereas the IgG1 level began to rise between days 7 to 14. Although the IgG2a serum levels increased initially after the *M. corti* and *T. crassiceps* infections, they usually decreased or remained unchanged 10 days after infection. These data show that the polyclonal IgE and IgG1 responses of mice to helminth infection are linked.

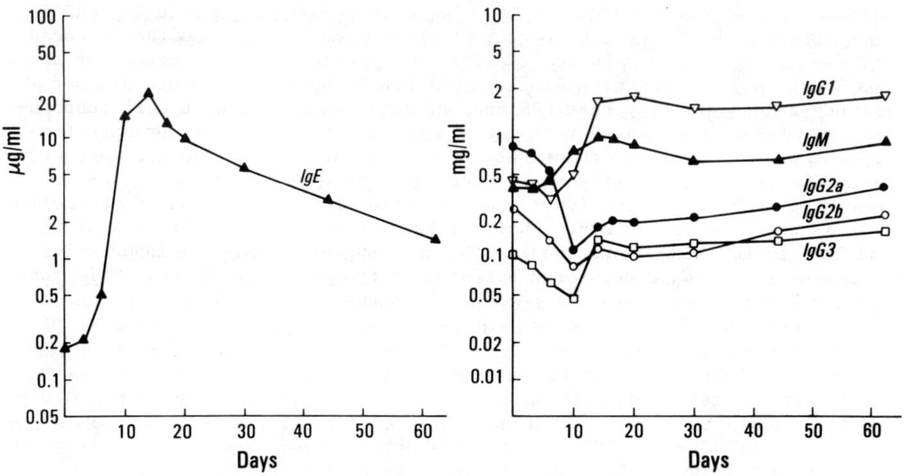

Fig. 1. Changes of serum levels of different isotypes after infection of BALB/c mice with 400 *N. brasiliensis* larvae.

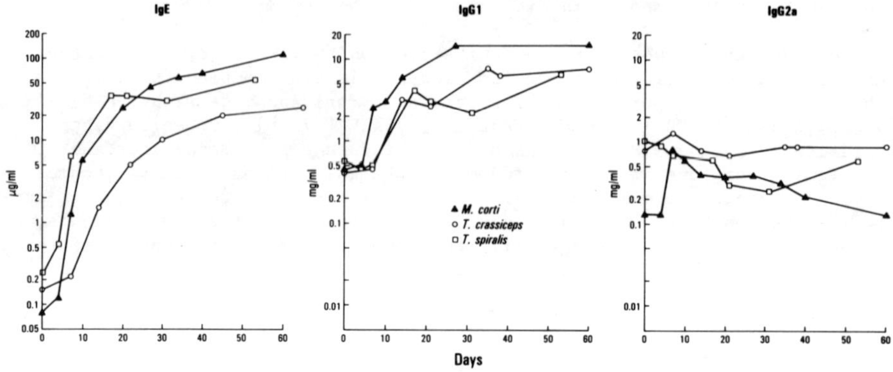

Fig. 2. Changes of the IgE, IgG1 and IgG2a serum levels in BALB/c after infection with *M. corti* (100 larvae), *T. crassiceps* (10 larvae) or *T. spiralis*. Mean of 3 mice.

As shown in Table 2, the antibody response to parasite extracts was predominantly of IgM and IgG1 subclass, which reflect the quantitatively much larger increase of the IgM (0.5 mg/ml) and IgG1 (2 mg/ml) than IgE serum levels (20 µg/ml). However, significant activities in the IgG2a subclass were also detected in *M. corti* and particularly *T. crassiceps* infected mice despite a decrease of the IgG2a serum levels. We do not know the % of the IgG1 and IgE serum increase that is antibody to the parasite; however, it is believed that a large portion is non-antiparasite IgG1 or IgE. The reason for the polyclonal increase of IgE and IgG1 is unknown. It may be the result of release of IL-4 into the circulation that affect antibody producing cells which are at the stage susceptible to lymphokine induced isotype switching. In contrast to the polyclonal response, B cells forming anti-parasite antibodies presumably obtain the lymphokine signal directly from the T helper cells, which would explain why IgG2a antibodies were formed to *M. corti* and *T. crassiceps* despite the decrease of the IgG2a and increase of the IgG1 serum levels. These data indicate that the regulation of the polyclonal response is not necessarily linked to the regulation of the antigen specific response.

The mechanism by which helminth parasites induce the restricted IgE/IgG1 response is not understood. Because IL-4 induces IgE and IgG1 secretion *in vitro*, it is likely that the parasites stimulate preferentially T helper cells secreting IL-4. It is presently unknown whether two classes of T helper cells exist, one forming IL-4 and the other IFN-γ, or whether pluripotential T helper cells differentiate into either IL-4 or IFN-γ secreting cells after interaction with antigen. The latter is more likely the case, because only T cell clones that have been cultured for a long period of time secrete either IL-4 or IFN-γ, whereas short-term cultured murine [12] or human [22] T helper cells secrete both. Still the question remains why the parasites stimulate mainly IL-4 secreting T helper cells. It is unlikely the nature of the parasite antigens. As shown by Teale and colleagues [20, 21] and confirmed in our laboratory, injections of dead *M. corti* parasites elicit an anti-parasite response that is evenly distributed among different isotypes, in particular equal amounts of IgG1 and IgG2a antibodies to the

Table 2. Isotypes of antibodies reacting with parasite extracts in sera obtained 14 days after infection

Parasite	cts/min at 1:50 serum dilution*					
	IgM	IgG3	IgG1	IgG2b	IgG2a	IgE
N. brasiliensis	14,400	30	6,500	200	90	70
M. corti	41,000	1,500	48,000	7,200	9,500	1,600
T. crassiceps	21,000	2,000	30,000	5,600	23,000	500

*Determined by a previously described radioimmunoassay [7]

parasite are formed. The polyclonal increase of the serum levels also affect all IgG subclasses with IgG2a showing the largest increase. The fact that both IgG1 and IgG2a anti-parasite antibodies are formed by mice injected with dead parasites demonstrates that the parasite antigens can activate both IL-4 and IFN-γ producing T helper cells. One possibility that could explain the predominant IgE/IgG1 response to live helminth parasites could be the secretion by the parasite of a substance that either suppresses IFN-γ secreting cells or enhances IL-4 forming T helper cells. Another explanation for the isotype restricted immune response could be the restricted anatomic location of parasite infection analogous to IgA responses occurring characteristically in Peyer's Patches. Whereas the localization of live parasites in the lung, peritoneum, and/or liver may produce IgE/IgG1 responses, dead parasites may be processed at many different sites leading to formation of antibodies of all isotypes. To investigate which of these two possibilities is more likely, we compared the effect of a subcutaneous (s.c.) with the intraperitoneal (i.p.) infection with live parasites. In contrast to *M. corti* parasites which did not remain at the s.c. injection site, *T. crassiceps* larvae multiplied subcutaneously and persisted at the injection site for 30 to 60 days. As can be seen in Fig. 3, the s.c. infection induced a different polyclonal response from the i.p. infection. After s.c. infection, the IgG2a serum levels increased to a greater extent than the IgG1 levels, the opposite of the changes in IgG1 and IgG2a levels after i.p. infection. Furthermore, mice s.c. infected with *T. crassiceps* formed about 10 fold less IgE than the i.p. infected mice. This experiment makes it unlikely that the parasites secrete a substance that enhance or suppress certain T helper cells. It favors the possibility that accessory cells at different anatomic sites are responsible for the IgG1/IgE isotype restriction. Experiments reported by Spalding and Griffin [23] regarding IgA isotype restriction support this suspicion. These authors cultured dendritic cells isolated from spleens or Peyer's Patches with T cells and pre-B cells and quantitated the IgA and IgG secretion of the *in vitro* differentiated B cells. They showed that the source of the dendritic cells and not that of the T cells determined whether the B cells formed IgA or IgG. Dendritic cells from Peyer's Patches induced IgA, whereas splenic dendritic cells induced IgG secretion. These findings suggest that T helper cells may differentiate under the influence of accessory cells to secrete either IL-4 or IFN-γ, a possibility that we are presently investigating. If proven correct, it would suggest that the abnormality in atopic patients leading to increased IgE formation may be at the level of accessory cells and not or only partially a T cell defect as is presently suspected. If a cytokine

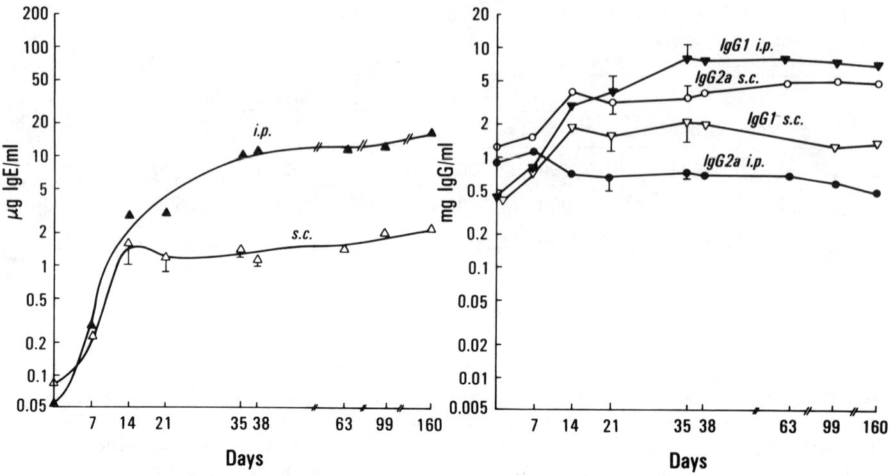

Fig. 3. Changes of IgE, IgG1 and IgG2a serum levels in BALB/c mice infected either peritoneally (i.p.) or subcutaneously (s.c.) with 10 *T. crassiceps* larvae.

produced by accessory cells is found that induces T cells to differentiate into IL-4 secreting cells, it may be possible to interfere with its action on T cells. This would offer a new therapeutic intervention for preventing allergic diseases.

CONCLUSIONS

In the mouse, it has clearly been demonstrated that IL-4 induces and IFN-γ suppresses LPS activated B cells to secrete IgE and IgG1. Although it is unknown whether IgE and IgG1 formation is always linked *in vivo*, mice infected with helminth parasites show a concomitant rise of the IgE and IgG1 serum levels. However, it remains to be shown whether the mechanism of IL-4 induced IgE and IgG1 formation is identical and whether factors in addition to IL-4 are necessary for the switch from IgM to IgE secretion to occur. Also, little is known of the role of human IL-4 in IgE formation. Furthermore, it is not resolved whether two types of T helper cells exist, one secreting IL-4 and the other IFN-γ, or whether T helper cells are pluripotential with regard to interleukin formation and differentiate after interaction with antigen into either IL-4 or IFN-γ secreting cells. We propose that accessory cells differing in cytokine production at different anatomic sites may be responsible for the differentiation of T helper cells to become IL-4 or IFN-γ secreting cells. This could explain the isotype restricted response to helminth parasites and allergens because the immune response to these agents occurs at restricted anatomic sites.

ACKNOWLEDGEMENTS

Supported by NIH grant AI-10734. Biomedical Research Support Program grant RRO 5514, and a grant from Eli Lilly Company. This is publication no. 5341-IMM from the Research Institute of Scripps Clinic, La Jolla, CA, USA.

REFERENCES

1. Kishimoto, T. and Ishizaka, K. (1973). Regulation of antibody response *in vitro*. VI. Carrier-specific helper cells for IgG and IgE antibody response. J. Immunol., 111, 920.
2. Isakson, P.C., Pare, E., Vitetta, E.S. and Krammer, P.H. (1982). T cell derived B cell differentiation factor(s): Effect on the isotype switch of murine B cells. J. Exp. Med., 155, 734.
3. Coffman, R.L., Ohara, J., Bond, M.W., Carty, J., Zlotnik, A. and Paul, W.E. (1986). B cell stimulatory factor-1 enhances the IgE response of lipopolysaccharide-activated B cells. J. Immunol., 136, 4638.
4. Snapper, C.M. and Paul, W.E. (1987). Interferon-γ and B cell stimulatory factor-1 reciprocally regulate Ig isotype production. Science, 236, 944.
5. Coffman, R.L. and Carty, J. (1986). A T cell activity that enhances polyclonal IgE production and its inhibition by interferon-γ. J. Immunol., 136, 949.
6. Mosmann, T.R., Cherwinski, H., Bond, M.W., Giedlin, M.A. and Coffman, R.L. (1986). Two types of murine helper T cell clones. I. Definition according to profiles of lymphokine activities and secreted proteins. J. Immunol., 136, 2348.
7. Lebrun, P. and Spiegelberg, H.L. (1987). Concomitant immunoglobulin E and immunoglobulin G1 formation in *Nippostrongylus brasiliensis*-infected mice. J. Immunol., 139, 1459.
8. Zakroff, S.G.H., Beck, L. and Spiegelberg, H.L. (1988). Concomitant increase of IgE and IgG1 and decrease of IgG2a serum levels in mice infected with helminth parasite. Fed. Proc., 47, A1250.
9. Coutelier, J-P., van der Logt, J.T.M., Neessen, F.W.A., Warnier, G. and van Snick, J. (1987). IgG2a restriction of murine antibodies elicited by viral infections. J. Exp. Med., 165, 64.
10. Finkelman, F.D., Katona, I.M., Urban, J.F., Snapper, C.M., Ohara, J. and Paul, W.E. (1986). Suppression of *in vivo* polyclonal IgE responses by monoclonal antibody to the lymphokine B-cell stimulatory factor-1. Proc. Natl. Acad. Sci. USA, 83, 9675.
11. Finkelman, F.D., Katona, I.M., Mosmann, T.R. and Coffman, R.L. (1988). IFN-γ regulates the isotypes of Ig secreted during *in vivo* humoral immune responses. J. Immunol., 140, 1022.
12. Glasebrook, A., Roeder, W., Laxer, J., Hom, J. and Firestein, G. (1988). Characterization of murine T cell clones secreting both IL-2 and IL-4. Fed. Proc., 47, A474.
13. Nusslein, H.G. and Spiegelberg, H.L. (1988). Secretion of Ig of different classes and subclasses by human B cells stimulated with different polyclonal activators and lymphokines. Fed. Proc., 47, A1250.
14. Lucas, A. and Asser, S. (1986). The capsular carbohydrate of *Hemophilus influenza* type b is a potent mitogen for murine B lymphocytes. J. Immunol., 137, 3130.
15. McKenzie, D.T., Filutowicz, H.I., Swain, S.L. and Dutton, R.W. (1987). Purification and partial sequence analysis of murine B cell growth factor II (interleukin 5). J. Immunol., 139, 2661.
16. Snapper, C.M., Finkelman, F.D. and Paul, W.E. (1988). Differential regulation of IgG1 and IgE synthesis by interleukin-4. J. Exp. Med., 167, 183.

17. Jarrett, E.E.E. and Bazin, H. (1974). Elevation of total serum IgE in rats following helminth parasite infection. Nature (London), 251, 613.
18. Ishizaka, T., Urban, J.F. and Ishizaka, K. (1976). IgE formation in the rat following infection with *Nippostrongylus brasiliensis*. I. Proliferation and differentiation of IgE bearing cells. Cell. Immunol., 22, 248.
19. Chapman, C.B., Knopf, P.M., Hicks, J.D. and Mitchell, G.F. (1979). IgG1 hypergammaglobulinaemia in chronic parasitic infections in mice: Magnitude of the response in mice infected with various parasites. Aust. J. Exp. Biol. Med. Sci., 57, 389.
20. Abraham, K.M. and Teale, J.M. (1987). Isotype restriction during infection of mice with the cestode *Mesocestoides corti* role of immune suppression. J. Immunol., 138, 1699.
21. Teale, J.M. and Abraham, K.M. (1987). The regulation of antibody class expression. Immunol. Today, 8, 122.
22. Romagnani, S., Del Prete, G.F., Maggi, E., Parrouchi, P., Tin, A., Macchia, D. and Ricci, M. (1988). Interleukin-4 is an essential factor for IgE synthesis induced *in vitro* by human T cell clones and their supernatants. Fed. Proc., 47, A1250.
23. Spalding, D.M. and Griffin, J.A. (1986). Different pathways of differentiation of pre-B cell lines are induced by dendritic cells and T cells from different lymphoid tissues. Cell, 44, 507.

Human Natural Killer Cells and their Fc epsilon Receptors in the Regulation of IgE and Other Immunoglobulin Isotypes

A. Saxon and H. Kimata

Division of Clinical Immunology and Allergy, Department of Medicine, UCLA School of Medicine Los Angeles, California 90024, USA

ABSTRACT

The induction and immunoglobulin (Ig) regulatory function of human natural killer (NK) cells expressing Fc receptors (FcR) for IgE (FcER-II) and IgA (FcAR) were investigated. Enriched human NK (NKH-1+) cells incubated for 16 hours with immune complexes of defined Ig isotype composition displayed FcER-II and FcAR receptors in an isotype specific fashion. This FcR induction was dose and time dependent, being maximal at 16-24 hours and requiring protein and RNA synthesis but being independent of cell division. The NKH-1+ cells were non T cells as CD3 depleted cells expressed FcER or FcAR while CD3 positive NKH-1 positive cells were not inducible. The isotype specific FcR cells also showed the ability to specifically enhance ongoing IgE or IgA production from lymphoblastoid B cell lines. Supernatants from the NK cells similarly enhanced ongoing Ig synthesis. This enhancement was shown to primarily result from B cell differentiation as Ig production increased with little or no increase in thymidine incorporation. The NK cell supernatants had the ability to enhance ongoing Ig synthesis from large bone marrow or blood B cells while it could not induce Ig production from either small resting B cells or small resting B cells that were co-activated with Staph aureus Cowan I. The NK driven differentiation activity functioned after the known B cell differentiation factors and could not be replaced by any of the presently known interleukins, interferons or cytokines. Finally, we showed that supernatants derived from immune complex induced FcER or FcAR NK cells mimicked the isotype specific effects on IgE or IgA production respectively with the FcR NK cells themselves. Thus it appears that a subset of human NK cells can potentially be both responsive to the Ig isotype in their environment and participate in the control of Ig isotypes as part of regulatory circuits involved in humoral immune responsiveness.

KEYWORDS

IgE production; Immunoglobulin Isotype Control Natural Killer Cells; Fc Receptors; Fc Epsilon Receptors; FcER-II.

INTRODUCTION

There is a rapidly expanding appreciation of the role for cells expressing Fc receptors (FcR) for specific isotypes in the regulation of isotype specific immunoglobulin (Ig) production [1-4]. We originally reported that T cells, when exposed to defined IgE immune complexes (IgE-IC) were activated, presumably through the low affinity Fc epsilon receptor (FcER), to specifically inhibit ongoing IgE synthesis in an isotype specific fashion [5]. Subsequently T cells expressing receptors for IgE, IgA as well as other isotypes have been detected and induced on human peripheral blood lymphocytes and virally transformed human T cell lines [6-8]. The role and number of T cells with this low affinity FcER (FcER-II) remain controversial.

We have focused our attention on FcR cells copurifying with T cells, cells which are, in fact, natural killer (NK) cells. We have investigated their expression of FcR for IgE and IgA to determine the potential *in vitro* and *in vivo* role such cells may play in isotype regulation. It has been known for some time that human NK cells, be they are freshly derived, induced or as IL-2 dependent cloned cell lines, produce a variety of B cell regulatory factors [9-10] though the ability of NK cells themselves had not been specifically addressed. We studied the induction of FcER or FcAR on purified human NK cells and subsets thereof. Additionally we have looked at the ability of NK cells from patients with disturbances in circulating levels of Ig to be induced for various FcR to determine the influence of the environment that the cells are derived from [4]. By inducing and purifying FcR cells, we have been able to show that such cells can produce IgE and IgA specific regulatory activity. This activity functions as late acting differentiation factors on both *in vivo* derived immunoglobulin producing cells and *in vitro* immunoglobulin producing cell lines and that this activity could not be ascribed to any of the presently known or described lymphokines [11].

MATERIALS AND METHODS

Cell Separation. Human mononuclear cells were isolated from heparinized venous blood or bone marrow using buoyant density centrifugation and T cells were purified by the AET sheep red blood cells rosetting technique [12]. Large and small B cells were separated by unit gravity separation as described [13]. NK cells were obtained from the T cells (AET- sheep red blood cell binding population) by immune adherence on NKH-1 monoclonal antibody coated plates [14]. This yielded more than 95% NKH-1 positive cells by indirect immunofluorescence and less than 5% CD3 positive cells, less than 1% CD20 positive cells and less than 0.5% monocytes. These cells were further separated into CD3 plus and minus cells by immune adherence [15] using anti-Leu 4 (CD3 antibody). Cells depleted of CD3 in this fashion were less than 1% CD3 positive while the CD3 positive populations were more than 95% CD3 positive.

Induction and Detection of FcER or FcAR NK Cells. NK cells were incubated with defined immune complexes of monoclonal anti-IgE plus purified myeloma IgE (IgE-IC) or monoclonal anti-IgA plus myeloma IgA (IgA-IC). Concentrations were used which were found to induce FcER and FcAR as previously published [4,14]. FcER induction was maximal at 24 hours. FcER cells were detected by an immunohistochemical stain using IgE-IC as previously described [14] or by staining with an anti-FcER (anti-CD23) monoclonal antibody (Mab 135) that was kindly provided by Dr Guy Delespesse, Montreal, Canada). FcAR positive cells were detected by immunohistochemical techniques using IgA-IC as described [16]. Fc gamma receptor cells were detected by immunofluorescence using the Leu 11 monoclonal antibody (Becton-Dickinson Monoclonal Center, Mountain View, CA).

Indicator Cells Lines. Human Ig producing cell lines synthesizing various isotypes were employed *in vitro*. Cell lines AF-10, GM 1500, CESS, and GM 1056 produced IgE, IgG, IgG and IgA respectively as reported [17]. The cell lines were maintained in culture continuously and seeded at 5000 cells per well for individual experiments.

Production of NK Cell Supernatants. Production of supernatants (NKSN) from the initially purified NK cells or subfractions of these cells (FcER plus, FcAR plus or CD3 plus or minus) were obtained by extensively washing the cells and then culturing for 24 hours in complete RPMI. The supernatants of the cells was then harvested. For T cell supernatants containing IgE enhancing activity (TCSN-A), T cells from atopic individuals were cultured overnight without additional stimulation. The supernatants were then removed, filtered through ultrafiltration membrane cones with a cut off at 50,000 and then filtered through a 0.22 μm millipore filter. This TCSN-A did not contain detectable amounts of IgE and has been shown to enhance IgE synthesis specifically from small resting B cells [18].

Cytokines and Measurement of Interleukins. Interleukin 1 beta was kindly provided by Dr S. Gillis from Immunex Corporation, Seattle, Washington. Interleukin 2 was purchased from Amgen, Thousand Oaks, California. Interleukin 4 was provided by Dr T. Yokota and Ms A.V. Kimmenade, DNAX Research Institute of Molecular and Cellular Biology, Palo Alto, California. Interleukin 6 (IL-6) was kindly provided by Drs T. Hirano and T. Kishimoto, Institute for Molecular and Cellular Biology, Osaka, Japan. Interferon beta was provided by Triton Bioscience Incorporated, Alameda, California. Interferon gamma was purchased from Biogen Incorporated, Cambridge, Mass. A partially purified preparation of low molecular weight B cell growth factor was purchased from Cellular Products Inc., Buffalo, NY. Tumor necrosis factor-alpha and granulocyte macrophage colony stimulating factor were the kind gift of Dr David W. Golde, Division of Hematology, UCLA. Interleukin 2 activity was measured using the CTLL T cells as described [19] while IL-6 activity is measured by using an IL-6 mouse hybridoma cell lines MH60, an IL-6 (BSF2) dependent cell line kindly provided by Drs T. Hirano and T. Kishimoto, Osaka, Japan.

Immunoglobulin Measurements. Isotype specific immunoglobulin production in the culture supernatants was assayed by isotype specific ELISAs as previously described [18].

RESULTS

Induction of FcER and FcAR by Isotype Specific Immune Complexes. Freshly isolated NKH-1 positive cells from normal donors displayed no FcER positive cells and only few FcAR positive cells (0.8 ± 0.5%). When these cells were incubated with IgE-IC or IgA-IC, there was a dose and time dependent increase in the display of the Fc receptors [7]. Enhanced FcR cells were detected by 6 hours, reached a maximum at 24 hours and declined thereafter. These effects were isotype specific as FcER was induced by IgE-IC but not

IgA-IC and FcAR were induced by IgA-IC only (Fig. 1). Following induction for 16 hours, FcER receptor display averaged 4.8 ± 2.3% while FcAR positive csells had a mean of 10.1 ± 4.7%. At the same time Fc gamma receptor (CD11)

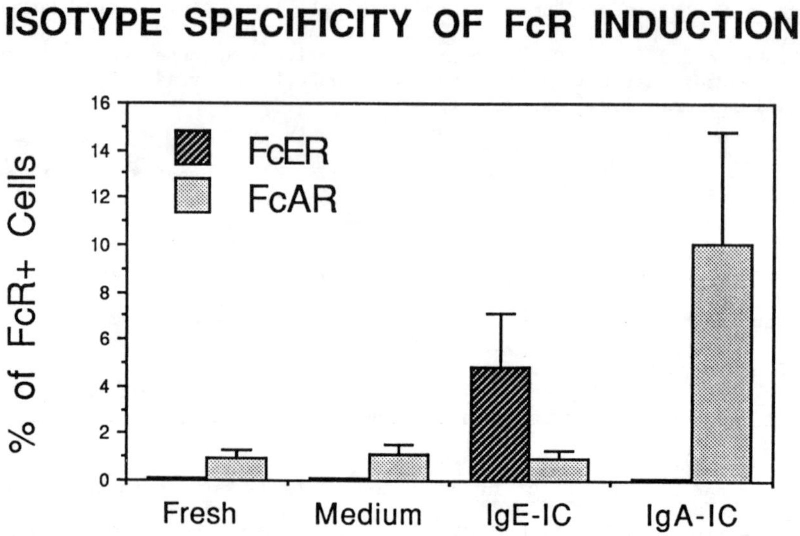

Fig. 1. Isotype specific induction of FcER or FcAR by IgE-IC or IgA-IC. Purified human NKH-1+ cells (2×10^6) were incubated 16 hrs. with IC (0.1 μg/ml) or medium alone as noted, washed extensively and tested for their expression of FcER or FcAR by immunohistochemical or immunofluorescence staining.

display was not altered under any of these conditions with approximately 90% of the cells remaining positive. Induction of FcAR and FcER required protein and RNA synthesis as they could be inhibited completely by treatment with cyclohexamide (50 μg/ml) or actinomycin D (0.4 μg/ml) while irradiation (3000 rads) did not alter FcER induction. To exclude the possibility that contaminating CD3 positive cells (the NKH-1 positive cell population was approximately 3-5% CD3+) were the cells being induced for Fc receptor, NKH-1 positive cells were further separated into CD3 plus and minus populations. IgE-IC or IgA-IC induced FcER and FcAR respectively on the CD3 minus but not on the CD3 plus populations (Table 1). CD3 positive cells (either NKH-1 plus or minus) could not be induced to express FcER or FcAR by the immune complexes.

<u>Effect of Immune Complex Stimulated NK Cells Ig Synthesis by B Cell Lines.</u> When NKH-1+ cells were incubated with IgE or IgA-IC to induce FcER or FcAR respectively, the resulting cells demonstrated the ability to regulate ongoing IgE production from cell lines in an isotype specific fashion. Thus there was a specific increase in IgE synthesis when IgE-IC incubated NK cells were added to microtiter wells containing 5000 cells each producing IgE, IgA

Table 1. Induction of FcR on CD3 positive or negative NKH-1+ cells

Cells + immune complexes	% of FcR
(A)	FcER
NKH-1+ + IgE-IC	4.0 ± 0.7*
NKH-1+/CD3- + IgE-IC	4.1 ± 0.6*
NKH-1+/CD3+ + IgE-IC	0
NKH-1-/CD3+ + IgE-IC	0
(B)	FcAR
NKH-1+ + IgA-IC	6.2 ± 0.7*
NKH-1+/CD3- + IgA-IC	7.2 ± 0.7*
NKH-1+/CD3+ + IgA-IC	1.0 ± 0.3
NKH-1-/CD3+ + IgA-IC	0.5 ± 0.5

* P<0.01 compared to control cultures.

and IgG (Table 2). Similarly FcAR induced NK cells showed preferential enhancement of IgA synthesis in these cultures. These results were not due to carryover of Ig from the immune complexes used to treat the NK cell populations as these populations showed no Ig in their supernatant when cultured alone. Additionally, NK cells from patients with either very high or very low levels of specific isotypes failed to show FcR induction for that isotype or isotype Ig enhancement following incubation with IC containing the isotype(s) in question (Table 2).

NK Supernatants (NKSN) Enhance Ongoing Ig Synthesis from Human Cell Lines. When supernatant was obtained from the uninduced natural killer cells (NKSN), it was able to induce enhanced production of IgE from AF10, IgG from GM-1500 and CESS, IgA from GM-1056 and IgM from CBL and 3332 cell lines. This occurred in a dose dependent fashion with effects being evident at 5% NKSN and being optimal at 50% (volume/volume) of NKSN in 200 microliter culture wells with the various cell lines. While Ig production was increased from 140-180% by the NKSN, there was no increased proliferation of the B cell lines as evidence by our failure to see increased thymidine incorporation and by the fact that there was no change in the cell number in the cultures at day 3. It is worth noting that one of the cell lines was CESS which is a known BCGF responsive cell line. When NKH-1 positive and negative cells were again separated into CD3 minus and CD3 plus populations as described earlier, the NKSN differentiation activity was found to be produced by the NKH-1 plus CD3 minus subset of cells [14].

NKSN Acts to Enhance Ongoing Spontaneous Ig Synthesis. To investigate where in B cell differentiation NKSN activity was effective, freshly obtained human peripheral blood or bone marrow cells were separated by unit gravity fractionation into large activated or small resting populations [13]. The

Table 2. Effect of Immune Complex Stimulated NK Cells on Ig Synthesis from B Cell Lines

Donors	Incubation of NK cells with	% change of Ig synthesis		
		IgE	IgA	IgG
Normal donors				
(N=3)	IgE-IC	22 ± 4	3 ± 6	3 ± 6
	IgA-IC	2 ± 2	54 ± 9	4 ± 2
Hyper IgE immunodeficiency				
(N=3)	IgE-IC	-2 ± 3*	0 ± 6	1 ± 6
	IgA-IC	1 ± 1	42 ± 6	-1 ± 1
Atopic dermatitis				
(N=3)	IgE-IC	-1 ± 3*	3 ± 3	-3 ± 4
	IgA-IC	-1 ± 1	47 ± 6	-4 ± 6
IgA myeloma				
(N=2)	IgE-IC	22 ± 2	-4 ± 0	2 ± 2
	IgA-IC	2 ± 2	0 ± 2*	2 ± 1
Selective IgA deficiency				
(N=2)	IgE-IC	26 ± 3	-3 ± 1	1 ± 1
	IgA-IC	-2 ± 2	0 ± 3*	3 ± 1
Dysgammaglobulinemia type I				
(N=1)	IgE-IC	-6 ± 1*	-1 ± 1	-4 ± 1
	IgA-IC	-4 ± 2	-2 ± 1*	0 ± 1
Common variable immunodeficiency				
(N=3)	IgE-IC	0 ± 2*	0 ± 5	2 ± 2
	IgA-IC	-2 ± 2	0 ± 6*	-2 ± 2

$$\% \text{ change} = \frac{\text{Ig with IC stimulated NK} - \text{Ig with medium treated NK cells}}{\text{Ig in control cultures}}$$

* p<0.01 for decreased Ig enhancement compared to normal donors.

large activated cells fractions contained cells that spontaneously produce Ig *in vitro* in contrast to the small resting cells which require activation and differentiation signals *in vitro* in order to synthesize Ig. NKSN enhanced the IgG, IgM and IgA synthesis between 190 and 350% from the large B cells from blood or bone marrow while NKSN was ineffective in inducing Ig synthesis from either small resting B cells alone or small resting B cells that had been co-stimulated with SAC for cells derived from bone marrow and blood. These results suggested that NKSN acts after to early differentiation factors such as IL-2 and possibly even after IL-6. Therefore small resting B cells from normal individuals were stimualted with the supernatant of atopic individuals T cells (TCSN-A) which has been shown to induce IgE synthesis from such cells by functioning as an isotype differentiation factor [18,20,21]. When NKSN was added with TCSN-A, there was increased enhancement of IgE synthesis and this was most effective when the NKSN was added 5 days after the initiation of cultures with TCSN-A (Table 3). Controls of NKSN alone failed to induce IgE synthesis in this system.

Table 3. NKSN Enhances IgE Synthesis from Peripheral Small B Cells as a Late Acting BCDF

Addition of Supernatants to B cells on: Day 0	Day 5	IgE (pg/ml)
Medium	Medium	<80
NKSN (50%:vol/vol)	Medium	<80
Medium	NKSN	<80
TCSN-A (50%:vol/vol)	Medium	287 ± 50
TCSN-A+NKSN	Medium	416 ± 48 (45%)*
TCSN-A	NKSN	583 ± 52 (103%)*
TCSN-A+Cycloheximide	NKSN	<80

*% enhancement compared to TCSN-A alone.

Supernatants from NK Cells Induced to Display Fc Receptors Alter Ig Synthesis in an Isotype Preferential Fashion

FcER or FcAR positive and negative cells were enriched by immune adherence as described in the Materials and Methods. These cells were then utilised to generate NKSN. As shown in Figure 2, NKSN from the FcER positive NKH-1+ cells enhanced IgE synthesis from the IgE producing cell line significantly while NKSN derived from the FcER negative cells did not. Similarly NKSN derived from FcAR induced NK cells enhanced IgA synthesis while that from the FcAR negative cells did not. These effects demonstrated isotype specificity in that the FcAR+ cell NKSN was without effect on IgE synthesis and the FcER+ cell NKSN did not alter IgA synthesis. All supernatants had an enhancing effect on IgG synthesis which was not altered by the various manipulations undertaken. This enhancement in Ig synthesis (IgE or IgA) was a differentiation effect as proliferation of the IgE or IgA producing cells was not increased (3 ± 3%) in spite of the fact that there was between 36 and 44% enhancement of Ig synthesis. This B cell differentiation activity was also shown to be the product of the NKH-1 positive, CD3 negative cells (Fig. 3).

Natural Killer Cell Ig Differentiation Activity can not be Substituted for by Known Cytokines

As it is known that NK cells produce a variety of cytokines such as IL-1, IL-2, interferons, BCGF, GM-CSF [9,10,22,23], we tested the effects of various recombinant or purified cytokines in our assay of B cell differentiation of human cell lines to determine if we could reproduce the effects of NKSN. Specifically IL-1 beta, IL-2, IL-4, interferon-beta 1, interferon-gamma, GM-CSF, tumor necrosis factor-alpha and BCGF failed to induce Ig synthesis in our assays. The interleukins were used at up to 100 U/ml, the interferons at 1000 U/ml, GM-CSF at up to 1000 picomoles/liter, TNF-alpha at

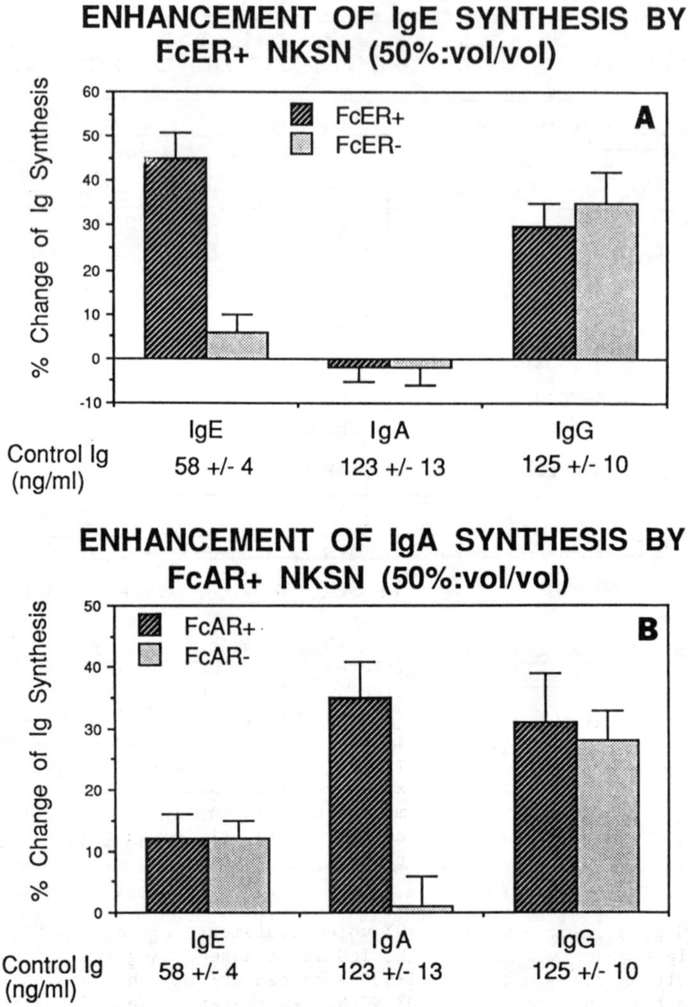

Fig. 2. Supernatants from FcER or FcAR induced NK cells demonstrate an isotype specific enhancement of Ig production. Supernatants (50% vol:vol) from FcER positive or negative (A) or FcAR positive or negative (B) enriched NK cells were added to mixtures of 5×10^3 each of AF-10, GM-1500 and GM-1056 cells. Three days later, the amount of each Ig produced in the wells was determined. Results are expressed as the mean (±1 SD) % change of control Ig from 3 experiments. Net absolute IgE, IgA and IgG production in the controls were 58 ± 4 ng/ml, 123 ± 13 ng/ml and 125 ± 10 ng/ml respectively. The FcER populations had less than 1% FcAR+ cells while the FcAR populations had no detectable FcER+ cells.

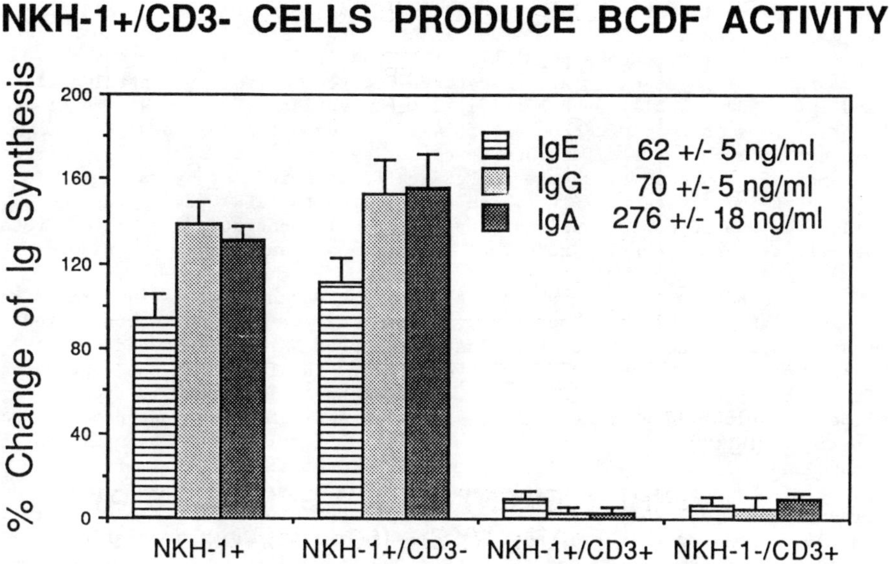

Fig. 3. NKSN B cell differentiation activity is produced by NKH-1 positive and CD3 negative cells. Supernatants (50% vol:vol) from various cell populations shown were tested for their ability to induce Ig synthesis from B cell lines in a 3 day culture. Results are expressed as the percent change (+1 SD) from control medium treated cultures in triplicate. Net absolute amounts of Ig (±1 SD) in the control cultures are shown.

up to 100 pg/ml, and BCGF at a 10% volume:volume. IL-6 at 100 U/ml did enhance Ig synthesis from the IgE, IgG and IgA secreting cell lines which was clearly above background (21-43% enhancement) but this was still 4-5 fold less than that seen with 50% NKSN. We therefore tested the NKSN for the amount of AL-6 as well as IL-2 it contained. NKSN contained only 0.1 unit per ml of IL-6. Furthermore IL-6 induced proliferation (between 30 and 100%) of the B cell lines, an effect not seen with the NKSN. No IL-2 activity was detected in the NKSN.

DISCUSSION

We have found that human NK cells (NKH-1 and high affinity sheep red blood cell receptor positive) from normal donors can be induced by defined immune complexes, comprised of a specific Ig isotype, to express Fc receptors for that isotype. This induction appeared to be isotype specific as IgE-IC induced only IgE FcER while IGA-1 induced FcAR. Neither of these manoeuvres altered Fc gamma receptor display and co-cultivation experiments with IgE-IC plus IgA-IC suggested that some NK cells express FcER and FcAR at the same time (data not shown). This is in accordance with the reported multiple FcR induction in an isotype specific fashion on a T cell hybridoma cell line [24]. The fact that we could induce a small per cent (5-10%) of NKH+1 cells to display FcER or FcAR is not surprising in that NKH-1 positive/CD3 negative cells are a heterogeneous population [25] and it is likely only a minor proportion of these cells are the subset related to isotype control of IgE and IgA. Additionally even specific T cell hybridoma lines that are cloned express less than 10% FcER or FcAR [24]. The detection of these Fc receptors was not a technical artifact due to either cytophilic Ig or a non-specific binding of immune complexes as no FcER cells were induced by using IgA-IC or with heat inactivated IgE-IC (which cannot bind to FcER) [26]. The FcR induction was IC concentration as well as time dependent and required protein as well as RNA synthesis indicating induction was an active mechanism.

Whereas we have previously shown that IgE-IC treated NK cells in normals can enhance IgE in an isotype specific fashion [14], we have now demonstrated that this works equally well for IgA Fc receptor NK cells induced by IgA-IC [4]. To extend these studies we have utilized the fact that the supernatant from such NK cells (NKSN) also show enhancement of Ig synthesis from cell lines constitutively producing IgE, IgA, IgG and IgM. Supernatants from both unstimulated and IgE-IC and IgA-IC stimulated NK cells could be utilized to induce and enhance ongoing Ig synthesis. We demonstrated this effect was primarily on cell differentiation as there was a marked increase in Ig production with absolutely no alteration in thymidine incorporation or cell number at the end of three days. To be sure that the differentiation activity we detected for was not one of the recognized human cytokines, we tested a battery of cytokines to see if they could mimic the effects of NKSN. IL-1, IL-2, interferon-gamma, interferon-beta 1, GM-CSF, tumor necrosis factor-alpha, BCGF as well as interleukin 6, were shown not to be able to replace NKSN activity. Though NKSN did control miniscule amounts of IL-6, one thousand fold greater doses of recombinant IL-6 failed to replace the activity of NKSN. In addition, IL-6 showed a distinctive proliferative effect on the cell lines that was not seen with the NKSN. We did not test IL-5 directly. It has been reported to enhance Ig synthesis from SAC activated B cells [27] while our NKSN failed to do so.

Our results using large activated versus small resting B cells from both blood and bone marrow showed that the NK differentiation activity only worked on cell already producing Ig (in this sense, analogous to the *in vitro* maintained B cell lines). Thus NKSN effect appears to function after the signal of IL-2 or IL-6 as SAC activated B cells did not respond to NKSN and these cells are IL-2 and IL-6 reactive [28]. This result is extended by the observation that NKSN, while unable to induce IgE synthesis from small blood B cells on its own, was able to synergize the TCSN-A to further enhance TCSN-As IgE differentiation effect and that NKSN was most effective when added subsequent to TCSN-As action. Thus is appears that NK cells may make a late acting differentiation activity similar to that which has been recently described for (BSF2-4) by [29]. In fact, Brenner and colleagues have described the situation where activated NK cells appear to provide help for B cells in patients who have received a T cell depleted allogeneic bone marrow graft [30-31] while Bonagura and coworkers have recently reported a patient with panhypogammaglobulinemia who had normal *in vitro* B and T helper cell function but who lacked detectable NK cells [32].

The question still remains as to whether the isotype specific effects we observed with NK cells incubated with IC of defined isotype composition or NKSN derived from such cells are different activities from that observed with the "unstimulated/uninduced" NK cells. This can not be definatively answered at present but will require purification and characterization of the individual factors involved. So far we have not been able to locate or derive a source of such material to provide the quantities necessary for such an undertaking. However, we feel it is likely that the unstimulated NK cells and their supernatants actually represent a mixture of the isotype specific activities and our IC induction simply enhances the activity of these cells for a particular isotype. This is in accord with the data that NK cells can express more than one class of FcR which is in response to the Ig isotype in their mileau. Additionally, we have reported that the NK cells from patients with either absent IgE or IA were unable to be induced to express FcER or FcAR respectively nor were the NKSN derived from such cells able to drive differentiation for the missing isotype [4]. In fact, we also found that patients with very elevated levels of these same Igs (hyper IgE syndrome or IgA myeloma) also were defective in their FcR inducability. This strongly supports the notion that the NK FcR expression is part of isotype regulatory circuits in which the cells both respond to and influence the expression of levels of isotype involved in the human antibody response.

Acknowledgements. We would like to thank Drs. Fergus Shanahan, Stephan Targan, Jon Kagan and Elliott Sherr for their helpful discussions and collaboration in various aspects of the studies reported herein.

REFERENCES

1. Neauport-Sautes, C., Rabourdin-Combe, C. and Friedman, W.H. (1979). T cell hybrids bear FcG receptors and secrete suppressor immunoglobulin binding factor. Nature, 277, 656-659.
2. Yodoi, J. and Ischizaka, K. (1980). Lymphocyte bearing Fc receptors for IgE. IV. Formation of IgE-binding factor by rat T lymphocytes. J. Immunol., 124, 1322-1329.
3. Kiyono, H., Mosteller-Barnum, L.M., Pitts, A.M., Williamson, S.I., Michalek, S.M. and McGhee, J.R. (1985). Isotype-specific immunoregulation. IgA-binding factors produced by FcA receptor-positive T cell hybridoma regulate IgA responses. J. Exp. Med., 161, 731-747.
4. Kimata, H. and Saxon, A. (1988). A subset of natural killer cells are induced by immune complexes to idsplay Fc receptors for IgE and IgA and demonstrate isotype regulatory function. J. Clin. Invest., in press.
5. Hassner, A. and Saxon, A. (1984). Isotype-specific human suppressor T cells for IgE synthesis activated by IgE-anti-IgE immune complexes. J. Immunol., 132, 2844-2849.
6. Yodoi, J. and Ishizaka, K. (1980). Induction of FcE-receptor bearing cells *in vitro* in human peripheral lymphocytes. J. Immunol., 124, 934-938.
7. Young, M.C., Leung, D.Y.M. and Geha, R.S. (1984). Production of IgE-potentiating factor in man by T cell lines bearing Fc receptor for IgE. Eur. J. Immunol., 14, 871-878.
8. Nutman, T.B., Delespesse, G., Sarfati, M. and Vilkman, D.J. (1987). T cell derived IgE-binding factors I. Cloned and transformed T cells produce IgE-binding factors. J. Immunol., 139, 4049-4054.
9. Kasahara, T.B., Djeu, J.Y., Dougherty, S.F. and Oppenheim, J.J. (1983). Capacity of human large granular lymphocytes (LGL) to produce multiple lymphokines: interleukin 2, interferon and colony stimulating factor. J. Immunol., 131, 2379-2385.
10. Vyakarnum, A., Brenner, M.K., Reittie, J.E., Houlker, C.H. and Lachmann, P.J. (1985). Human clones with natural killer functions can activate

B cells and secrete B cell differentiation factors. Eur. J. Immunol., 15, 606-610.

11. Kimata, H., Sherr, E.H. and Saxon, A. (1988). Human NK cells produce a late acting B cell differentiation activity. J. Clin. Immunol., in press.
12. Saxon, A., Feldhaus, J.L. and Robins, R.A. (1976). Single step separation of human T and B cells using AET-treated SRBC rosette. J. Immunol. Methods, 12, 285-288.
13. Sherr, E., Saxon, A. and Wells, J.R. (1988). Functional and phenotypic characterization of human B lymphocyte subsets isolated by unit gravity sedimentation. Int. Arch. Allergy appl. Immunol., 85, 154-160.
14. Kimata, H., Shanahan, F., Brogan, M., Targan, S. and Saxon, A. (1987). Modulation of ongoing human immunoglobulin synthesis by natural killer cells. Cell. Immunol., 107, 74-88.
15. Wysocki, L.J. and Sato, U.L. (1978). "Panning" for lymphocytes. A method for cell selection. Proc. Natl. Acad. Sci. US, 75, 2844-2848.
16. Kanowith-Klein, S. and Saxon, A. (1985). Fc epsilon receptor on human cell lines and peripheral blood lymphocytes detected by binding of IgE immune complexes. J. Clin. Immunol., 5, 38-45.
17. Hassner, A. and Saxon, A. (1983). Inhibition of ongoing myeloma IgE synthesis *in vitro* by activated human T cells. J. Immunol., 130(4), 1567-1572.
18. Sherr, E.H., Stein, L.D., Dosch, H. and Saxon, A. (1987). IgE-enhancing activity directly and selectively affects activated B cells: evidence for a human IgE differentiation factor. J. Immunol., 138(11), 3836-3843.
19. Merrill, J.E., Mohldtrom, C., Uittenbogaart, C., Kermani-Arab, V., Ellison, G.W. and Myers, L.W. (1984). Response to and production of interleukin 2 by peripheral blood and cerebrospinal fluid lymphocytes of patients with multiple sclerosis. J. Immunol., 133, 1931-1937.
20. Romagnani, S., Maggi, E., Del Prete, G.F. and Ricci, M. (1983). IgE synthesis *in vitro* induced by T cell factors from patients with elevated serum IgE levels. Clin. Exp. Immunol., 52, 85.
21. Saryan, J., Leung, D.Y.M. and Geha, R.S. (1983). Induction of human IgE synthesis by a factor derived from T cells of patients with hyper-IgE states. J. Immunol., 130, 242.
22. Scala, G., Allavena, P., Djeu, J.Y., Kasahara, T., Ortaldo, J.R., Herberman, R.B. and Oppenheim, J.J. (1984). Human large granular lymphocytes are potent producers of interleukin-1. Nature, 309, 56-59.
23. Postoia, V., Cozzolino, F., Torcia, M., Castigli, E. and Ferrarini, M. (1985). Production of B cell growth factor by a Leu-7+, OKM1+ non-T cell with the features of larte granular lymphocytes (LGL). J. Immunol., 134, 3179-3184.
24. Daeron, M., Neauport-Sautes, C., Yodoi, Y., Moncuit, J. and Fridman, W.H. (1985). Receptors for immunoglobulin isotypes (FcR) on murine T cells. II. Multiple FcR induction on hybridoma T cell clones. Eur. J. Immunol., 15, 668-674.
25. Lanier, L.L., Le, A.M., Civin, C.I., Loken, M.R. and Philips, J.H. (1986). The relationship of CD16 (Leu-11) and Leu-19 (NKH-1) antigen expression on human peripheral blood N cells and cytotoxic T lymphocytes. J. Immunol., 136, 4480-4486.
26. Ishizaka, K., Ishizaka, T. and Menzel, A. (1967). Physiochemical properties of reaginic activity. VI. Effect of heat on gamma E, gamma G, and gamma A antibodies in the sera of ragweed sensitive petients. J. Immunol., 99, 610-618.
27. Azuma, C., Tanabe, T., Konoshi, T., Noma, T., Matsuda, F., Yaoita, Y., Takasu, K., Hammarstrom, L., Smith, C.I.E., Saverinson, E. and Honjo,

T. (1986). Cloning of cDNA for human T-cell replacing factor (interleukin 5) and comparison with the murine homologue. Nucleic Acid Res., 14, 9149-9158.

28. Hirano, T., Taga, T., Nakano, N., Yasukawa, K., Kashiwamura, S., Shimizu, K., Nakajima, K., Pyun, K.H. and Kishimoto, T. (1985). Purification to homogeneity and characterization of huamn B-cell differentiation factor (BCDF or BSFp-2). Proc. Natl. Acad. Sci., 82, 5490-5494.

29. Noma, T., Misuta, T., Rosen, A., Hirano, T., Kishimoto, T. and Honjo, T. (1987). Enhancement of the interleukin 2 receptor expression on T cells by multiple B-lymphotropic lymphokines. Immunology Letters, 15, 249-253.

30. Brenner, M.K., Reittie, J.E., Grob, J-P., Wimperis, J.Z., Stepeens, S., Patterson, J., Hoffbrand, A.V. and Prentice, H.G. (1986). The contribution of large granular lymphocytes to B cell activation and differentiation after T cell depleted allogeneic bone marrow transplantation. Transplantation, 42, 257-261.

31. Brenner, M.K., Vyajarnam, A., Reittie, J.E., Wimperis, J.Z., Grob, J.P., Hoffbrand, V. and Prentice, H.G. (1987). Human large granular lymphocyte induce immunoglobulin synthesis after bone marrow transplantation. Eur. J. Immunol., 17, 443-447.

32. Bonagura, V.R., Cunningham-Rundles, S., Edwards, B. and Valacer, D.J. (1988). Common variable hypogammaglobulinemia recurrent pneumocystis carinii pneumonia on IV gammaglobulin and natural killer deficiency. J. Allergy Clin. Immunol., 81, 288 (Abstr.).

Discussion: Session Seven

Chairman: Jack Pepys

1. Question to Alec Sehon

A.A. Ansari. Is there any possibility that these subjects produce anti-PEG antibodies?

Alec Sehon. It is neither easy to produce anti-PEG antibodies in animals, unless you use Freund's complete adjuvant, nor to test for PEG antibodies. However, Pharmacia have conducted clinical trials of the PEG-modified allergens in at least 350 patients and found that some may have a low level of IgM class antibody to PEG. It is worth pointing out that enzyme-PEG conjugates have been approved by the FDA for treating children with enzyme deficiencies, e.g., a severe combined immunodeficiency disease due to a lack of adenosine deominase.

2. Questions to Larry Schwartz

Herbert Mansmann. Larry, that was an excellent discussion of something that has been missing so far from this Symposium, namely that allergy is the study of diseases associated with mast cell activation. My question refers to your slide which showed the different types of mast cell and the response to stimulation by substance P and morphine. I missed the point of where the different mast cells might be. Are they in all tissues or is there any distribution?

Larry Schwartz. In normal tissues the T (i.e., tryptase-containing) or mucosal mast cell is the predominant type in the mucosa of the small intestine and in the alveoli of the lung. The TC (i.e., tryptase- and chymase-containing) mast cell predominates in the skin - it is analogous to the rodent connective tissue mast cells. However, in both cases you see a mixture of the two types as indeed you do when you look in other areas of the small intestine, the lung (bronchi), nasal mucosa and conjunctiva. This is in normal patients, so we are now looking at the distribution of the two types of mast cell in various diseases to see whether it may be related to pathogenesis.

Advances in the Biosciences Vol. 74

Printed in Great Britain

Sheldon Spector. I have two questions. The first one relates to your challenge experiments: what medications must you withhold because of their ability to block mast cell degranulation? The second relates to tartrazine challenge: there is much controversy about mechanisms, but some people reckon that a tartrazine challenge can provoke asthma; monitoring tryptase levels could presumably resolve the issue.

Jack Pepys. You might answer that last question for aspirin and non-steroidal anti-inflammatory drugs as well!

Larry Schwartz. As to what medications to withhold, that has not yet been looked at systematically. The patients undergoing nasal and bronchial challenge were off medication, because the rhinitics were challenged out of season and the asthmatics had mild well-controlled symptoms.

We have not looked at tartrazine challenge, but have observed three different patients each with an adverse reaction to one of three cyclooxygenase inhibitors: aspirin, motrin and indocin. The first two patients presented with hypotension and hives and both had tryptase levels above the normal value of 5 ng/ml - of 12 and over 100 ng/ml respectively. Our third, the one with the indocin reaction, presented with bronchospasm only; her tryptase level, obtained about 5 hours after the episode, was about 80 ng/ml and then gradually decreased. We hope to look at this more systematically with the collaboration of Don Stevenson, but at first sight we have 3/3 patients with evidence of mast cell activation after drug challenge.

Robert Hamburger. Do you observe any cross-reactivity with any serum enzymes when you are measuring tryptase?

Larry Schwartz. We cannot find any cross-reactivity, yet. I think, in part, that is because we have two antibody preparations which give us a double specificity in this sandwich assay.

Herbert Mansmann. Do you have any thoughts about late-phase reactions?

Larry Schwartz. We haven't looked at the lung yet, but with Bert Zweiman's group we have looked at the cutaneous late-phase reaction. It is complicated (and controversial). Our experiments involved challenge with allergen in a skin test chamber for a limited time period, followed by replacement with buffer before monitoring for the release of mediators into the chamber. There is an early appearance of both tryptase and histamine which goes away with time, and we never see a late-phase reaction either clinically or with mediators under those conditions. If you leave allergen in the chamber, you see that tryptase has the same kinetics as if you had removed it: it peaks and then goes back down within a few hours to baseline. In contrast histamine levels peak, go back down to a plateau but remain elevated for the entire time. There is also a clinical second phase reaction, but not a second peak of histamine - it is a continuous release. In an attempt to understand this behaviour, the experiment was repeated, i.e., the patient was re-challenged with allergen for a second time and the release of mediators released into buffer was then monitored. Histamine was released but not tryptase. However, the tryptase content has not been depleted, because stimulation at this point with codeine releases both tryptase and histamine. So, what is the explanation for this? My interpretation depends on the very low levels of tryptase which can be released from basophils. We can see basophils arriving on the scene, albeit in small numbers, and think they are responsible for the persistent release of histamine and the ability of an allergen to release histamine but not tryptase after an intermittent challenge. But there are alternative explanations, so the issue has not yet been completely resolved.

3. Questions to Hans Spiegelberg

Aftab Ansari. I forget the order of the mouse constant region genes in the mouse, but was wondering why it is in man that IgE and IgG4 are stimulated at the same time, while in the mouse it is IgE and IgG1.

Hans Spiegelberg. Of course all the different classes of immunoglobulins use the same set of variable region genes, but immunoglobulin constant region genes do not have the same sequence in mouse as in man. In mouse the order is mu, delta, gamma 3, gamma 1, gamma 2b, gamma 2a, epsilon and alpha. While in man it is mu, delta, gamma 3, gamma 1, alpha 1, gamma 2, gamma 4, epsilon and alpha 2; our *in vitro* experiments suggest that the stimulation is in the order IgG1, IgG2, IgG3 and very little IgG4. However, unlike IgE, the increase in IgG4 can still be measured even with a poor stimulator like LPS. People who claim that LPS stimulates the mouse in gene order always conveniently omit IgA, because it is the last in sequence yet always gives more than they want to have.

Wayne Thomas. I was thinking along similar lines about how this fits in with the actual transcripts that you take during evolution of the classes. What sort of data exists at the clonal level? Do these factors like LPS tend to make them switch from IgG1 to IgE?

Hans Spiegelberg. I don't know whether they make first IgG1 and then IgE, but I think such a study has been reported.

Marianne Brüggemann. No, that has not been shown. It seems to be that they go either from mu to G1, or from mu to E. Referring back to the gene order: it does not correspond in rat, or mouse, or in man.

Hans Spiegelberg. So you do not believe that there is class selection?

Marianne Brüggemann. No, nor do I believe there is any indication that it is due to the length or strength of the switch region.

Hans Spiegelberg. Coffman did some clonal analysis and found IgM, IgG1 and IgE all in the same culture supernate.

Jim Braun. I do not wish to exactly disagree with these things, but would like to modify slightly Marianne's statement just to say that there are clear-cut examples of sequential switching of B-cells. Sequencing around the switch region will show that when a cell is making IgE there are residuals of other switch regions, for example one of the gammas. Although this is not a common occurrence, and often it is indeterminate, it nevertheless shows that the cell can go through a couple of switch events.

Robert Hamburger. A quick question Hans about those studies with the live versus dead parasites. Was the route of entry the same?

Hans Spiegelberg. Yes, the intra-peritoneal route was used for both live and dead parasites. Of course the dead ones were injected every week, so their concentration rose and fell as they were broken down and digested; the live ones remained alive and went to the liver.

Robert Hamburger. The route of entry was always thought to be important for parasites in terms of entry through the gut wall as against being processed in other places.

Hans Spiegelberg. Right.

4. Questions to Andrew Saxon

Hans Spiegelberg. I have difficulty in understanding the last part of your talk. Do you mean that IgE immune complexes can turn off any B-cell immunoglobulin secretion?

Andrew Saxon. No, you can turn off IgE secretion with binding at CD23. It is not an effect on proliferation, our immediate question is if CD23 is a growth factor how do you cut down on the proliferation? Neither the cell number nor the cell cycle changes, and the percentages of G1 are identical (60% under all conditions).

Hans Spiegelberg. But all normal B-cells have CD23, so IgE-immune-complexes would turn off all normal B-cells whether they make G or E. This I cannot understand.

Andrew Saxon. No, this is interesting. We have other cell lines making other immunoglobulins, and although they have CD23 they do not turn off. I want to be honest with you though, those cells are infected with EBV, unlike this cell, and EBV maybe driving them through constitutive expression of CD23 and acting differently on them. I don't want to go to the nap on that saying that I have the perfect negative control. We have not yet gone back to look at early populative cells, we have looked at atopic patients' B-cell complexes and they are also turned off in an isotype specific fashion by IgE-immune complexes. We have not yet tested them with functional antibodies against CD23 complex, we wanted to do so and also to go back to these early cells with EBV and see how CD23 comes into the picture, although I'm not sure that this is a part of the real world.

Larry Schwartz. I am curious about the Fc receptors expressed on the NK cells (FcER-II) - they peak at 24 hours and are gone by 48 hours. What happens to it? With its disappearance does IgE enhancing activity also disappear? If you re-present it after 3 days with immune complexes is it re-stimulated, or have lost the response, i.e., is it only an early response?

Andrew Saxon. That is a good question. What we had to do was thoroughly wash out the massive dose of IgE (24 μg), but it does disappear. I did not show you the time-kinetics, but the peak of expression is around the peak of the supernatant activity. We did not try the re-stimulation experiment, and I do not know whether the FcER-II would be re-expressed; *in vivo* I guess it would be, because they seem to be just listening to what is going on around them.

Alec Sehon. Is there a definite ratio of IgE to monoclonal anti-IgE in your immune complexes? In addition, do you also need the Fc portion of the monoclonal anti-IgE?

Andrew Saxon. Yes, the complexes involve a definitely defined ratio, and one of the things which held us up was the poor reproducibility of making these complexes. We made them by mixing the IgE and monoclonal anti-IgE, leaving them forever, but that way you make only very small complexes, aggregates of two, three and four - absolutely no gigantic ones. Your second question about the Fc portion of the mouse monoclonal antibody - the Fab portion works just as well so there is no binding of the Fc region.

Birgit Helm. As far as I am aware from reading the literature you only get CD23 expression on cells secreting and having membrane-bound mu. It was suggested, I think by Kishimoto, and also by Gordon, that on switching to either epsilon, gamma or alpha you lose CD23.

Andrew Saxon. It was only Kishimoto; the French and the rest of the world that I read about don't mention that. He said that it was constitutively expressed. You don't see CD23 expressed on most resting cells. It is induced by IL-4 and seems to be an activation factor as it is actually the receptor for growth factor.

Birgit Helm. Yes, of course, this is what you suggested. What slightly confuses me is that with the U6266 cell line you have CD23 expression with IgE, which is contrary to what one would expect from Kishimoto's experience.

Andrew Saxon. After stimulation of B-cells we find from a few % to 23% have surface CD23. At any rate I think this is one of the very few occasions when Kishimoto may be wrong.

Birgit Helm. I see, so you have observed CD23 expression together with say gamma or epsilon or .....

Andrew Saxon. Well, to be honest we didn't actually look at those 23% of B-cells as to what is on their surface because of difficulties with limiting numbers in 3-color flow cytometry.

Birgit Helm. Well, the myeloma cell line may be an exception in that it expresses both and is not really representative of the serological conditions.

Andrew Saxon. I don't really agree with that - the myeloma cells may exaggerate, but they don't lie.

Session Eight

COMPUTER ASSISTED MEDICAL DECISION MAKING

Chairman: A. Said El Shami

Building Computer Aids for Medical Decisions: a Review and Discussion of the Techniques Available

J. Wyatt

Charing Cross and Westminster Medical School,
London SW1P 2AR, UK

ABSTRACT

After an analysis of how doctors make decisions, the areas are identified where support by computer aids might be useful. A variety of methods that have been used to construct medical decision-aids are then described by reference to examples, and general principles for the building of a successful system proposed. Particular emphasis is placed on ensuring that the performance of the system is evaluated, and that it is easy to use by the intended community.

The conclusions drawn are that suitable roles for computer decision-aids exist in medicine, that some are already being filled by systems built using a variety of methods, and that computers are of proven benefit in this application. Nevertheless, scientific evaluation is vital to a successful outcome.

KEYWORDS

Medical decisions; computer decision-aids; expert systems; statistics; evaluation; usability.

1.0. INTRODUCTION: DO WE NEED COMPUTER DECISION-AIDS?

Many doctors are familiar with the concept of using a computer to process patient data to produce either diagnostic or therapeutic advice. These "computer decision-aids" are a recurring theme in debates about improving the quality of medical decisions, but until recently, so few were in use and the legal issues surrounding them (Zeide, 1986) so difficult, that Croft's prediction (Croft, 1972) remained a distant prospect:

> "*If dependable mathematical models for disease diagnosis could be constructed, they would have a vast impact on the costs and availability of medical care in this country ... and even greater in the world's underdeveloped countries.*"

I hope to demonstrate that such systems do now have a useful, if restricted, role in medicine, that there are a variety of well-documented methods that can be used to construct them, and that at least a dozen such systems of proven efficacy are in everyday use. The paper is structured as follows. Firstly, in the remainder of this section I discuss how doctors make decisions, and demonstrate that while they are proficient in some areas, they would benefit from support in others. In Section 2, I review available techniques for building computer decision-aids, with examples of systems that fill the niches identified in Section 1. In Section 3, I will discuss some of the general issues raised by these systems, including the questions of evaluation and of ensuring that such systems are easy to use. Finally, in Section 4, I will attempt to synthesise these insights into some general principles for building a successful decision-aid.

1.1. What are the Components of Medical Decision-making?

The question of how doctors make their decisions has considerable implications for not only the quality but also the cost of health care. It has, therefore, attracted the interest of doctors themselves, cognitive and experimental psychologists, sociologists, economists and even anthropologists over the last few years. We therefore have a range of detailed, if sometimes conflicting, descriptions of the medical decision-making process (eg. Pauker *et al.*, 1976; Blois, 1980; O'Neil *et al.*, 1988). I have attempted to produce a simplified diagram (Fig. 1) which identifies the major components of a medical decision.

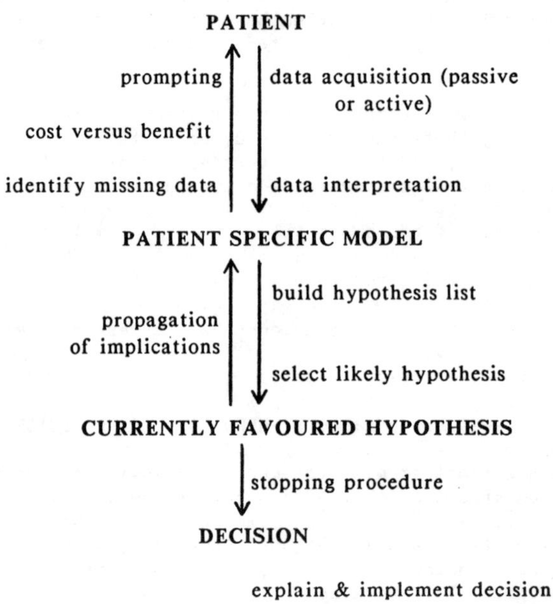

Fig. 1. The components of a medical decision

Decision-making usually starts with the patient, or someone on their behalf, volunteering a handful of facts about their current problem. The doctor then interprets these data (a stage which is more obvious if the data are laboratory results), and uses them to build a model of the patient's problem; this "patient-specific model" (PSM) is similar to Weed's "problem list". The PSM is then compared both with the doctor's stored medical knowledge and his experience (facts about common diseases, similar problems recently encountered etc.) to build a list of hypotheses that could explain most or all of the features of the PSM; this we usually call the differential diagnosis. It is often a list of four or five conditions; rarely, a feature volunteered by the patient is pathognomonic of just one disease, and the diagnosis is reached at this point.

Usually, however, the hypothesis list consists of more than one item, and further data-gathering is necessary to determine which is the most likely. The process of refining the differential diagnosis consists of a complex cycle of activities. The first stage is the "propagation" of the implications of each disease on the list to suggest what other features might be present in this patient. Then we decide which further data would differentiate between the rival hypotheses, and check on which it is reasonable to seek by analysis of the cost of the data, in terms of discomfort or morbidity for the patient, and the benefit, in terms of its contribution to the decision. Thirdly, we go back to the patient, elicit the required data, and compare it with our PSM. We then rebuild the hypothesis list, which leads on to further propagation and data-gathering. This iteration may proceed for several cycles before some "stopping procedure" is activated, such as a reduction in the number of hypotheses to one, or the realization that there is no benefit in further differentiating between two close candidates. Only then do we "make up our mind" and both implement the decision and explain it to the patient.

1.2. How Well do Doctors Perform at these Tasks?

From this introduction, it must be clear that every decision combines a wide range of tasks, from the "creative" building of the patient-specific model, to the routine of searching through lists of features of each disease on the differential diagnosis for those that are most useful in differentiating one from another. It must also be clear that, when the decision is being made by junior doctors working round the clock with the inadequate technical and other backup facilities common in the NHS, mistakes can occur. Where do these mistakes happen, in terms of the components outlined above?

Starting with the collection of data, many patients do not volunteer a complete list of their symptoms to the doctor, so that missing details need to be requested. If he forgets, vital details may be omitted, leading to diagnostic errors. It has been shown that simply providing doctors with a checklist of questions to ask and signs to elicit will improve diagnostic accuracy by 10% (Adams *et al.*, 1986; Wyatt *et al.*, 1986).

The next difficulty occurs in the interpretation of a patient's symptoms, signs and laboratory data. We may think we know what patients mean when they refer to their "stomach", but a study has shown that patients' comcepts of the site and function of their organs differs widely from ours (Pearson *et al.*, 1982). The problem is similar for physical signs: some doctors do not seem to know how to elicit them or what they mean, as any clinical examiner's report will show. Laboratory test results ought to be immune from subjective bias; but studies have shown that prior knowledge influences our interpretation of ECGs [Gjorup *et al.*, 1986) and other tests, and that even experienced doctors disagree about the interpretation of chest X-rays in about 10% of cases. In some cases, there is not even this degree of concensus as to what the test results show (Tweeddale *et al.*, 1983). In short, the eliciting and interpretation of both patient and laboratory data is of variable quality, with surprising variations between and even within observers.

The question of how well doctors build a patient-specific model, and then how they use it and their medical knowledge to identify the "correct" list of hypothesis, is more difficult to answer. However, it is known that doctors' diagnostic accuracy in the accident and emergency department in conditions such as acute abdominal pain (de Dombal *et al.*, 1974) and chest pain (Goldman *et al.*, 1982) is of the order of 50-80%. Patients who die in hospital and come to post-mortem examination have undiagnosed conditions in 53% of cases, with serious conditions such as TB missed in 13% (Mercer *et al.*, 1985). Experimental studies of decision-making have suggested that humans cannot consider more than four or five parameters simultaneously when making decisions, and that we ignore any which exceed this number without sufficient consideration of their relevance. There is also evidence that humans tend to place undue emphasis on recent experience: this is seen, for example, when laboratories witness a rush of tests for a rare disease following the diagnosis of a case in the hospital. This evidence appears to suggest that doctors find both the hypothesis selection and propagation stages of decision-making particularly difficult.

1.3. Which Components might Computers perform better than Doctors?

Computers are machines which store and manipulate data. The advantage of using them to assist in medical decisions stems from their impressive storage capacity, rapidity of data retrieval and manipulation, and their obsessive accuracy. Medicine is a complex area of human endeavour, with the core knowledge estimated at one million "facts" and the total at double or triple this figure (Pauker *et al.*, 1976). Although experienced doctors clearly have an overall framework available to them, their knowledge is often more detailed in certain areas, and they perform better on some days than on others. A computer, correctly programmed, could potentially "know" all that there is to know about medicine, and demonstrate the same performance 24 hours a day, seven days per week. The problem is that many of the million or so facts only "make sense" in the context of our knowledge about the real and imaginary worlds outside of medicine; and we still know very little about representing these in the computer.

In practical terms, then, it is those components of medical decision-making requiring fast, accurate access to large amount of data that are most suitable for computer assistance. These include the interviewing of patients (or the prompting of doctors acquiring the data), some data interpretation, and the process of building and refining the hypothesis list, especially by propagation of implications back to patient data. These areas are discussed in relation to examples of the techniques that have been used to implement them, in Sections 2.1 to 2.5 below.

2.0. TECHNIQUES AVAILABLE FOR BUILDING MEDICAL DECISION-AIDS

Although the distinctions are arbitrary, for reasons that should be clear from the discussion of the medical decision-making process above, I shall discuss the techniques that have been used to build medical decision-aids in three sections. In Section 2.1, I review some systems which assist in the collection of patient data, in Section 2.2, systems which assist in the interpretation of patient data, before going on to review a range of approaches to assist doctors in constructing or refining the hypothesis list in Section 2.3.

2.1. Techniques to Assist in Data Collection

As discussed earlier, there is evidence that doctors do not always obtain a complete or accurate history from patients, and this, combined with the time taken to elicit a complex history in areas such as psychiatry or the diagnosis of allergic diseases, has led to the construction of patient interviewing programs. Another possibility, which also assists by improving the quality and completeness of the patient data, is to provide checklists that are tailored to the patient's problems. A final possibility, of building computerised "clinical robots" to examine patients, has not yet become a reality, unless one considers such machines as echocardiographs their precursors.

2.1.1. Computerised patient interviewing systems. Early work on patient interviewing programs, using 35 mm slides of questions and full QWERTY keyboards, soon gave way to more sophisticated approaches (Lucas *et al.*, 1976). These use branching questionnaires, a variety of input devices including a special five key keyboard, and programs that can process the replies to add-up a simple score, such as that used by some psychiatric history-taking programs, or "weights of evidence", as in GLADYS (Spiegelhalter, 1983).

An example of a straightforward interviewing program is ARCHIE, which is written in BASIC and runs on a small microcomputer in an allergy clinic (Tanser *et al.*, 1984). It was built using questions culled from a variety of allergy pro-formas and questionnaires, and was tested in the clinic as a paper questionnaire, then coded as a program once the wording was improved. Patients enter answers using "yes". "no" and "don't know" buttons, and 96% of a sample of 50 found the process "easy". It is interesting that 82% preferred giving their history to the computer than to the doctor; the authors commented that this was probably because of the lack of time pressure, and there being less embarrassment if they did not know the answer. Nearly all patients completed the questionnaire in less than 20 minutes; the machine then prints out a summary of the history, as well as suggestions about which skin tests to perform. A benefit of the system is that it automatically builds up a data-base of symptoms which can be subsequently used as the basis for a true diagnostic program.

The major advantages of computerised patient interviews are the more standardised and more complete acquisition of data (Quaark *et al.*, 1987); a further point is that many patients will admit more to a computer than to a doctor (Lucas *et al.*, 1977). Such systems are therefore useful for sensitive areas such as obtaining alcohol and sexual histories. Possible problems include the inflexibility of some programs, which require a major re-write if one question is altered, and the difficulties some patients have with using even a simplified keyboard. Obviously, patients with reading difficulties or other physical handicaps may find such systems impossible to use. Their ideal role appears to be in the acquisition of historical data in an area where lengthy histories and sensitive questions are a problem.

2.1.2. Computerised Checklists. The ability of a computer to store large amounts of textual material with rapid, accurate retrieval has led to the building of decision-aids which exploit this. Essentially, an expert creates an "online textbook", with an index as a series of lists or "menus". The user inputs one key item, such as a diagnosis or problem, the computer searches its database and then prints out its stored advice. This may take the form of suggestions about further questions to ask the patient, signs to elicit or tests to be performed, or it may discuss treatment options.

Two examples of this approach merit description. The House Officers' Information System, HOIS, is a database which contains advice about the diagnosis and treatment of some eighty common medical problems for use by house officers (Young, 1980). It has been shown to improve the completeness of history recording and to reduce the number of unnecessary tests performed out of hours. The Edinburgh Poisons Information System (Proudfoot *et al*, 1984) is a similar database, linked up to a Prestel node, and is accessible by telephone from anywhere in the UK by using a suitable viewdata terminal. It deals with the clinical and laboratory features of accidental and self-poisoning, and gives advice about treatment. Since its introduction in 1983, the number of enquiries has doubled every year, and now over 90% of enquiries to the Edinburgh centre are made by viewdata.

Such systems have considerable benefits, despite their simple structure. They are relatively simple to set up, can run on cheap machines, are fast in operation, and some are of proven efficacy. Against this, they are limited to the display of "canned text" in response to a query, lacking a patient-specific model, so cannot cope with unusual or conflicting situations, or reason intelligently about problems which were not foreseen by their designers. Thus, their ideal role is in the giving of advice about common problems to middle grade staff, especially in urgent cases with limited data available.

2.2. Data Interpretation Systems

We have already discussed the difficulties doctors have in the timely, accurate, unbiased interpreting of laboratory tests. This is particularly acute in certain areas, such as ECGs and lung function tests, and has stimulated work on computer interpretation.

The most widespread of these programs are incorporated in automated ECG interpreting machines, such as those manufactured by Marquette and Hewlett Packard. These convert the signals from the instrument into a digital form before identifying the components of interest, such as the P, QRS and T waves. The program then uses one of a variety of methods to measure certain parameters, such as the PR interval, and to classify them as normal or otherwise. Pattern recognition procedures, principal component analysis and other statistical techniques have all been used in such systems. Another example is the use of programs to interpret lung function tests; these may be written in either conventional (Pack *et al*, 1977) or rule-based (Aikens *et al*, 1983) languages (see later discussion of knowledge-based systems).

The advantages of such systems are that they free doctors from performing a routine task, ensuring accuracy and consistancy by the use of an explicit set of criteria. Their disadvantages are the theoretical risk of "de-skilling", and the costs of installation and maintenance. Maintenance costs may be considerable in domains where agreement is not universal, or progress is rapid; the ideal role is therefore in repetitive test analysis using stable criteria.

2.3. Diagnostic Assistants

This aspect of decision-support, although the most complex, has attracted most interest, judging by the number of different approaches that have been adopted (Fig. 2). Interest in this area was stimulated in the early sixties by the building of large clinical databases and the "rediscovery" of Bayes' Theorem, continued in the seventies with a variety of statistical approaches, and has moved on in the eighties with the application of knowledge-based techniques. Current opinion seems to agree that it is only by combining several differing approaches that we will succeed in building the robust, accurate, usable systems that have been promised for over 20 years.

1. Clinical algorithms
2. Database matching
3. Bayesian statistical approach
4. Decision theory
5. Knowledge-based "expert" systems

Fig. 2. Techniques for building decision-aids

2.3.1. Clinical Algorithms. A clinical algorithm consists of a set of instructions about data collection, "decision points" where the data is compared with some criterion, and "terminal nodes". These are stopping points, and consist either of advice about a diagnosis or treatment, or the suggestion that further help is necessary as the case is too complex for this approach. The algorithm is usually laid out as a structured flowchart on paper, but may easily be computerised (eg. Auvert *et al.*, 1986), for use by either the patient themselves or by an inexperienced health worker on their behalf.

Most algorithms are written for a naive user to allow him/her to filter patients at some primary contact point: this is a particular need of developing countries, where the few trained staff are overwhelmed with simple problems requiring modest expertise. Perhaps the best known series are for use in East Africa (Essex, 1975), to assist primary health care workers to cope with the simpler needs of their clients. In experiments carried out in a hospital, second year medical students using these algorithms were as accurate and seven times as fast as doctors using the conventional techniques.

The advantages of algorithms are that they assist in both data collection and in the decision itself, are fast, portable and cheap, especially if produced as a book of charts, and are surprisingly accurate on common problems. However, as the user has only "yes" and "no" options to select at decision points, they rely on all the clinical data being available, so cannot be used with patients who don't understand all the questions, if a certain test is not available, or retrospectively with incomplete data. Similarly, they are limited in their coverage, and soon become unwieldy to build and use if attempts are made to extend them. This means that they rely on expert help being available. Nevertheless, they are a useful screening tool for paramedics, and provide a graphic way of making a decision tree explicit for teaching purposes.

2.3.2. Database Matching. One of the earliest approaches to providing computerised support to a doctor with a patient whose clinical features are known but the diagnosis is not, is by attempting to match this patient with one or more from a database. The database consists of a large number of patients with known clinical features, each of whom has been carefully followed up to determine the correct diagnosis and outcome. If one can accurately match the new case with one from the past, then it is not unreasonable to assume that the diagnosis and outcome of this new case will resemble that of the previous one. Various mathematical tests of "exactness of matching" allow uncertainty about the results to be expressed, and if no match is found, then the "nearest neighbours" can be located by dimension-reduction calculations.

The HELP system (Warner *et al.*, 1972), built at the Latter Day Saints Hospital, Salt Lake City, uses this principle. It started as a database of laboratory results and a hospital-wide network; clinicians were asked to write 'HELP sectors', initially to flag abnormal test results, but later to provide advice about new patients based on the experience gained with previous similar cases. A range of experiments have shown that the HELP system is effective in improving both the quality of care given and the final outcome.

Clearly, this type of approach has much to recommend it, providing that the critical element, a large accurate database, is available. Sometimes the results of testing of such decision-aids have been disappointing, as several problems can beset them. Firstly, the database must be large enough to include examples of both typical and atypical presentations of all the diseases to be covered by the system. Secondly, the definition of each of the symptoms, signs and laboratory results to be recorded must not change during the life of the system. This is usually a quite unrealistic requirement, as to build such a database may take years or even decades. Thirdly, the method adopted for determination of "nearest neighbours" in the frequent case of there being no exact match must always find the correct cases - which in multidimensional space may not always be so simple!

2.3.3. Bayesian Statistical Approach. A technique which has been more widely used, but resembles the database matching technique in some ways, is the use of Bayes' Theorem. This states that it is possible to calculate the probability of some outcome, O, from the "prior probability" of that outcome in the population, p(O), and the "conditional probabilities" of each of the features observed, f1 - fn, given that the outcome is present, p(f1/O) - p(fn/O). In medical terms, the outcome of interest is usually a diagnosis, and the features or "indicants" may be symptoms, signs or test results. Both the prior and the conditional probabilities can be obtained from database of cases, each of which is exhaustively followed-up and investigated until the true diagnosis is known. Once the probabilities are calculated, it is a simple matter to calculate the probability of a new patient with an unknown diagnosis but known features f1 - fn having diagnosis X, Y or Z, and ranking the resulting list in descending order of probability. The program to perform this calculation is fast and can run on a hand calculator.

An excellent example of the evolution of a decision-aid over a period of 15 years is one built by de Dombal's team using this approach, the Leeds abdominal pain program (Horrocks *et al.*, 1972). This is designed for use by junior doctors in casualty departments, and relies on the collection of some 50 items of patient data, including simple test results, on a structured form. It has been shown to improve the diagnostic accuracy of doctors by

20%, although evidence from a recent multi-centre trial (Adams *et al.*, 1986) suggested that the "checklist effect", of doctors obtaining more accurate and complete profiles of the patient by filling out the form, was partly responsible for the improvements seen (Wyatt *et al.*, 1986). However, the halving of perforated appendix rates and 60% reduction in negative laparotomies during this trial must rate as one of the most impressive examples of the benefits of computer decision-aids.

However, there are problems. Firstly, as in our previous example, we need a database of cases with known diagnoses, which should ideally include at least five cases of each diagnosis <u>per item of data used</u> (Wasson *et al.*, 1985) - thus, if ten indicants are used, the database should contain over 50 cases of each disease. Secondly, there are doubts about the stability, over both time and space, of the probabilities obtained from such databases, despite some work from de Dombal's group (Horrocks *et al.*, 1976). Thirdly, the Reverend Bayes made it clear that he assumed both independence of indicants and that all outcomes were mutually exclusive and exhaustive (ie. never occurred together, and accounted between them for 100% of the cases) - neither of these assumptions hold in clinical medicine. Although the computer can be programmed to display the relevant conditional probabilities used in the calculations, and even perform "what if?" revisions, the explanations provided by such systems are poor. Finally, as there is no representation in the program of the relationships between indicants and outcomes in other than "heuristic" or associational terms, there is no capacity to cope with the unexpected or atypical case: which is often the one with which the doctor most needs help.

Despite these criticisms, the Bayesian technique has a proven role in the building of decision-aids, particularly in those areas where a database is available which includes all the likely diagnoses, which are mutually exclusive, and where independence of indicants is likely. To avoid possible problems due to variations in prior and conditional probabilities, it is helpful if the training set is collected in the same place as, and just prior to, the use of the decision-aid.

<u>2.3.4. Decision Theoretical Approaches</u>. Decision theory is widely used in industry (Lindley, 1971), and attempts have been made to apply it to the rational solution of medical problems. The theory is again simple: an expert draws up a decision tree of all possible outcomes of a certain situation, then assesses both the probabilities at the "chance nodes" (often by reference to a database) and the "utility" of each branch. The user then chooses a path that maximise the expected utility for the patient.

This approach has been used to decide which patients with a deep venous thrombosis (DVT) should be heparinised (Emerson, 1974). After identifying all possible outcomes of a DVT - namely, fatal and non-fatal pulmonary embolus, no complication, and death from bleeding due to over-anticoagulation, he attached probabilities to these, which depended on both age and smoking history, and concluded that heparin treatment was only justified in certain categories of patients.

The benefit of this approach is that it makes explicit otherwise obscure value judgements, and is thus an ideal way of teaching rational decision-making to trainee doctors. However, to use this approach in clinical practice is impractical, for it relies on the vexed concept of "utility". For over two decades, argument has raged over how to value patient outcomes on a uniform scale, and despite some promising inventions such as the scale of valuations of disease states (Rosser *et al.*, 1978) and "quality adjusted life years" (Office of Technology Assessment, 1979), the debate continues. A second criticism is that such an approach is unduly mechanistic, relying on so-called objective measures when many patients feel that only their subjective opinion matters to them; finally, some argue that the decision chart itself is only applicable to the specific clinical situation for which it was constructed, and that it is unreasonable to build charts for every decision in the whole of medicine (Feinstein, 1977).

2.3.5. Knowledge-based Systems. Reflection about how people make decisions led workers in the early seventies to build computer models of human reasoning, which became known as "knowledge-based" or "expert" systems. The development of these systems is inextricably linked with attempts to apply them to medical problems, so that their history and current state is well covered by two books of papers about the development of medical expert systems (Clancey *et al.*, 1984; Buchanan *et al.*, 1984).

The building of these systems relies on the close cooperation of a "domain expert" who contributes his knowledge and insight. In its simplest form, the expert introspects, often with the help of a "knowledge engineer" and some example cases, and produces a series of rules of the form **If** {one or more conditions} **Then** {one or more conclusions}. These are then coded in a programming language such as Lisp, and then processed together with items of patient data to draw conclusions (Hayes-Roth *et al.*, 1983). Further sophistication, such as the incorporation of ad-hoc "certainty factors", rudimentary explanation facilities (display of the rule that produced the conclusion), and a simple control structure (set up a goal; find subgoals which lead to it; solve the subgoals first, if necessary by querying the user) were added, and the first medical expert system, MYCIN (Shortliffe *et al.*, 1975) resulted (see Fig. 3). A major advantage of this approach was that, for the first time, the knowledge embedded in the system was transparent to the expert, who could verify it; this led to a more rapid cycle of development, and the achievement of impressive accuracy in the processing of bacteriological and clinical data to produce advice about the likely cause of a septicaemia or meningitis (Yu *et al.*, 1979).

IF the infection is primary-bacteria
AND the site of the culture is one of the sterile sites
AND the suspected portal of entry is the GI tract
THEN there is suggestive evidence (0.7) that the identity of the organism is bacteroides

Fig. 3. Sample MYCIN Rule

The MYCIN system was important in the development of medical knowledge-based systems not only because of what it did do, but also because of what it did not do. It spawned many offspring, and now second generation systems, such as ONCOCIN (Shortliffe *et al.*, 1984) are becoming available. Some of these use not only rules but also facts and "semantic networks" to make explicit our knowledge about the relationships between parts of the body, time etc. and to synthesise a "deep model" of the clinical problem (Szolovits, 1988). This approach potentially leads to more robust problem-solving behaviour and more comprehensive explanations than are possible with simpler systems. Unfortunately, such advanced systems take several man-years to build, require expensive computers, and still run unacceptably slowly. Nevertheless, they are beginning to overcome some of the disadvantages of the early systems, specifically by allowing easier extension and modification of the "knowledge -base", better explanations, less use of ad-hoc heuristics, and more sympathetic and adaptable control strategies.

It has been unkindly suggested that knowledge-based systems were a solution looking for a problem. Perhaps this was the case a decade ago, although at least two systems, PUFF (Aikens *et al.*, 1983) and SPE/Expert (Weiss *et al.*, 1981) were in use by 1982. The most suitable role for these early systems was that of data interpretation, with limited clinical exposure. The second generation systems are now coming in to clinical use for more complex decision-support such as the implementation of oncology protocols (ONCOCIN), the "critiqueing" of plans for treatment (ATTENDING, Miller, 1983) and the suggestion of differential diagnoses from lists of clinical features (DXplain, Barnett *et al.*, 1987).

3.0. GENERAL COMMENTS

3.1. Which Systems are in Use Today?

Although doctors perform many tasks that computers will never take over from them, there is a need for decision-aids in certain areas, and there are effective methods available for building them. Some systems are already in use and are of proven efficacy (see Fig. 4).

System name	Number of sites
ECG interpretation	1000's
PUFF	1
Cliniscan / SPE	? dozens
Edinburgh poisons viewdata	UK network
HELP	4
Leeds abdominal pain	20
CFAP's allergy system	4
ONCOCIN	1
DXplain	USA network

Fig. 4. Which systems are in use today?

3.2. The Need for Evaluation

A recurring theme throughout the development of these systems is the need for, and difficulty of, evaluation (Spiegelhalter, 1983; Gaschnig *et al.*, 1983; Miller, 1986; Wyatt, 1987a). All decision-aids should be evaluated, like any new medical technology. This is most necessary in the development phase, when it is frequently unclear whether modifications are improving the decision-aid or not. However, it is also a requirement before any system is released for general use, as, if the usual legal definition of "reasonable care" is taken in cases of alleged professional negligence following the use of a decision-aid, the doctor must be able to point to a published scientific trial demonstrating that the aid is effective.

Apart from preliminary evaluations of performance under laboratory conditions, this means that a randomised trial comparing the effect of the system on patient outcome against a control group should be performed in the field (see Fig. 5). However, for some systems, it is unreasonable to expect the system to exert an effect on patient outcome, and evaluation of intermediate parameters such as accuracy of diagnosis, yield of positive tests, acceptability or quality of reasoning is justified.

3.3. What Makes These Systems Usable?

A further problem that is becoming more obvious as these systems penetrate into the clinical arena is that, despite excellent performance, the users find them difficult to exploit. Although much work has been done on the human-computer interface, ease of use extends beyond this (Wyatt, 1987b; Wyatt, 1988).

Firstly, the system must be appropriate, in terms of fulfilling a useful clinical role (ie. there must be a real problem), while the interface, explanations, and advice itself must match the users and their working environment. Secondly, the system must be of proven effectiveness, in both laboratory and field trials, and its side-effects must be insignificant. Thirdly, it must be reliable, not only in the sense of the hardware which is used, but also in its method of reasoning, manner of coping with uncertainty, and in its behaviour when it reaches the bounds of its knowledge.

Accessibility is a further key requirement of users: this means that the system must be cheap, with minimal learning required, and should preferably be implemented on a fast, portable machine. Finally, medical practice varies widely from place to place and over a period, so systems must be adaptable easily. The transparency of the "knowledge-base", the separation of control strategies from facts, and other features of second generation knowledge-based system will probably allow them to best meet this criterion.

4.0. SUMMARY AND CONCLUSIONS

4.1. Stages in the Building of a Successful Decision-aid

From all the discussion of problems above, one might wonder where to begin in constructing a computer decision-aid. I suggest an eight stage procedure, starting with a careful and complete definition of the problem the system is supposed to solve, of the prospective users and how to obtain the "correct" answer to the problem. Then collect some sample data and find out the correct solution in the cases to which it relates. This "training set" can then be studied and, according to one's preferences and the discussion above, an appropriate "reasoning" technique can be selected. It is usually a fairly simple matter to build the first prototype, and test it on some data. Much more difficult is to reject it, when it proves unsuitable, and try another technique or even gather fresh data. Eventually, according to the performance and likely ease of use of the prototypes, the best can be selected and further developed. When it is performing well, it can be "frozen", and a formal evaluation performed with fresh data in the laboratory. As long as the performance is still satisfactory, and ethical

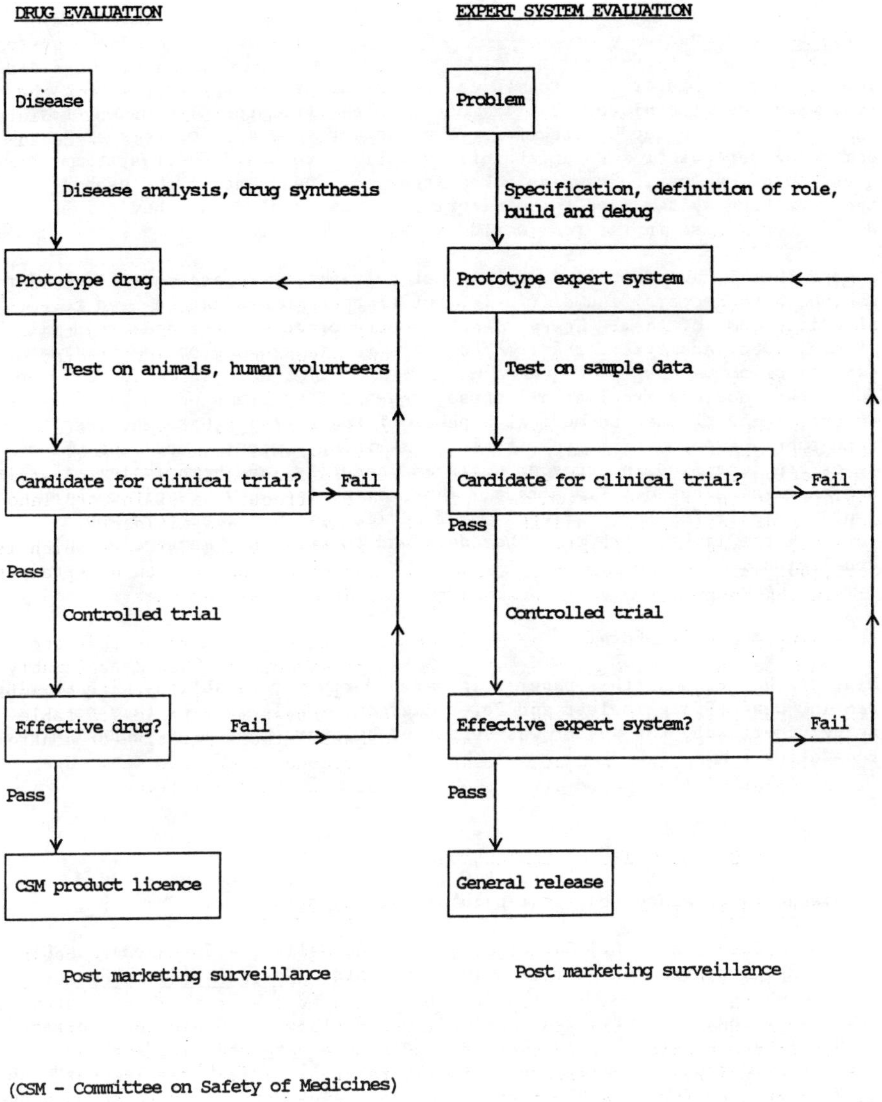

Fig. 5. The evaluation of drugs contrasted with the proposed evaluation of medical expert systems.

permission can be obtained, the system can then be evaluated in a field trial, and measurements made of usability and its effect on patient outcome. The final, often neglected stage, consists of long-term maintenance.

4.2. Conclusions

The field of computer aids to medical decisions has been ploughed up by successive new techniques, but it must be clear from the discussion above that no one technique by itself is "the answer", and most workers are now combining methods from a variety of disciplines to build their systems. The question of evaluation remains a key issue, as does that of the usability of the resulting system, and the monitoring of new problems as they arise during system use in the real world.

I hope that I have demonstrated that not only are there aspects of decision-making which doctors find difficult, but that computers can be used to alleviate some of them. However, an advocate of computer-aided diagnosis (Blois, 1980) has stated that: *"The most important question is not 'Where can we use computers?' but 'Where must we use human beings?'"* There is no doubt that doctors excel at all those open-ended problems which require understanding of the non-medical aspects of the world; these include the more subtle aspects of history taking, examining patients, deciding which tests are justified in terms of costs and benefits, and both implementing and explaining the decisions made. Thus, even if adequate decision-aids could be built for the entire range of medical problems, doctors would remain vital links, both to gather data and to explain a machine's conclusions.

ACKNOWLEDGEMENTS

Many of the ideas in this paper originated during conversations with Charles Pantin, David Spiegelhalter and Peter Emerson. While some of this work was being undertaken, the author was supported by a grant from the BUPA Medical Foundation.

REFERENCES

Adams, I.D., Chan, M., Clifford, P.C., *et al.* (1986). Computer aided diagnosis of acute abdominal pain: a multicentre study. BMJ, 293, 800-804.

Aikens, J.S., Kunz, J.C., Shortliffe, E.H. and Fallat, R.J. (1983). PUFF, an expert system for pulmonary function data. Comp. & Biomed. Res., 16, 199-208.

Auvert, B., Aegerter, P., van Look, F., *et al.* (1986). A hand-held system designed for rural health workers. Comp. & Biomed. Res., 19, 80-89.

Barnett, G.O., Cimono, J.J., Hupp, J.A., *et al.* (1987). DXplain: an evolving diagnostic decision-support system. JAMA, 258, 67-74.

Blois, M.S. (1980). Clinical judgement and computers. NEJMed., 303, 192-197.

Buchanan, B.G. and Shortliffe, E.H. (1984). Rule-based expert systems: the MYCIN experiments. Pub. Addison-Wesley, Reading, Mass.

Clancey, W.J. and Shortliffe, E.H. (eds.) (1984). Readings in medical AI. Addison Wesley, Reading, Mass.

Croft, D.J. (1972). Is computerised diagnosis possible? Comp. & Biomed. Res., 5, 351-367.

de Dombal, F.T., Leaper, D.J., Horrocks, J.C., *et al*. (1974). Human and computer-aided diagnosis of acute abdominal pain: further report with emphasis on performance of clinicians. BMJ, 1, 376-380.

Emerson, P.A. (1974). The application of decision theory to the prevention of DVT following myocardial infarction. Bull. Inst. Maths., 10, 33-36.

Essex, B.J. (1975). Approach to rapid problem solving in medicine. BMJ, 3, 34-36.

Feinstein, A.R. (1977). The haze of Bayes, the aerial palaces of decision theory, and the computerised Ouija board. Clin. Pharm. & Therapeutics, 21, 482-496.

Gaschnig, J., Klahr, P., Pople, H., Shortliffe, E., Terry, A. (1983). Evaluation of expert systems: issues and case studies. In: Building Expert Systems, (eds. F. Hayes-Roth, D.A. Waterman & D. Lenat), Addison-Wesley.

Gjorup, T., Kelbaek, H., Stenbygard, L., *et al*. (1986). Effect of knowledge of serum enzyme concentrations on doctors' interpretation of ECGs. BMJ, 292, 27.

Goldman, L., Weinburg, M., Weisburg, M., *et al*. (1982). A computer-derived protocol to assist in the diagnosis of emergency room patients with acute chest pain. NEJMed., 307, 588-596.

Hayes-Roth, F., *et al*. (1983). Building expert systems. Pub. Addison-Wesley, Reading, Mass.

Horrocks, J.C., McCann, A.P., Staniland, J.R., *et al*. (1972). Computer-aided diagnosis of acute abdominal pain. BMJ, 2, 5-9.

Horrocks, J.C., Lambert, D.E., McAdam, W.A.F., *et al*. (1976). Transfer of computer-aided diagnosis of dyspepsia from one geographical area to another. Gut, 17, 640-644.

Lindley, D.V. (1971). Making decisions. Pub. Wiley Interscience, London.

Lucas, R.W., Card, W.I., Knill-Jones, R.P., *et al*. (1976). Computer interrogation of patients. BMJ, 2, 623-625.

Lucas, R.W., Mullin, P.J., Luna, C.B., *et al*. (1977). Psychiatrists and a computer as interrogators of patients with alcohol-related illness. BJPsych., 131, 160.

Mercer, J., Talbot, I.C. (1985). Clinical diagnosis: a post-mortem assessment of accuracy in the eighties. Postgrad. Med. J., 61, 713-716.

Miller, P.L. (1983). Medical plan-analysis: the ATTENDING system. Proc. IJCAI-83, 239-241.

Miller, P.L. (1986). The evaluation of AI systems in medicine. Comp. Methods & Programs in Biomedicine, 22, 5-11.

Office of Technology Assessment (1979). A review of selected federal vaccine and immunization practices. Pub. US Congress, Washington DC.

O'Neill, M.F., Fox, J., Glowinski, A. (1988). Diagnostic support in The Oxford System of Medicine. Technical Report, Biomedical Computing Unit, Imperial Cancer Research Fund Labs., Lincoln's Inn Fields, London.

Pack, A.I., McClusker, R., Moran, F. (1977). A computer system for processing data from routine pulmonary function tests. Thorax, 32, 333-341.

Pauker, S.G., Gorry, G.A., Kassirer, J.P., Schwartz, W.B. (1976). Towards the simulation of clinical cognition: taking a present illness by computer. Am. J. Med., 60, 981-996.

Pearson, J. and Dudley, H.A.F. (1982). Bodily perceptions in surgical patients. BMJ, 284, 1545-1546.

Politakis, P., Weiss, S.M. (1982). A system for empirical experimentation with expert knowledge. Proc. 15th Hawaii Int. Conf. on Systems Science, 2, 649-657.

Proudfoot, A.T. and Davidson, W.S.M. (1984). Acceptance of Viewdata for poisons information. BMJ, 289, 1420-1421.

Quaak, M.J., Westerman, R.F., van Bemmel, J.H. (1987). Comparison between

written and computerised patient histories. BMJ, 295, 184-190.
Rosser, R. and Kind, P. (1978). A scale of valuations of states of illness: is there s concensus? Int. J. of Epidemiol., 7, 347-358.
Shortliffe, E.H., Scott, A.C., Bischoff, M.B., *et al*. (1984). An expert system for oncology protocol management. In: Rule-based expert systems: the MYCIN experiments (eds. B.G. Buchanan, E.H. Shortliffe), Addison-Wesley, Reading, Mass., Chapter 35.
Shortliffe, E.H., Davis, R., Axline, S.G., *et al*. (1975). Computer based consultations in clinical therapeutics... the MYCIN system. Computers & Biomed. Res., 8, 303-320.
Shortliffe, E.H. (1987). Computer programs to support medical decisions. JAMA, 258, 61-66.
Spiegelhalter, D.J. (1983). Evaluation of clinical decision aids, with an application to a system for dyspepsia. Statistics in Med., 2, 207-216.
Szolovits, P. (ed.) (1982). Artificial intelligence in medicine. Pub. Westview, for AAAS, Washington.
Szolovits, P., Patil, R.S. and Schwartz, W.B. (1988). Artificial intelligence in medical diagnosis. Annals Int. Med., 108, 80-87.
Tanser, A.R. and Davies, A. (1984). ARCHIE: the use of a microcomputer in an allergy clinic. Br. J. Dis. Chest, 78, 135-139.
Tweeddale, P.M., Sudlow, M.F., Crompton, G.K., *et al*. (1983). So its easy to interpret pulmonary function tests? Thorax, 38, 238.
Warner, H.R., Olmsted, C.M. and Rutherford, B.D. (1972). HELP: a program for medical decision-making. Comp. Biomed. Res., 5, 65-74.
Wasson, J.H., Sox, H.C., Neff, R.K. and Goldman, L. (1985). Clinical prediction rules: applications and methodological standards. NEJMed., 313, 793-799.
Weiss, S.M., Kulikowski, C.A., Galen, R.S. (1981). Developing microprocessor-based expert models for instrument interpretation. In: Proc. Seventh IJCAI, Kaufmann Inc., Los Altos, pp.853-855.
Wyatt, J.C., Crichton, N.J., Emerson, P.A. (1986). Computer assisted diagnosis of acute abdominal pain. BMJ, 293, 1305.
Wyatt, J.C. (1987a). The evaluation of clinical decision support systems: a discussion of the methodology ised in the ACORN project. Lect. notes in Med. Inform., Vol.33, pp.15-24 (Proc. AIME'87, Marseilles).
Wyatt, J.C. (1987b). Improving the usability of knowledge-based medical decision-aids: proposal for a European research programme. Presented at the EEC AIM Workshop 6, 25/11/87, Brussels.
Wyatt, J. (1988). Lessons learned from the field evaluation of ACORN. Seminar presented at Massachusetts Institute of Technology, Boston, 20/4/88.
Young, D.W. (1980). An aid to reducing unnecessary investigations. BMJ, 281, 1610-1611.
Yu, V.L., Fagan, L.M., Wraith, S.M., *et al*. (1979). Antimicrobial selection by computer: a clinded evaluation by infectious disease experts. JAMA, 242, 1279-1282.
Zeide, J.S., Liebowitz, J. (1987). Using expert system: the legal perspective. IEEE Expert, Spring 1987, pp.19-22.

Predicting Allergen Involvement with Symptoms

Charles Pantin

ABSTRACT

The place of decision-aids in science and the criteria required to build and maintain one are discussed. The development, evolution and use of the atopy diagnostic system used at Benenden Chest Hospital is evaluated against the above criteria. The problems and opportunities of decision-aids, especially in the area of quality control, are described.

KEYWORDS

Medical decisions; computer decision-aids; expert systems; statistics; evaluation; usability; allergy.

INTRODUCTION

Mathematics and Information Technology are servants, not masters. They are essential in evaluating any scientific hypothesis including those on allergy [1,2] but do not by themselves predict or make decisions. The knowledge required must come from outside. They can only model that knowledge and use rules from that knowledge to predict or make decisions. Dr Wyatt has shown that their value is their consistency and infallible memory. A decision-aid needs to be treated as any other test, such as Immunoglobulin-E. Thus it must be built in a rigorous and well-evaluated way. following the steps mentioned by Dr Wyatt (Table 1), against which criteria I shall discuss the Benenden Chest Hospital atopy screening system.

PROBLEM

Why allergy prediction by computer?

In the United Kingdom (U.K.) approximately 15% of the population is atopic [3]. Thus eight million people may be sensitized to common environmental allergens, although not all will have related symptoms. Only a small pro-

Advances in the Biosciences Vol. 74

Printed in Great Britain

Table 1. Stages in building a decision-aid

1. Define problems, patients, users.
2. Collect sample database.
3. Select appropriate mathematical technique.
4. Build prototype and redesign, using sample data.
5. Collect test database.
6. Evaluate system with test database.
7. Evaluate system in use.
8. Continual maintenance and quality control.

portion can attend the small number of Allergy clinics in the United Kingdom. Can we identify those patients in most need of specialist advice? The value of knowing the atopic status of the patients is clinically a controversial subject outside the scope of this article, but some would argue that elimination [4] and desensitization [5] are of increasing use.

The allergy unit at Benenden Chest Hospital offers a laboratory based service to general practitioners and hospital consultants for measurement of total Immunoglobulin-E (IgE) and specific IgE antibodies to common environmental allergens in blood samples. In 1978, the service was running into problems with no history arriving with the blood samples (Fig. 1). The laboratory was having to test targe panels of allergens. Many tests were negative and money was being wasted.

A review of the service decided that it must answer the two questions in Table 2. The allergy unit needed to offer a system which was able to obtain the patient's history, analyse it, reduce the number of tests on the blood sample to a rational few, and guide the referring doctors (Table 3).

DEVELOPMENT OF THE SYSTEM [6]

The questionnaire was developed through many variations to ask the questions an allergy specialist would use to elucidate whether symptoms might be caused by a specific allergen (Table 4). Pilot studies on patient acceptability were carried out until the present layout was established.

A database of 1500 patients, aged over two years, were referred by their doctors for suspected atopic symptoms, was gathered. All patients or their parents filled in the questionnaire and had total IgE and specific IgE measured to the thirteen commonest U.K. inhalant allergens (Table 5), plus six common foods (egg, milk, codfish, peanut, wheat, hazel nut).

The database was used to choose an appropriate method to answer the questions in Table 2. The probability of finding a positive specific IgE antibody at different levels of total IgE was evaluated. A cluster analysis [7] of the allergens to which antibodies were found showed that antibodies grouped into nine groups, eight of which contained antibodies to one of grass, house dust mite or cat (Table 6). The likelihood of each answer to the individual questions were calculated for the mutually exclusive and exhaustive four

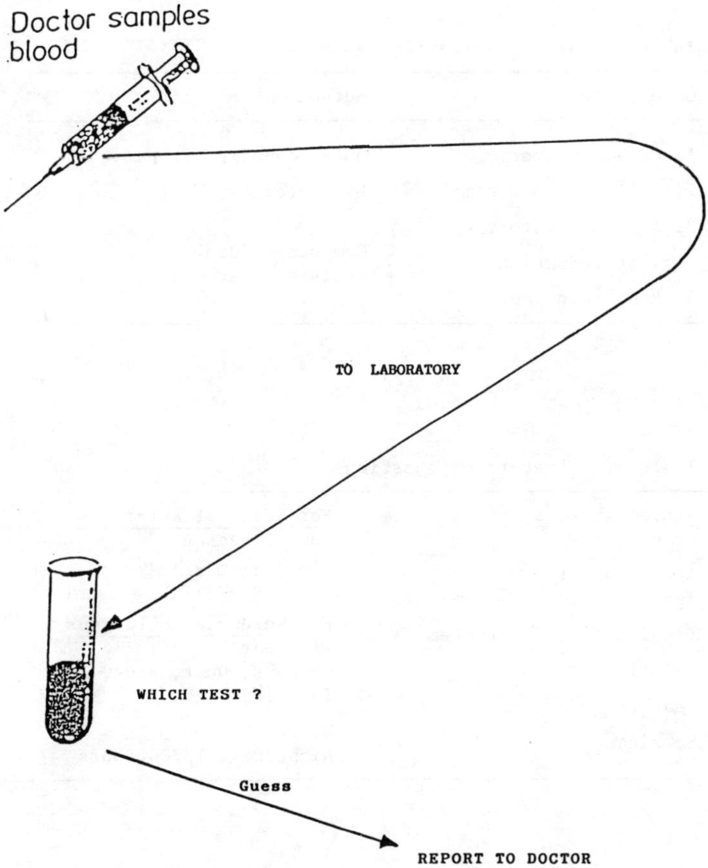

Fig. 1. Analysis of blood samples before the decision-aid was introduced.

Table 2. Aim of atopic screening

1. Is the patient atopic?
2. Are the patient's symptoms likely to be related to those allergens to which specific IgE antibodies are found?

Table 3. Atopic screening system

Objectives	Method
1. Obtain history	Questionnaire for patient
2. Collect blood samples	By doctor
3. Analyse history 4. Test reduction 5. Report to doctor	Computer-aided decision system

Table 4. Clusters of questions

1. System where symptoms occur	7. For Seasonal Allergens Summer Season Grass, Tree, Weed, Outside, Day
2. Age of onset Age	
3. Previous atopic problem	8. For Perennial Allergens All year Animals, Dust, Beds, Indoors, Night
4. Relatives	
5. Weather	
6. Smoking	9. Winter Season Night/Day, In/Outdoors

Table 5. Common British allergens tested

Perennial	Seasonal
House dust mite	Grass (meadow fescue)
Cat	Mugwort
Dog	Plantain
Penicillium notatum	*Alternaria tenuis*
Aspergillus fumigatus	*Cladosporium herbarum*
Horse	Silver birch
	London plane tree

Table 6. Cluster analysis of 19 allergens

267 Patients RAST positive to at least one allergen

1. GRASS
2. *Alternaria*, *Aspergillus*, *Penicillium*, *Cladosporium*
3. MITE, Egg, Milk.
4. Plantain, GRASS, MITE, Plane
5. Codfish, CAT, God, Horse, MITE

---

6. Horse, Dog, *Penicillium*, CAT, *Cladosporium*, *Aspergillus*, Egg, MITE, GRASS, Birch
7. Peanut, Wheat, Mugwort, Plane, Hazel Nut, Plantain, Birch, GRASS, Milk, Egg, *Penicillium*, Horse.
8. CAT, Mugwort, GRASS, *Alternaria*
9. All 19 Allergens

groups of patients; those having antibodies to seasonal inhalant allergens only, perennial allergens only, both or none.

This data was used to select the appropriate decision-aid techniques to answer the aims of the project (Table 2); Bayes theorem with independent clusters [8] and rules. From the likelihood ratios of answers to the individual questions for the four groups obtained from the developmental database, algorithms were constructed to predict the probability of a patient having antibodies to seasonal and/or perennial inhalant allergens in their blood samples from the pattern of answers given in their questionnaire, using Bayes theorem. These probabilities were combined with the results of the total IgE and specific IgE antibody tests to grass, mite and cat to form the screen for atopy. This screen determines how diligently the search for specific allergens should be conducted.

Antibodies to further allergens were looked for if the system felt the probability of finding them was high, based on the following rules held in the microcomputer;

1) More perennial allergens if specific IgE to mite (*D. pteronyssinus*) and/or cat were found OR total IgE was over 80 kU/1 and the perennial history algorithm gives a more than 20% probability of finding antibodies to a perennial allergen.

2) More seasonal allergens if specific IgE to grass (meadow fescue) pollen was found OR total IgE was over 80 kU/1 and the seasonal history algorithm gives a more than 20% probability of finding antibodies to a seasonal allergen.

3) If the total IgE was over 200 kU/1, even if the probability of any

positive RAST was low, then the following moulds were to be tested: *Alternaria Tenuis*, *Aspergillus fumigatus*, *Cladosporium herbarum* and *Penicillium notatum*.

4) Similarly, irrespective of the history analysis, if the total IgE was over 2000 kU/l then all the remaining allergens in Table 5 were to be tested.

Rules for which individual allergen to test were built into the system based on the knowledge of the time and place of their occurrence, related to where and when the patients symptoms occurred, found by the questionnaire.

Therefore, the guessing game of Fig. 1 had been replaced by the structured analysis of the history obtained from the questionnaire and the results of total IgE and specific IgE antibodies measured in the blood samples as shown in Fig. 2.

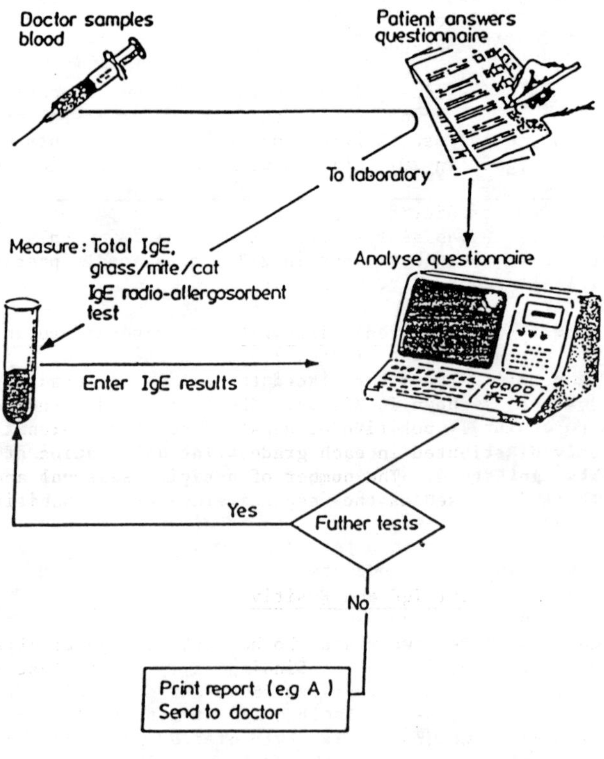

Fig. 2. Analysis of blood samples and questionnaires after decision-aid was introduced.

EVALUATION WITH FRESH DATA [6]

All 554 patients who were referred during a 12 month period (to avoid seasonal variations) entered the study and were investigated by the new system.

The system's predictions were compared with the results of testing their sera for antibodies to all of the panel of thirteen common U.K. allergens (Table 5). Excluded from the full analysis were half of those patients who were very unlikely to have specific IgE antibodies to any of the panel allergens. Therefore, of the 216 patients with a total IgE under 40 kU/l, no grass, mite or cat specific antibodies and whose history predicted under a 20% probability of finding specific IgE antibodies to any common U.K. allergen, only 108 were tested against the full panel. Of 554 sera tested, 298 (54%) had specific IgE antibodies to at least one common environmental allergen with 83 cases of seasonal allergens alone, 93 perennial only and 122 both seasonal and perennial.

Atopy Screening System Compared with Test Panel Results

The system selected only 3861 allergen tests compared with the 7202 necessary with the panel of 13 allergens. Of the suggested tests, 842 (22%) were positive and 945 (24%) were borderline. Altogether 3341 tests were predicted as unnecessary, and of these, 34 (1%) were positive and 19 (0.6%) borderline results. Dividing into seasonal and perennial groups.

Seasonal:- The system assigned 298 patients as probably negative and 256 as probably positive. When compared with the results of the allergen panel, positive seasonal tests results were found in only 4 patients with negative predictions (1.3%), but in 201 patients with positive predictions (79%).

Perennial:- 256 were predicted as probably negative and 298 as probably positive. Positive perennial test results were found in only two patients with negative predictions (0.8%) but in 213 with positive predictions (71%).

Relation of History Symptom Predictions with Test Panel Results

Seasonal and perennial algorithms placed the 554 questionnaires among the history predictions as shown in Fig. 3. The distribution showed that seasonal histories were usually positive or negative but that perennial histories were more evenly distributed in each grade. The calibration of the probabilities are shown in Fig. 4. The number of positive seasonal and perennial test results both increased as the respective history probability predictions increased.

Relationship Between Total IgE and Positive Test Results

The 418 patients aged over two years who had the full panel allergens tested were analysed. The probabilities of finding any positive test results were calculated for total IgE concentrations above and below 80 kU/l and again above and below 200 kU/l, using the result that 54% of patients had at least one positive result. Knowledge of both total IgE and grass/mite/cat specific IgE antibody concentrations was a better predictor of a positive test result than was total IgE alone (Fig. 5). Furthermore, negative grass/mite/cat test results and total IgE less than 80 kU/l made any other positive test result most improbable (Fig. 5c)

Grass/mite/cat test results were more discriminating than total IgE in identifying the atopic person. Of the 298 patients with any positive result 283 were positive to grass/mite/cat. Thus, the probability of a positive test result in a patient who was grass/mite/cat negative was 0.05.

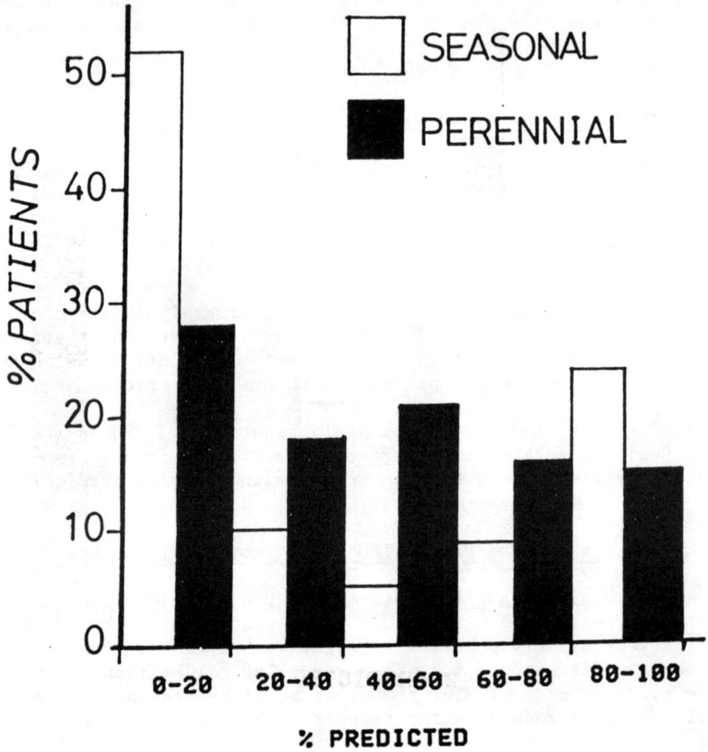

Fig. 3. Distribution of probabilities calculated from the clinical history of finding any positive antibodies to seasonal and perennial allergens.

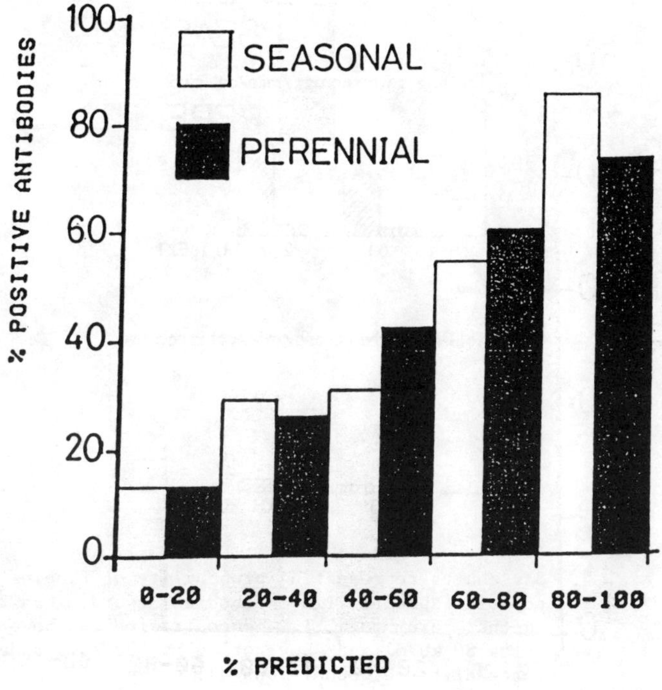

Fig. 4. Correlation of seasonal and perennial history probabilities and the percentage of patients with any positive antibodies to seasonal and perennial allergens in each 20% range of probabilities.

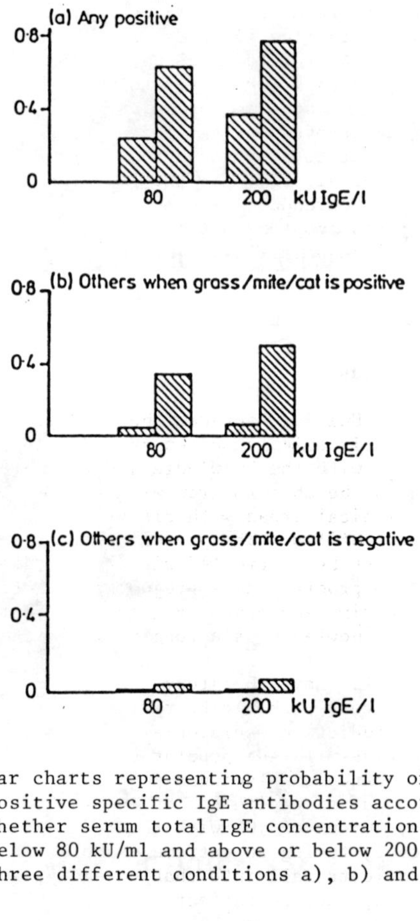

Fig. 5. Bar charts representing probability of finding any positive specific IgE antibodies according as to whether serum total IgE concentration is above or below 80 kU/ml and above or below 200 kU/l under the three different conditions a), b) and c).

CLINICAL USE

The system has been in regular use in the Allergy Unit at Benenden Chest Hospital since 1982. A contradiction scoring system checks the incoming questionnaires and only 5-8% are rejected. The others are entered into the computer and the results of the algorithms integrated with the blood tests as in Fig. 2. The baseline tests - total IgE and grass/mite/cat specific IgE - are measured. The results are entered into the computer, which checks if the four criteria detailed in the development section are satisfied. If so, it suggests further specific allergens as possible causes of the patients symptoms. Finally reports of the forms A, B and C are printed to send to the referring doctor. These forms consist of three parts, firstly administrative details, then integrated blood test results and comments, and finally a resumé of the clinical history on which the comments are based.

History predictions are a necessary control on the validity of the IgE results. Allergens to which specific IgE antibodies are discovered may be irrelevant to the presenting complaint. In report A, the specific allergens to which antibodies are found are suggested by the history. In Report B, negative test results are reinforced by a non-atopic history. However, in Report C the positive antibodies to seasonal allergens found agree with history, but despite a 94% prediction of perennial allergens, none were found. This may indicate an allergen yet to be discovered, possibility occupational, a non-IgE mediated mechanism or technical problems with the tests. Similarly, the discovery of antibodies to grass pollen in a case of perennial asthma may refer to a previous seasonal hayfever, or perhaps is an indication of wheat flour allergy. In the last two weeks, referral to an allergy specialist is recommended. Indeed, referral should be considered when either the patient's symptoms are not fully explained by the report or advice on management is required.

CONCLUSION

Problems and Opportunities

At present, the Allergy Unit at Benenden Chest Hospital is running an atopy diagnostic system which is able to screen an individual patient for common U.K. inhalant allergens with the knowledge and technology of 1982. However, a diagnostic system must be able to grow with the new knowledge in allergy, to adapt to new geographical areas with different common enviromental allergens and patient profiles. The present system lacks this flexibility. However, this growth must be controlled with repeated evaluation and quality control. It is easy to program one's prejudices! Setting up a diagnostic system at one point in time and place is good, but continual quality control and upgrading with new knowledge is a long-term commitment.

Quality control is vital. Any laboratory test - and a decision-aid should be regarded as another test - should be rigorously controlled. The Calibration (Fig. 4) of any predictive decision-aid or test such as the atopy diagnostic system, varies with the population on which it is used. If the test predicts an 80% chance of placing a patient in a category based on calculations from a population with 20% of people in that category, and the population is stable, 80 out of 100 with the 80% prediction will fall into that category. However, taking an extreme example, if the population changes so that no patients fall into that category, and this shift in population is not noted, then no patients with the 80% prediction will fall in that category. There will be no 80% predictions if the test is 100% sensitive and specific but the less sensitive and specific the test, the less robust the predictions of that test are to the shifts in the population on which it is used. Thus from Fig. 3, if is obvious that seasonal history predictions are more robust than the perennial ones. Thus shifts in population must be checked and the necessary adjustments made to the probabilities on which the algorithms are based when significant shifts are noted.

However, these are opportunities, not just problems. To extend the system to new knowledge - new questions, allergens, new tests - the methods used to build and evaluate the atopy diagnostic system can be automated so that, if a laboratory gathers a developmental database, the probabilities required to develop a similar system are generated as the database is collected. Similarly, the evaluation phase could be part of the system, not only initially, but also continually running as quality control of the working system. For example the profile of patients processed through the system can be constantly monitored for deviations, as in any other laboratory tests (Fig. 6). If a trend is observed, as in Fig. 6, with an increase in patients with antibodies to seasonal allergens at expense of those with no antibodies

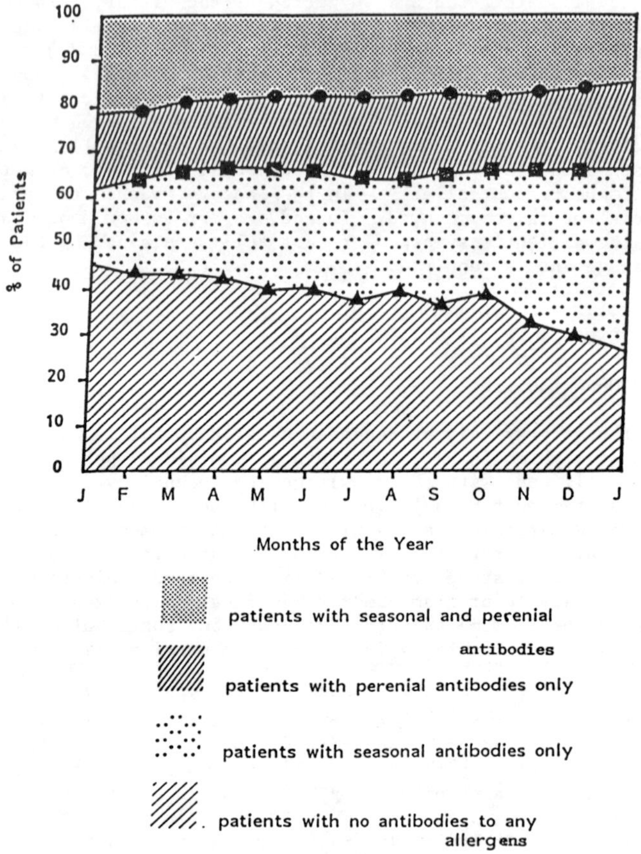

Fig. 6. Quality control: Patient profile monthly.

found, the system can adjust its probabilities to minimise the number of false negatives and positives. Quality control of this type is required independently of the mathematical technique used to build the decision-aid.

Extensions to an allergy diagnostic system are possible in several areas. New database handling techniques can increase the use made of the database collected of many patients' clinical history and screening tests. By integrating the system into an automatic laboratory sample analyser, the laboratory's assay protocol would be generated direct from the patient's history, which might be directly entered by the patient via a keyboard in the allergy clinic [9].

All these clever uses of information technology depend on the quality of the knowledge on which the system is built. A decision-aid is a tool or test which enables a human being to make better use of knowledge about a subject such as allergy. However, the decision-aid can only reflect the knowledge built into it and, as the knowledge changes, so must the decision-aid.

REPORT FROM: RAST Allergy Unit, Benenden Hospital, Cranbrook, Kent TN17 4AX (Tel.0580-240495)

ALLERGY INVESTIGATION PATIENT NO: 10 James KIRK IS A P.O. & C.S.S.S. MEMBER

DATE OF BIRTH : 16/11/51
DATE OF QUESTIONNAIRE : 11/02/87

TO:

Dr Bones
Sick Bay
Starship Enterprise
Galactica

TODAYS DATE : 01-03-1988

Fold

SYMPTOMS : Involve-EYES-NOSE-CHEST

BLOOD TEST RESULTS AND COMMENTS TOTAL IgE = 384 units/ml.

The total IgE is above most common adult non-allergic levels - atopy likely.

PHADEBAS IGE-RAST RESULTS FOR SPECIFIC ALLERGENS

(RAST results are graded :0,1,2,3,4, where 0 IS NEGATIVÉ and 4 IS STRONGLY POSITIVE,
1 IS BORDERLINE EXCEPT for MOULDS when it is regarded as PROBABLY clinically significant.)

Patient has multiple positive specific IgE antibodies to common allergens and so is atopic
The HISTORY
STRONGLY SUGGESTS that symptoms may be related to seasonal allergens
and specific IgE antibodies were found to:
GRASSES--TIMOTHY GRASS(4)
WEEDS--PLANTAIN(2) but not MUGWORT(1),NETTLE(1)

but no specific IgE antibodies were found to:
EARLY SUMMER MOULD--CLADOSPORIUM HERBARUM(0)
LATE SUMMER MOULD--ALTERNARIA TENUIS(0)

The HISTORY
WEAKLY SUGGESTS that symptoms may be related to perennial allergens
and specific IgE antibodies were found to:
DUST--HOUSE DUST MITE PTERONYSSINUS(4)

but no specific IgE antibodies were found to:
CATS--CAT EPITHELIUM(0)
DOGS--DOG DANDER(0)
INDOOR MOULD--PENICILLIUM NOTATUM(0)
IN/OUTDOOR MOULD--ASPERGILLUS FUMIGATUS(0)

SUMMARY OF GRASS, DUST-MITE AND COMMON ANIMAL ANALYSES

** PROBABILITIES of OUTDOOR and INDOOR ALLERGENS contributing to symptoms are 97 % and 55 %. **
GRASS pollens, DUST-MITES and common PETS cause most U.K. ALLERGIES,
and it is VERY LIKELY that GRASS * MITE * contribute(s) to this patient's symptoms.

REPORT A

```
The above comments are based on the following Clinical History:-

CLINICAL HISTORY ASSESSED BY QUESTIONNAIRE
------------------------------------------
     3  relation(s) have a possible ATOPIC history
    Age of onset was  10

SYMPTOMS ARE SEASONAL  AND  SYMPTOMS ARE PERENNIAL

            Months occur..... 5   6   7   8
            Months worst..... 6   7
                              OCCUR IN :
                                4  FROM  5  MONTHS of the GRASS season
                                3  FROM  4  MONTHS of the WEED season
                                3  FROM  5  MONTHS of the CLADOSPORIUM HERBARUM season
                                3  FROM  5  MONTHS of the ALTERNARIA TENUIS season
                                2  FROM  4  MONTHS of the TREE season
                              WORST IN:
                                2  FROM  3  MONTHS at height of the GRASS season
                                1  FROM  3  MONTHS at height of the CLADOSPORIUM HERBARUM season

   WORST NEAR GRASS, TREES, WEEDS
   Aggravated by DUSTING &/or making BEDS
    REGULARLY IN CONTACT WITH :-CATS  REGULARLY IN CONTACT WITH :-DOGS  REGULARLY
    IN CONTACT WITH :-FISH
   OCCUR BOTH INDOORS & OUTDOORS
   OCCUR IN 2 SETS, ONE AT NIGHT, ONE IN THE DAY
   IMPROVE ON WET SUMMER DAYS(suggests weeds more than outdoor moulds)
    ERROR  ** PERENNIAL/WINTER SYMPTOMS NOTED IN 'TIME' QUESTION BUT NOT IN 'WEATHER' QUESTION

   SMOKING :- NEVER (Less than 400 cigarettes in patients lifetime.)

   NO SKIN TESTS PERFORMED

FOOD:
-----
    Patient believes NOT AFFECTED by food.
* Factors POSSIBLY indicating food allergy are listed below :
 ----------------------------------------------------------
     Age of onset less than 30 years, IgE over 200u/ml


MISCELLANEOUS ALLERGIES:
-----------------------
    None mentioned

--------------------------------------------------------------------------------------------
```

REPORT A

REPORT FROM: RAST Allergy Unit, Benenden Hospital, Cranbrook, Kent TN17 4AX (Tel.0580-240495)

ALLERGY INVESTIGATION PATIENT NO: 8946 D. SCOTTIE

DATE OF BIRTH : 09/03/82
DATE OF QUESTIONNAIRE : 12/02/88

TO:

Dr Bones
Sick Bay
Starship Enterprise
Galactica

TODAYS DATE : 17-02-1988

Fold

SYMPTOMS : Involve-CHEST

BL(TEST RESULTS AND COMMENTS TOTAL IgE = 21 units/ml.

The total IgE is in normal range - atopy possible.

PHADEBAS IGE-RAST RESULTS FOR SPECIFIC ALLERGENS

(RAST results are graded :0,1,2,3,4, where 0 IS NEGATIVE and 4 IS STRONGLY POSITIVE,
1 IS BORDERLINE EXCEPT for MOULDS when it is regarded as PROBABLY clinically significant.)

Patient had no positive specific IgE antibodies to common inhalant allergens discovered
The HISTORY
STRONGLY SUGGESTS that symptoms may be related to a non-IgE cause

The HISTORY
does NOT SUGGEST that symptoms may be related to seasonal allergens
and no specific IgE antibodies were found to:
GRASSES--TIMOTHY GRASS(0)

The HISTORY
as NOT SUGGEST that symptoms may be related to perennial allergens
and no specific IgE antibodies were found to:
DUST--HOUSE DUST MITE PTERONYSSINUS(0)
CATS--CAT EPITHELIUM(0)

**** no specific IgE antibodies were found to the following foods:
FISH(COD)(0) WHEAT(0) COCOA(0) CHEDDAR CHEESE(0)
The history suggests that the patient may be allergic to :
coloured drugs
SUMMARY OF GRASS, DUST-MITE AND COMMON ANIMAL ANALYSES

** PROBABILITIES of OUTDOOR and INDOOR ALLERGENS contributing to symptoms are 1 % and 13 %. **
GRASS pollens, DUST-MITES and common PETS cause most U.K. ALLERGIES,
but NONE of these common allergens are CLEARLY identified as triggering this patient's symptoms.

REPORT B

```
. comments are based on the following Clinical History:-

 AL HISTORY ASSESSED BY QUESTIONNAIRE
 -----------------------------------
 2  relation(s) have a possible ATOPIC history
Age of onset was  2

SYMPTOMS ARE PERENNIAL

   HOUSE has DAMP/MOULDS
   OCCUR BOTH INDOORS & OUTDOORS
   OCCUR BOTH NIGHT & DAY

   SMOKING :- NEVER (Less than 400 cigarettes in patients lifetime.)

SKIN TESTS PERFORMED
--------------------
     Positive to..foods.

FOOD:
-----
   Patient believes the following foods affect his/her symptoms:
   T.HEAT. FLOUR.
* Factors POSSIBLY indicating food allergy are listed below :
 -----------------------------------------------------------
   Symptoms start within 0-1 HOUR
   This history and atopic profile
   WEAKLY SUGGESTS that symptoms may be related to food
   but no specific IgE antibodies were found to:
            FISH(COD)(0)  WHEAT(0)  COCOA(0)  CHEDDAR CHEESE(0)

MISCELLANEOUS ALLERGIES:
------------------------
     coloured drugs.
--------------------------------------------------------------------------------
```

REPORT B

REPORT FROM: RAST Allergy Unit, Benenden Hospital, Cranbrook, Kent TN17 4AX (Tel.0580-240495)

ALLERGY INVESTIGATION PATIENT NO: 8943

Mr Vulcan SPOCK

DATE OF BIRTH : 05/03/43
DATE OF QUESTIONNAIRE : 08/02/88

TO:

TODAYS DATE : 17-02-1988

Dr Bones
Sick Bay
Starship Enterprise
Galactica

Fold

SYMPTOMS : Involve-EYES-NOSE-CHEST-SKIN-GUT

BLOOD TEST RESULTS AND COMMENTS TOTAL IgE = 628 units/ml.

The total IgE is above most common adult non-allergic levels - atopy likely.

PHADEBAS IGE-RAST RESULTS FOR SPECIFIC ALLERGENS

(RAST results are graded :0,1,2,3,4, where 0 IS NEGATIVE and 4 IS STRONGLY POSITIVE,
1 IS BORDERLINE EXCEPT for MOULDS when it is regarded as PROBABLY clinically significant.)

Patient has positive specific IgE antibodies to at least one common allergen and so is atopic
The HISTORY
STRONGLY SUGGESTS that symptoms may be related to seasonal allergens
and specific IgE antibodies were found to:
GRASSES--TIMOTHY GRASS(4) MIXED TREES (0)
MIXED WEEDS (1)
The HISTORY
STRONGLY SUGGESTS that symptoms may be related to perennial allergens
but no specific IgE antibodies were found to:
DUST--HOUSE DUST MITE PTERONYSSINUS(0)
CATS--CAT EPITHELIUM(0)
DOGS--DOG DANDER(0) MIXED MOULDS (0)
BIRD--BUDGERIGAR FEATHERS(0)
RABBIT--RABBIT EPITHELIUM(0)
OTHER ANIMAL--HAMSTER EPITHELIUM(0),RAT EPITHELIUM(0)

**** no specific IgE antibodies were found to the following foods:
MILK(0) CRAB(0)

SUMMARY OF GRASS, DUST-MITE AND COMMON ANIMAL ANALYSES

** PROBABILITIES of OUTDOOR and INDOOR ALLERGENS contributing to symptoms are 98 % and 94 %. **
GRASS pollens, DUST-MITES and common PETS cause most U.K. ALLERGIES,
and it is VERY LIKELY that GRASS * contribute(s) to this patient's symptoms.

REPORT C

```
.........ents are based on the following Clinical History:-

  _AL HISTORY ASSESSED BY QUESTIONNAIRE
 -----------------------------------------
   3  relation(s) have a possible ATOPIC history
  Age of onset was  42

SYMPTOMS ARE SEASONAL  AND  SYMPTOMS ARE PERENNIAL

           Months occur..... 1   2   5   6   11   12
           Months worst..... 1   2   6
                             OCCUR IN :
                              2  FROM  5  MONTHS of the GRASS season
                              1  FROM  4  MONTHS of the WEED season
                              1  FROM  5  MONTHS of the CLADOSPORIUM HERBARUM season
                              1  FROM  5  MONTHS of the ALTERNARIA TENUIS season
                              4  FROM  5  MONTHS of the ASPERGILLUS FUMIGATUS season
                              2  FROM  4  MONTHS of the TREE season
                             WORST IN:
                              1  FROM  3  MONTHS at height of the GRASS season
                              2  FROM  3  MONTHS at height of the ASPERGILLUS FUMIGATUS season

   WORST NEAR GRASS, TREES, WEEDS
   Aggravated by DUSTING &/or making BEDS
    CAUSED BY :-CATS  CAUSED BY :-DOGS  REGULARLY IN CONTACT WITH :-BIRDS  REGULARLY
    IN CONTACT WITH :-RABBITS  REGULARLY IN CONTACT WITH :-FISH  REGULARLY IN CONTACT
    WITH :-BUDGIE RAT HAMSTER AND GERBILS
   OCCUR IN 2 SETS, ONE INDOORS, ONE OUTDOORS
   OCCUR IN 2 SETS, ONE AT NIGHT, ONE IN THE DAY
   IMPROVE ON WET SUMMER DAYS(suggests weeds more than outdoor moulds)

   SMOKING :- NEVER (Less than 400 cigarettes in patients lifetime.)

 SKIN TESTS PERFORMED
 --------------------
   Positive to..animals, H.D.M., grass, others especially DOGS CATS CEREALS
   POLLENS

FOOD:
-----
   Patient believes the following foods affect his/her symptoms:
   SPIRITS, MILK, CRAB,
* Factors POSSIBLY indicating food allergy are listed below :
 -----------------------------------------------------------
   Skin & Chest symptoms, Gut symptoms
  Symptoms start within 0-1 HOUR
   Skin tests positive to one of house-dust mite,animal or grass, IgE over 200u/ml
  This history and atopic profile
  WEAKLY SUGGESTS that symptoms may be related to food
  but no specific IgE antibodies were found to:
               MILK(0)  CRAB(0)

MISCELLANEOUS ALLERGIES:
-----------------------
  None mentioned

-----------------------------------------------------------------------------------------
```

REPORT C

REFERENCES

1. Fife, D., Twarog, F.J. and Geha, R.S. (1983). Evaluation of clinical data in childhood asthma. Am. J. Dis. Child, 137, 945-8.
2. Osvath, P., Kovacs, Z., Godo, B. and Markus, V. (1972). Analysis of skin test results and case history data for asthmatic children using a computerised punched-card method. Acta Allergol., 27, 287-301.
3. Merrett, T.G., Burr, M.L., St. Leger, A.S. and Merrett, J. (1980) Circulating IgE levels in the over seventies. Clin. Allergy, 10, 433-9.
4. Platts-Mills, T.A.E. and Chapman, M.D. (1987). Dust mites: Immunology, allergic disease, and environmental control. J. Allergy Clin. Immunol., 80, 755-75.
5. Warner, J.O., Price, J.F., Soothill, T.F. and Hey, E.N. (1978). Controlled trial of hyposensitization to *Dermatophagoides pteronyssinus* in children with asthma. Lancet, 2, 912.
6. Pantin, C.F.A. and Merrett, T.G. (1982). Allergy screening using a microcomputer. BMJ., 285, 483-7.
7. Everitt, B. (1974). In: Cluster Analysis, London, Heinemann.
8. Norusis, M.J. and Jacquez, J.A. (1975). Diagnosis 11. Diagnostic models based on attributable clusters: a proposal and comparisons. Comput. Biomed. Res., 8, 173-88.
9. Tanser, A.R. and Davies, A. (1984). Archie: The use of a microcomputer in an allergy clinic. Br. J. Dis. Chest, 78, 135-9.

Clinical Usage of Computers in Allergy Diagnosis

Jack Pavuk

Tech-Sys Data Systems, Inc., Richardson, Texas, USA

ABSTRACT

Continued advances in both economically priced Computer Hardware, and in more sophisticated Computer Software, argue for an increased role for Computer technology in the Clinical Diagnostic Process. Teaming a COMPUTER (with abilities to quickly gather, organize, and graphically represent data) with a HUMAN (capable of providing rational and deductive analysis of that data) can provide the most efficient means to a fast, efficient, and accurate medical diagnosis. An application in Computer-assisted Allergy Diagnosis is presented in support of such a premise.

KEYWORDS

Computer; software; evaluation; diagnosis; allergy.

Most clinicians approach a medical diagnosis in a structured manner, even though they may not be completely aware that a formal structure is being applied. The experienced diagnostician, especially, feels no sense of a formal problem-solving structure - he instinctively probes, perceives, and rationalizes his way to a final diagnosis. In actuality, however, he DOES follow a "subliminal" pattern in reaching his final judgements.

If one were to analyze the steps taken (subliminally, or otherwise) in problem-solving, the sequence would resemble the following:

Initial Information Gathering
Organization of the Gathered Information
Initial Hypothesis (based on the preliminary information)
Probing for Additional Information to Support/Refute Hypothesis
Re-Organization of ALL Information
Final Deductive Analysis based on ALL Information

Advances in the Biosciences Vol. 74

Printed in Great Britain

As magnificent as the Human Brain is, there are limitations in man's UTILIZATION of the brain's capacity and capabilities. We often know we have been exposed to a piece of information, but can't quite "put our finger on it". The information was permanently stored in our memory banks, but it was temporarily misplaced among the myriad of total brain memory cells. We often "have too much on our minds" to consider other than the current situation at hand.

COMPUTER QUALITIES

The Good News Is...	The Bad News Is...
Blinding Speed of Reports	Slow Data Entry ***
Cross—Referencing	Rigidity of Spelling
Graphing of Data	Data Overwhelming
Look—up Tables	No 'REASONING' Power
Absolute Duplicability	Impersonal
Objectivity	Inability to 'PROBE'
Huge Data Capacity	Garbage In ... Garbage Out

Computers also have storage cells, filled with information concerning diverse topics. Some of its cells may contain Spreadsheet information regarding accounting functions. Another memory area may store and process our correspondence, utilizing a word processor. Still other areas house information which may be charted graphically for better visualization of interrelationships between data pieces.

Computers are blessed, however, with having no emotional, physical, or other impediments to their performing at top efficiency at all times. With a properly functioning, properly devised "Program" to direct that Operating System, all information in its memory cells is instantly available all the time.

It would seem then, that a computer could be used in concert with man's deductive and rationalization abilities to form an unbeatable team for making logical decisions and/or diagnoses, based on total recall of all stored data and on man's inherent logic and reasoning talents.

Such is the premise for having a computer-assisted approach to medical diagnosis.

If we were to look at the phases of the problem-solving outline which was previously described, we would find that the computer is actually better qualified to perform SOME of those functions than is the intellectually superior man. With any exposure to computers at all, one would have to agree that computers are better at the initial gathering and organizing functions of the problem-solving process. It can recall data from data bases with remarkable speed and accuracy. It can sort, graph, and interrelate that data in the blink of an eye.

Man, on the other hand, has inherent talents lacking in all computers - mostly in relation to his being able to reason, or think. A COMPUTER CANNOT

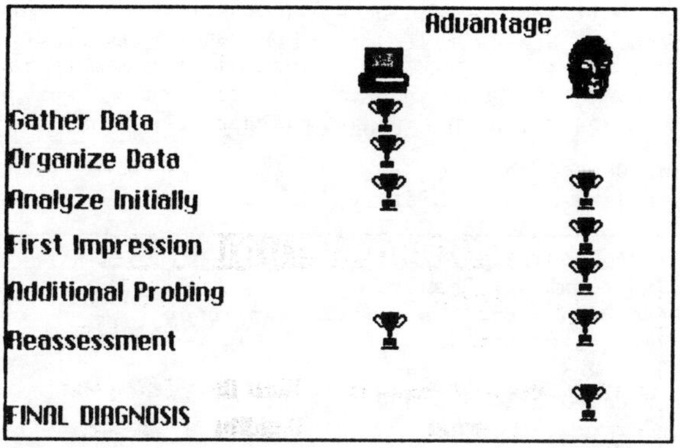

THINK! (It blindly follows instructions without regard to the rationale for those instructions.)

Therefore, the perfect marriage of man and computer would allow for the computer to carry the brunt of the gathering, processing, and presentation of data - while the man would then analyze, probe, re-analyze, and make the final decisions based on that data.

Aller-Dat is a computer-based tool for assisting physicians in Allergy Diagnosis. It is presented as an example of the Man-Computer team described above. Certain physical enhancements have been added to the computer program itself to facilitate Data Entry (Card Readers, Interfacing to Measuring Devices, etc.).

A great deal of effort was made to assure that the decisions reached by the "team" were decisions of the "brains" of the team, the man. The computer often suggests interrelationships based on raw data, but it cannot pass judgement on the quality of that data. Hence, the man is rightfully and logically tasked with the ultimate judgemental responsibility.

No attempt will be made to describe ALL of the features of the Aller-Dat Program. It has ancillary functions which are not germaine to the discussion at hand (Zip-Code Pollen Distribution, Patient Compliance with Injections,

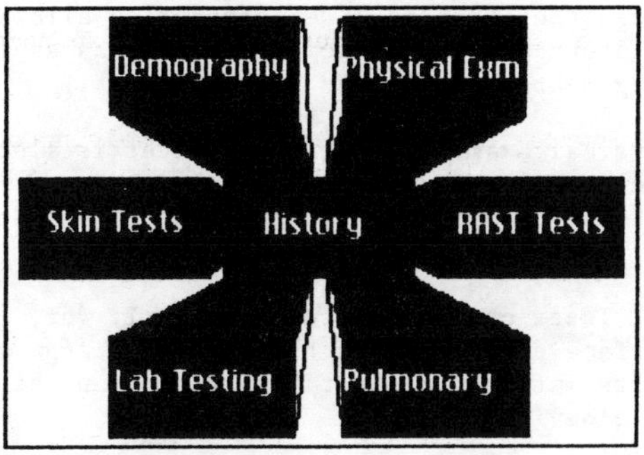

Cross-Reacting Foods, etc.). However, since this Panel is highly academically oriented, I will address the Research capabilities of the Patient Data Base (both Search and Statistical Analysis) at the conclusion of the presentation.

If one views a partial Aller-Dat Program MENU,

1- Patient Demographics
2- Computerized Allergy History
3- Physical Exam and Lab Work
4- Skin Test Results
5- RAST and Total IgE Results
6- Complete Data Summary and Physician Input
7- Test Results Only
8- Automatic Referral Letters
9- Extract Calculations and Dosage Schedules
etc.,

it can be seen that items 1 thru 5 represent Data to be "Input" into the computer. One drawback in entering data into a computer is in the method of entry, usually via the keyboard. In addition to being slow, keyboard entry is also subject to typing errors, etc. For this reason, we try, as much as possible, to eliminate keyboard entry. Except for demographic data (which must, of necessity, be typed), an optical card scanner is used, to a great extent, to translate pencil marks on a card into data input for the computer. These scanners are very accurate, and the data is entered very quickly.

As an example, the patient Allergy History Questionnaire is completed by the patients while they wait to see the physician. A typical allergy questionnaire (which can be customized to the individual Physician) has approximately 200-300 questions. The patient simply places a pencil mark next to the questions which are answered YES. The card which contains the pencil marks is then passed through the optical scanner, and all responses are recorded in approximately 4 seconds. The Program then analyzes those answers for approximately 20 seconds, and a two page History is printed.

Page one gives the patient history, in outline form. It is a verbal description of the patient's responses, which can be used as a permanent record for the patient's file, and as an interview guide from which to probe further into the background for the patient's responses.

Page two quantitates the answers (using a scoring matrix which can refer to previous answers and concurrently score various categories), and it graphs the results for: Atopy, non-Atopy, strength of the History for Trees, Grass, Weeds, etc. The resultant graphical representation allows for quick scanning of the patient's History by the physician, as well as saving time in making interrelational judgements by allergen class.

The physician's Physical Examination of the patient is also annotated on a card, with a simple pencil mark indicating only the area of abnormal observation. The default is for a normal finding - usually just a few pencil marks are required to record the complete Physical.

Skin Tests and RAST Tests can be recorded similarly (or, in the case of RAST Tests, we can interface directly with the measuring apparatus to automatically log the results into the patient Data Base when the blood sample is being counted by the device).

Once items 1 thru 5 of the menu are entered, items 6 and 7 "Output" the data to the Physician.

Item 6, the COMPLETE DATA SUMMARY, graphs and interrelates all of the data. An overall Allergic Index is calculated (using all information) and a list of suspected causative allergens is printed by having the computer correlate the RAST/Skin Test results with the patient History. Following the Summary printout is a Physician Worksheet which allows the physician to select: Diagnoses, Treatment Regimens, Allergens for Treatment, etc. These selections are made from multiple choice lists and entered by simply selecting numbers which are annotated on the worksheet. Once the worksheet is given to the person inputting the data to the computer, only a series of numbers (corresponding to those checked by the physician on the worksheet) need be typed to let the computer know the physician's desires.

Item 7, the TEST RESULTS ONLY, generates a one page summary of the pertinent information concerning the patient. The results also occupy only one screenful of data on the monitor, so that it can be viewed for a quick recap of the patient at a later date. The information contained on the page/screen includes: Total IgE, History Scores, RAST Test Results, Skin Test Results, Diagnoses, Allergens selected for Treatment, Immunoglobulins (G,A,M), Pulmonary Function, and selected responses from the patient History (Smoker, Insect Allergies, Atopic Family, Pets, etc.).

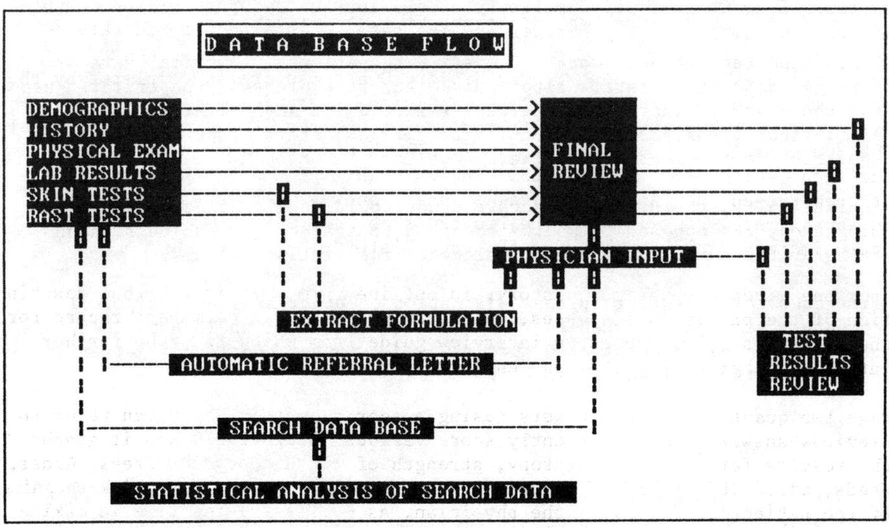

Items 8 and 9 are then available automatically, since the Computer now has the total picture (the initial data input, plus the Physician's desires). The referral letters are generated, without dictation, by searching both the patient's data base, and the Physicians desires, to create a letter of variable length and content, depending on the information in the data base. Similarly, the Extract Formulations and Dosage schedules are automatically generated since the computer now knows the results of the tests for each allergen, plus which allergens were selected by the physician for treatment.

Without going into the entire program, it should easily be recognized that the computer has made several important contributions to the overall diagnostic procedure:

1- Speed of Data Entry
2- Data Crunching and Graphic Presentation of the Data
3- Time Saving in Referral Letters and Extract Preparation
4- A Degree of Objectivity (computers have no biases)
5- Permanent Storage of Patient Data, Diagnosis, etc.

The last item (permanent storage) allows for an added benefit from the Aller-Dat Program - the ability to search the patient Data for ranges of any/all parameters which have been stored on any/all patients. Thus a set of criteria for a study can be input, the patients satisfying that criteria can be found and segregated for statistical analysis, and a built-in statistical module can do scattergrams, histograms, standard deviations, etc. on any of the patient parameters.

As an example of the time required for such a study to be done: from a total patient data base of 5000 patients, ranges for any/all stored patient parameters could be tagged for search (age, sex, test results, diagnosis, physical exam criteria, etc.). It would take the computer approximately 10 to 20 minutes to select the qualifiers and copy all of their numerical information to a separate floppy diskette. That diskette could then be analyzed statistically for any number of statistical procedures, with approximately 1 minute required to generate each scattergram, frequency distribution, descriptive statistic determination, etc.

It is important to note that ALL qualifying patients' numerical data is segregated to the separate floppy diskette, regardless of the criteria used for the Search. Therefore, patients selected for any criteria can be statistically analyzed RETROSPECTIVELY for parameters other than those specified in the Search Criteria.

Realistically, the data from such a study could easily be gathered in a single day (as opposed to months required to search file cabinets, hand transfer the data, and manually calculate the statistics).

Time does not allow for further discussion of other available types of computer-assistance in Medical Diagnosis. However, I would hope that this presentation has demonstrated that man and computer can live together in harmony, without either's ego being demeaned.

DISCUSSION

An Allergy Evaluation and Diagnosis is a complex procedure, its accuracy dependent on both the accuracy of the data gathered (by History, Lab and Physical Tests, etc.), and the acumen of the diagnostician in assessing that data.

Computers can play a vital role in assuring the completeness, accuracy, and visual representation of the data used in the Allergy Evaluation.

The physician plays the major role in applying sound logical and deductive reasoning in interpreting the data. Additionally, the physician is often blessed with instinctive talent which allows him to apply educated "gut feelings" as part of the diagnosis. No computer can supply that quality, which often proves to be the basis for a valid diagnosis.

It has been shown that the Man-Computer "team" is a viable option to Man, alone, in providing an expeditious, complete, and organized approach to Allergy Diagnosis.

Future developments in artificial intelligence and other, not yet discovered, computer advances may strengthen the arguments for such a liaison in all fields of decision making.

The Computer as Your Confidante

J. O. Warner and S. Shribman
Cardiothoracic Unit, Brompton Hospital, Fulham Road, London SW3 6HP, UK

ABSTRACT

The usefulness of computer-delivered questionnaires have been tested in a busy paediatric asthma clinic. Mothers were asked to respond on a standard computer keyboard to (1) an 18 frame psychiatric questionnaire, and (2) a more detailed clinical allergy questionnaire. The sessions were supervised by a clinic nurse and attended by mothers of 50 patients. The psychiatric investigation was judged to be very successful, akin to having a psychiatrist in attendance at the clinic, because of the better identification of the mother with depression and anxiety. Our data suggests that treatment for asthmatic children of these mothers should include family therapy. Unfortunately, the clinical allergy assessment via the computer was judged to be a disturbance in the medical process and many mothers failed to complete the questionnaire because of its complexity and length. We conclude that computers can be successfully used in an allergy clinic as an aid to evaluate emotional disturbance, but that interrogation directly related to the disease is best performed by a meaningful communication between the treatment provider and patients.

KEYWORDS

Clinical allergy; asthma; computing; questionnaire; psychiatry.

INTRODUCTION

Computers have been used with increasing frequency over the past 20 years as an aid to medical diagnosis [1]. Studies have indicated that patients are more likely to give honest answers about their more intimate problems if asked by a computer rather than a doctor [2]. The computer is less likely to raise an eyebrow or look disapprovingly upon sensitive issues revealed by a patient. As emotional problems can not only cause illness in children but also complicate the management of organic diseases, it is imperative that such problems are detected [3]. Unfortunately, this is difficult, time consuming and sometimes impossible. Psychiatrists are usually not immediate-

Advances in the Biosciences Vol. 74

Printed in Great Britain

ly available in busy paediatric clinics. We have utilized a computer-delivered questionnaire designed to identify depression and anxiety in mothers of children attending a paediatric asthma clinic.

METHODS

The method was developed by the clinical computer group at the Institute of Psychiatry in London. Its value was initially assessed in 125 patients individually diagnosed as suffering from depressive illnesses and 43 control subjects. The computer delivered 21 frames to each subject, 3 of instructions and 18 of questions from the Hamilton Depression Rating Scale [4], modified for self rating. The patients answered via a standard computer keyboard. The only supervision required was from a clinic nurse. The nurse initiated the procedure and then left the patients to run through the questionnaire on their own. Seventy-five of the patients were independently and blindly assessed for severity of depression by qualified psychiatrists using a global rating scale. The study showed that the system adequately discriminated the depressed from the non-depressed with 97% of the patients being correctly assigned. Furthermore, severity as assessed by the total score on the computer correlated well with the clinician's global assessment of severity ($R = 0.8$). Subjects and staff felt that the procedure gave useful additional information quickly and it did not disrupt the routine of a standard clinic [5]. An additional questionnaire was added to the original programme; the Salkind Self Rating Anxiety Inventory [6]. The system was automated so that the questionnaires ran sequentially and the answers were recorded without intervention of the operator. Subsequently, a printout of the scores for each question was produced for analysis.

At a later stage in the development of the procedure we added a complex questionnaire of birth history and allergy symptoms with the facility to print out a resumé of the relevant information. Whilst the psychiatric questionnaire surprisingly gained acceptance because of its simplicity, the clinical allergy history questionnaire was generally more difficult to apply because it was time consuming.

Fifty mothers were initially studied with the psychiatry questionnaire. They consulted the computer prior to seeing the paediatric chest physician and neither doctor nor mother were aware of the results. The doctor was asked to give a rating of the mother's emotional state blindly and also provided a rating of the severity of the child's asthma.

RESULTS

Fourteen of the fifty mothers were depressed by the computer rating but only four of these were identified by the doctor. Twenty two of the mothers were anxious but only ten of these were thought to be anxious clinically. The doctors also produced false positives in one for depression and twelve for anxiety. All but one of the depressed mothers were also anxious with a good correlation ($R = 0.72$) between anxiety and depression scores. Whilst the absolute severity of the child's asthma was not related to the occurrence of the depression or anxiety in the mother, those mothers that remained depressed on a subsequent assessment were more likely to have children whose asthma was unchanged or worse compared with mothers who either were not depressed or whose depression resolved [7]. No mother refused to complete the questionnaire and the only limitation was the ability to read English.

Unfortunately, many mothers failed to complete the clinical allergy questionnaire because of its complexity and the time required for completion. The resumé of the clinical history was found to be more intrusive into the subsequent consultation with the doctor and hindered the standard medical process. Its value was noted only in relation to the generation of consistent information for clinical trials.

DISCUSSION

We have shown that doctors, despite having an awareness that psychological difficulties may be present, fail to identify significant emotional disturbance in mothers of their patients. It is clear that the doctors were reviewing the child's progress and were not psychiatrists and therefore this result was not surprising. The pressure on time and manpower precludes the presence of psychiatrists in such clinics. The microcomputer technique to elicit such information is useful, not intrusive into the routine medical procedures, and identifies problems that would not otherwise have become apparent in the paediatric setting. Confidentiality of the system is maintained as the results are less accessible than ordinary hospital notes.

Computer techniques for administering such questionnaires are more efficient than pencil and paper methods. Furthermore, the subjects are unable to refer to their previous answers or the questions, which is known to affect the results. The high prevalence of depression in mothers of children with asthma certainly requires further investigation. A community study in southwest London has shown that poor mental health in the mother was strongly associated with undertreatment of the child's asthma [8]. Thus, our own observations that the children of mothers who remain depressed did less well than those of mothers who were not depressed or whose depression improved clearly indicates the need to detect this problem. It has been shown that family therapy used as an adjunct to conventional medical treatment is more effective than medical treatment alone in the management of childhood asthma [9]. Thus, the detection of an emotional problem can lead to alternative lines of management which may improve the situation.

Our computer administered questionnaire for gathering straightforward clinical information has not gained acceptance within our own clinic. We suspect that this may also be the case elsewhere. Whilst there have been a few publications recommending the use of computers in asthma clinics, such programmes have not been seen in action outside the original research setting [10]. The key to good management of a chronic disease such as asthma includes the need for meaningful communication between patients and the treatment provider in a non-threatening way which can promote open and honest communication and this is still best achieved in the standard clinical consultation setting [11] with perhaps the additional information provided by computer-aided assessments of emotional disturbance.

REFERENCES

1. Griest, J.H., Klein, M. and Erdman, M. (1976). Positive on-line diagnosis by computer. Am. J. Psychiat., 133, 1405-1408.
2. Lucas, R.W. *et al.* (1977). Psychiatrists and a computer as interrogators of patients with alcohol related illness: a comparison. Br. J. Psychiat., 131, 160-167.
3. Bingley, L. *et al.* (1980). A comprehensive management of children on a paediatric ward: a family approach. Arch. Dis. Child., 55, 558-561.

4. Hamilton, M. (1960). A rating scale for depression. J. Neurol. Neurosurg. Psychiatr., 23, 56-62.
5. Ancill, R.J. *et al.* (1981). Assessment of depression by microcomputer. Acta Psychiat.
6. Salkind, M. (1974). Self-rating scale for anxiety. PhD thesis, Univ. London.
7. Goldsworthy, S. *et al.* (1981). Computer-aided assessment of depression and anxiety. Prog. Respir. Research, 17, 277-281.
8. Anderson, H.R. *et al.* (1981). Influence of morbidity, illness label and social, family and health service factors on drug treatment of childhood asthma. Lancet, 2, 1030-1032.
9. Lask, B. and Matthew, D. (1979). Childhood asthma: a controlled trial of family psychotherapy. Arch. Dis. Childhood, 54, 116-119.
10. Tansor, A.R. and Davies, A. (1984). Archie: the use of a microcomputer in an allergy clinic. Br. J. Dis. Chest, 78, 135-139.
11. Hindi-Alexander, M. and Thorn, J. (1987). Compliance or non-compliance: that is the question. Amer. J. Health Promotion, S-11.

Discussion: Session Eight

Chairman: A. Said El Shami

1. Question to Jeremy Wyatt

Sheldon Spector. This is more of an addendum, because we together with a company in Denver have been using a microchip as an aid to medical decision-making in a rather unusual way. The microchip has been incorporated both in a nebuliser and a pill monitor, so that we can determine patient compliance with their prescribed treatment. During clinic attendance the "chip" was transferred to an "interpreter" which recorded the exact time (±4 minutes) that the patient had taken the medicine. This was compared with other monitoring devices which can only estimate over- or under-use of medication with the following interesting result. Our chip/interpreter discovered a sub-group of patients who used the correct amounts of medication, but in an erratic manner; that is to say sometimes they would over-medicate but would compensate later by using less of their medication. We were also able to check how well patients completed their diary cards; the answer was that all too often they lied! I guess they were just really trying to please the doctor.

Jeremy Wyatt. Were they aware of the fact that you were recording everything?

Sheldon Spector. In order to pass our human experimentation committee at UCLA we had to tell them that they were being 'monitored' but I think most patients did not know what that meant. In other studies they also were monitored, so we made it general enough so that most did not know what was happening. We concluded that we were probably looking at an "iceberg phenomenon", in that the people incorporated into our study had participated in other studies. Therefore presumably we had preselected the most compliant group of patients compared with those patients who either said that they were too busy etc., or who did not volunteer at all. Our next step is to do some intervention studies.

Jeremy Wyatt. That is very interesting. I have deliberately omitted all the applications of computers in patient monitoring systems instrumentation, and concentrated just on constructing clinical databases and diagnostic indices. But there is no doubt that all these applications can help with decision-making.

Advances in the Biosciences Vol. 74

Closing Remarks

Jack Pepys

We really have had an intensive two days; stimulating in every way, we have been given a wonderful opportunity of getting together and discussing things at a time in our specialty when we have reached an explosive revolutionary situation. The modern technologies, with sequencing, monoclonals, genetic understanding, are opening up a remarkable future for all of us, wherever our interests lie. This is the sort of meeting which might have been referred to in a story I'd like to tell you. There was a professor who came in on a Monday morning holding in his hand a letter, and he said to his secretary, please would you arrange for me to go to this meeting. She looked at it and said "But Professor, you can't go". And the professor asked why not, "Because this meeting was held last Friday". "Oh dear", he said, "I'm terribly upset". "But you needn't be", she said, "you went". Well that won't happen to us with this meeting, of that I am sure. This has been a memorable meeting in every way and I'm quite sure that when the publication appears, it will have a place a honour on our shelves and on many other shelves, as a point of reference at a remarkable time, and I think great credit is due to the foresight of DPC in taking advantage of this situation to bring us together.

Now very briefly what are just a few of the things which we have learnt in these two days. If we take allergens for example, there is the need for precision reagents to help us do our job properly; the monoclonals and the sequencing; the identification of epitopes. And there, I think, we have a problem, because no matter now precise our reagents are, they affect people who are not going to be so precise in the way they respond. I am a little perturbed when an epitope is described as a major epitope because it affects more than 50% of people, what of the remaining 50%? Its major only in the sense that you have some evidence which makes it a little more effective in a population than others, but the others are still very important. When it comes to diagnosis and treatment it may be that we will have to think beyond one major epitope and look at a pattern of epitopes.

The other thing is, that with the sequencing of allergens, it's possible to think of the next method of immunotherapy, better in many ways than injection treatment. And that is local immunotherapy. If we can sequence the relevant epitopes in small enough size, in monomeric form perhaps, so that they can then be applied locally, the chances are that we will get IgE blocking. We may reach the time when a few puffs of material in the nose or eyes or the bronchus may occupy the IgE sites without bridging them and make it possible to prevent reactions to the intact allergen.

Then when it comes to IgE itself we have a somewhat similar situation. This has been discussed across the board in many ways. The report of fragments of IgE was very interesting because there, along the lines of Bob Hamburger, is another method of achieving immunotherapy without having to inject. As you may know the history of injections of allergens has reached a sad end in the United Kingdom where we can no longer easily do this. So we have to have other methods and these are two promising avenues.

Then when it comes to methods of investigation, once again we need to have precision tools, and we have heard of DPC's method which seems to me to offer a very useful and flexible way of having a precision tool for our precision reagents. So we are now poised for the great advance, which will give us highly standardized and purified materials, this is a world-wide search and it has been greatly forwarded by what we have heard about. Just to pick up on one thing which I think we will all carry away with us, and that is the interest in the tryptase from the Mast cells which seems extremely interesting as a diagnostic tool in rather unusual situations.

I think that is all I would like to say without keeping you here very much longer. I would like to pay tribute to DPC for having us; tribute to Terry Merrett for organising a remarkable programme, wide ranging, covering many fields. I can't think of anything much which has been left out. I think we should all express our appreciation and thank you very much indeed.